THE OXFORD ENGINEERING SCIENCE SERIES

1 D. R. RHODES: *Synthesis of planar antenna sources*
2 L. C. WOODS: *Thermodynamics of fluid systems*
3 R. N. FRANKLIN: *Plasma phenomena in gas discharges*
4 G. A. BIRD: *Molecular gas dynamics*
5 H.-G. UNGER: *Planar optical waveguides and fibres*
6 J. HEYMAN: *Equilibrium of shell structures*
7 K. H. HUNT: *Kinematic geometry of mechanisms*
8 D. S. JONES: *Methods in electromagnetic wave propagation* (Two volumes)
9 W. D. MARSH: *Economics of electric utility power generation*
10 P. HAGEDORN: *Non-linear oscillations* (Second edition)
11 R. HILL: *Mathematical theory of plasticity*
12 D. J. DAWE: *Matrix and finite element displacement analysis of structures*
13 N. W. MURRAY: *Introduction to the theory of thin-walled structures*
14 R. I. TANNER: *Engineering rheology*
15 M. F. KANNINEN and C. H. POPELAR: *Advanced fracture mechanics*
16 R. H. T. BATES and M. J. McDONNELL: *Image restoration and reconstruction*
18 K. HUSEYIN: *Multiple-parameter stability theory and its applications*
19 R. N. BRACEWELL: *The Hartley transform*
20 J. WESSON: *Tokamaks*
21 P. B. WHALLEY: *Boiling, condensation, and gas–liquid flow*
22 C. SAMSON, M. LeBORGNE, and B. ESPIAU: *Robot control: the task function approach*
23 H. J. RAMM: *Fluid dynamics for the study of transonic flow*
24 R. R. A. SYMS: *Practical volume holography*
25 W. D. McCOMB: *The physics of fluid turbulence*
26 Z. P. BAZANT and L. CEDOLIN: *Stability of structures: principles of elastic, inelastic, and damage theories*
27 J. D. THORNTON: *Science and practice of liquid–liquid extraction* (Two volumes)
28 J. VAN BLADEL: *Singular electromagnetic fields and sources*
29 M. O. TOKHI and R. R. LEITCH: *Active noise control*
30 I. V. LINDELL: *Methods for electromagnetic field analysis*
31 J. A. C. KENTFIELD: *Nonsteady, one-dimensional, internal, compressible flows*
32 W. F. HOSFORD: *Mechanics of crystals and polycrystals*
33 G. S. H. LOCK: *The tubular thermosyphon*

The Tubular Thermosyphon

Variations on a Theme

G. S. H. LOCK
Department of Mechanical Engineering
University of Alberta

Oxford New York Tokyo
OXFORD UNIVERSITY PRESS
1992

Oxford University Press, Walton Street, Oxford OX2 6DP

Oxford New York Toronto
Delhi Bombay Calcutta Madras Karachi
Petaling Jaya Singapore Hong Kong Tokyo
Nairobi Dar es Salaam Cape Town
Melbourne Auckland

and associated companies in
Berlin Ibadan

Oxford is a trade mark of Oxford University Press

Published in the United States
by Oxford University Press, New York

A catalogue record for this book is available from the British Library

Library of Congress Cataloging in Publication Data
Lock, G. S. H.
The tubular thermosyphon: variations on a theme/by G. S. H. Lock.
Includes bibliographical references and index
1. Thermosyphons. I. Title.
TJ263.2.L63 1992 621.402'2—dc20 91–48199
ISBN 0–19–856247–0

Typeset by Cotswold Typesetting Ltd, Gloucester
Printed and bound in
Great Britain by Biddles Ltd,
Guildford and King's Lynn

To my wife who for many years has lived with
the tubular thermosyphon and me,
and cared for us both

PREFACE

I first became interested in the tubular thermosyphon as an undergraduate when I found myself utterly fascinated by the possibility of using the device to cool high-temperature gas turbine blades. It seemed to me that the harnessing of centrifugal accelerations of the order of 10^4 g to create a vigorous natural convection circulation had that stamp of ingenuity which marks the very best of engineering. At the same time, I was dimly aware that the convection pattern thus created presented a challenge to the imagination for it would be controlled by a complex force field operating in the confines of a tube. It was not known in advance how the device would perform. Intrigued by both the science and the technology, I began my Ph.D. thesis. The work was undertaken in the Stephenson Engineering Laboratories, then part of King's College of the University of Durham and now within the University of Newcastle upon Tyne. It was my good fortune to follow in the footsteps of three pioneers: H. Cohen, B. W. Martin, and F. J. Bayley, the last of whom supervised my thesis.

Like most graduate students, I started my work by assuming it to be a very specific task: to determine the heat transfer characteristics of a closed, tubular thermosyphon under single-phase conditions. Over a period of three years, I asked, what could be easier to fathom than natural convection inside a fluid-filled tube with one end hotter than the other, especially if the tube stands upright with gravity simulating centrifugal acceleration? But the simplicity was deceiving. After reading Lighthill's seminal paper of 1953, I suddenly realized that the internal circulation pattern was by no means obvious. Thus began my interest in the modelling of fluid flow systems. In the decades since, I have never ceased to be fascinated by the kaleidoscope of patterns which can occur in such limited circumstances; and seldom have I felt anything less than a sense of awe when confronted by the attendant complexities.

On a few occasions over the intervening years, my sense of adventure has got the better of me and I have felt compelled to suggest flow models in papers submitted to learned journals. Invariably, these suggestions succeeded in raising the ire of those reviewers who prefer to accept no statement that has not been independently checked and double-checked. I confess that it offends me to think that speculation, clearly identified as such, should not be allowed to play a role in the advancement of knowledge in a living literature; and it saddens me to think that the dead hand of such editorial policies should be allowed to remove so much of the delight of research, not to mention the timeliness. Accordingly, in dealing with various knowledge gaps discovered in the course of preparing this book I have not hesitated to speculate where I thought it would be helpful to the reader; if I have done so incorrectly, or unnecessarily, I hope the reader will be kind enough to favour me with an appropriate reference so that I may remove the deficiency.

But, returning to my graduate student days, I must also mention another awakening. This began as I searched for the full meaning of the word *thermosyphon*. Seeking generic linkages, and attempting to reconcile the many and varied forms and applications in which the device may be seen, I found that I had undertaken a much larger task, but one which provided me with a splendid education. It forced me to examine the structure of a definition, any definition, and to appreciate the balance between historical and current uses of terms, any terms; it showed me that while the same physical phenomenon may be found in different disciplines it is often treated differently, following the dictates of context and tradition; and it provided me with a glimpse of a pervasive system of links not only connecting one discipline with another but within a given discipline connecting concept to analysis, analysis to prediction, prediction to measurement, measurement to correlation, etc. This sense of interrelatedness and the unity that goes with it has been a valuable guide ever since.

Even so, in putting together this book I have become increasingly conscious that a thermosyphon means different things to different people. Those who work with natural convection, for example, are frequently unaware of the work that goes on in the two solitudes of boiling and condensation; authors of papers on loop thermosyphons seldom consult the literature on single tubes; and those who work with wick-lined tubes are often unfamiliar with buoyancy-driven systems. Increasing specialization and proliferation have made it increasingly difficult to stay abreast of developments in related fields, while the natural trend towards international conferences and networks has simply exacerbated the difficulty. It therefore seemed to me that a unified treatment of thermosyphons would be worth developing, if only to show the generic links and common principles.

To do this it was necessary to embrace all that might properly be concerned, or connected, with thermosyphon behaviour and usage. Excluding the loop thermosyphon, which I believe merits a separate treatment, the material presented here is intended to be a comprehensive introduction. Within reasonable space limitations, it is designed to offer a complete description of the tubular thermosyphon: mathematics and physics; theory and experiment; concept and application. Emphasis has been placed primarily on the fundamental principles of operation in order to provide not only the basis for assessing the performance of current devices but the insight necessary for the design of new devices in new circumstances. To that extent the book is forward looking.

In keeping with the central organizing principle of the work—variations on a simple theme—I have attempted to delineate the various categories into which thermosyphons fall. The book thus includes open and closed types, and deals with both gravitational and centrifugal force fields. It embraces single-phase and evaporative heat transfer in a variety of fluids and thus provides a thorough coverage of both forms of thermally induced circulation. It treats straight tubes and bent tubes, and deals with their orientation in the body force field. At appropriate points in the text, some ideas from thermodynamics and fluid dynamics are introduced and briefly discussed, partly to refresh the reader's mind and partly to emphasize the fundamental principles on which the *modus operandi* rests.

The structure of the book is clearly evident from the Contents list. For practical purposes, the chapters are organized more or less independently of each other, and this treatment within any given chapter may be sufficient for those readers who are interested in only one topic. However, I have also attempted to bind the whole together by making references between chapters, and between sections in a given chapter; thus, for example, the analogy between thermal buoyancy and Archimedean buoyancy is frequently mentioned so as to help bridge the gap between the single-phase and evaporative literature.

Finally, let me pay tribute to the many people who have helped me in this endeavour. To begin with, it is a pleasure to thank the staff of Oxford University Press, who facilitated the editorial work, not least by coaxing critical comments from reviewers. From among the latter, I would especially like to thank Dr Peter Whalley who read the entire manuscript and gave me the benefit of his experience. Several friends and colleagues were kind enough to read individual chapters: Dr M. Behnia, Dr C. A. Busse, Dr D. Chisholm, Professor E. Hahne, Dr B. S. Larkin, Dr S. Maezawa, Professor B. W. Martin, Professor W. Nakayama, Professor K. Negishi, Professor A. C. M. Sousa, Professor D. J. Steigmann, Professor C. L. Tien, Professor E. R. F. Winter, and Mr M Wu all provided me with critical comments which led to improvements in the manuscript. I am grateful for their advice and need only add that any remaining deficiencies are my responsibility alone. My students, R. Abdurahman, J. Fu, J.-C. Han, J. D. Kirchner, D. Ladoon, R. Natarajan, S. Park, G. A. Simpson, and L. Zhao, helped in numerous other ways and thus deserve my gratitude. The illustrations were prepared by Hiroshi Yokota of the Graphics Division of the University of Alberta, while the typing of successive drafts was done with care and patience by Helen Wozniuk, Gail Anderson, Betty-Ann Bloedorn, and Betty Saelhof of the Department of Mechanical Engineering. To all of the above I express by deep appreciation.

Edmonton G. S. H. L.
August 1991

ACKNOWLEDGEMENTS

Thanks are extended to the following organizations and individuals for their courtesy and kind permission to publish, in whole or in part, diagrams and photographs identified in the text and more fully in the Bibliography:

Academic Press; American Institute of Chemical Engineers; American Society of Mechanical Engineers; Arctic Foundations Inc. and Mr. E. L. Long; Cambridge University Press; Canadian Society for Mechanical Engineering; The Engineer; Furukawa Electric Company; Hemisphere Publishing Corporation; Her Majesty's Stationery Office; Institution of Mechanical Engineers; National Academy of Sciences; Oil and Gas Journal; Oxford University Press; Pergamon Press, Inc.; Royal Society; Professor J. Schenk; Springer-Verlag.

CONTENTS

NOMENCLATURE

a	radius
A	area, convective flux
Ar	Archimedes number
Bo	Bond number
c	specific heat
C	coefficient
d	diameter
D	diameter
Ek	Ekman number
Eo	Eötvös number
Eu	Euler number
f	function, fraction
F	area
Fm	figure of merit
Fr	Froude number
g	gravitational acceleration
G	acceleration
Gr	Grashof number
Gz	Graetz number
h	heat transfer coefficient
H	height, head
Ja	Jakob number
k	thermal conductivity, permeability
K	hydraulic conductivity, coefficient
Ku	Kutateladze number
l	length
L	length
m	mass, exponent
M	Mach number, mass
Mo	Morton number
n	number of moles
N	number
Nu	Nusselt number
$O()$	order of ()
P	pressure
Pe	Péclet number
Pr	Prandtl number
q	heat density (per unit area)
Q	heat
r	radius
R	radial coordinate, gas constant, resistance
Ra	Rayleigh number
Re	Reynolds number
$\mathfrak{R}$	eccentricity, universal gas constant
s	specific entropy
S	width
t	time, thickness, Rayleigh number divided by length–width ratio
T	temperature
u	normalized velocity in X direction
U	velocity (in X direction)
v	specific volume, filling ratio, normalized velocity in Y direction
V	velocity (in Y direction), volume
w	normalized velocity in Z direction
W	velocity (in Z direction), width
We	Weber number
x	normalized displacement in X direction, fraction
X	displacement coordinate
y	normalized displacement in Y direction
Y	displacement coordinate
z	normalized displacement in Z direction
Z	displacement coordinate

Greek letters

α	tilt angle
β	thermal expansion coefficient, angle
γ	ratio of specific heats, twist angle
Γ	mass flow rate per unit width
δ	thickness, displacement
Δ	increment length
ε	swirl angle
η	acceleration ratio
θ	temperature difference, angle
κ	thermal diffusivity
λ	latent heat
Λ	wavelength
μ	absolute viscosity, chemical potential

ν	momentum diffusivity
ξ	fraction convected
Π	osmotic pressure
ρ	density
σ	surface tension
Σ	summation
τ	shear stress
ϕ	normalized temperature difference
Φ	azimuthal coordinate
Ω	rotational speed

Subscripts

a	axial, air, annulus
a	radius a
A	adiabatic, annulus
b	bubble
B	base
cap	capillary
c	Coriolis, concentric
C	cooled, core
cr	critical
d	drag
D	diameter D
eff	effective
E	Ekman layer, evaporator
f	film
F	film
H	heated, height
I	liquid–vapour interface, intermediate
i	inner
in	inertial
l	laminar, liquid
L	liquid, length
m	mean, membrane
max	maximum
o	orifice, outer, operating
p	pore, isobaric
P	pool, pressure
r	radial, radius
s	sonic, surface, solid
sat	saturated
t	turbulent
T	temperature, thermosyphon
v	viscous, vapour
V	vapour, volume
w	wet
W	width, wall
X, Y, and Z	X, Y, and Z directions
δ	thickness δ
Λ	wavelength Λ
∞	bulk, terminal

Superscripts

$\bar{}$	average
$\dot{}$	per unit time
$'$	departure
c	characteristic
o	reference
α	exponent

1
INTRODUCTION

1.1 The nature of the tubular thermosyphon

At any given moment, huge amounts of thermal energy leave the earth's surface to be carried upwards by natural convection. In this process, energy is transported by bulk motion of the fluid within which it is temporarily stored. Typically, the process is much more effective than thermal conduction which relies on diffusive molecular interactions. The latter take place at molecular velocities which hover around the velocity of sound; but molecular motion is omnidirectional and thus creates no net energy flux in any particular direction unless the temperature, and hence the sonic velocity, varies in that direction. Convective velocities, on the other hand, are frequently much less than the sonic velocity but arise in circumstances where they are essentially unidirectional; moreover, they are capable of producing large vertical energy fluxes in the presence of small vertical temperature gradients. This distinction between the microscopic and macroscopic manifestations of molecular energy flux is at the root of many natural phenomena observed in the atmosphere, the oceans, and the core of the earth.

In the earth's gravitational field, the requirement for (upward) convection is twofold: the fluid density must vary, and the buoyancy forces thus produced must overcome any constraints. On the nominally horizontal surface of the earth, this commonly occurs when solar heating produces warmer air beneath cooler air. For small temperature gradients, heat is transmitted upwards by conduction, despite a small force imbalance. However, as the temperature gradient increases, conduction becomes increasingly inadequate as a transport mechanism, while the force imbalance grows. Ultimately, the imbalance becomes excessive and convective breakthrough occurs. The resulting cellular circulation pattern was first described in the pioneering studies of Thomson (1881) and Bénard (1900).

This example of natural convection draws attention to the essential feature of all such phenomena: namely, buoyancy or, to be more precise, thermally induced buoyancy. Buoyancy cannot exist if the fluid density is invariant, as may be demonstrated by considering the definition

$$\mathbf{F} = -\nabla P + \rho \mathbf{G} \qquad (1.1)$$

which combines the static pressure and body force terms in the equation of motion of a homogeneous fluid. The pressure gradient ∇P acting alone is most closely associated with forced convection, while the existence of a body force field creating an acceleration **G** gives rise to natural convection. The relation between the two terms becomes clearer when the pressure and density are both decomposed into two parts: a reference value corresponding to the fluid lying isothermally at rest; and the departure from this value. Thus, for example, for an incompressible fluid

$$\rho = \rho^{\circ} + \rho'$$

where ρ° is the density at a reference temperature T° and ρ' is the departure from ρ° caused by variations in temperature, while

$$P = P^{\circ} + P'$$

where P° is the hydrostatic pressure field at T° attributable to constraint in the presence of the acceleration field **G**, and P' is the departure caused by motion and variations in temperature. Since $\nabla P' = \rho'\mathbf{G} = 0$ at the reference condition, it follows that $\mathbf{F}^{\circ} = -\nabla P^{\circ} + \rho^{\circ}\mathbf{G} = 0$, and hence eqn (1.1) may be rewritten

$$\mathbf{F} = -\nabla P' + \rho'\mathbf{G} \tag{1.2}$$

in which the incompressible equation of state $\rho = \rho(T)$ enables the density departure to be approximated by

$$\rho' = \rho - \rho^{\circ} = -\rho\beta(T - T^{\circ})$$

where

$$\beta = -\frac{1}{\rho}\left(\frac{\partial \rho}{\partial T}\right)_{\mathrm{P}} \tag{1.3}$$

is the thermal expansion coefficient. From eqn (1.2) it follows that

$$\mathbf{F} = -\nabla P' - \rho\beta(T - T^{\circ})\mathbf{G} \tag{1.4}$$

which enables a new term $\beta\mathbf{G}\,(T - T^{\circ})$, an acceleration, to be defined. This is usually called the *thermal buoyancy*. Clearly, it vanishes if $\beta = 0$.

If the fluid is not homogeneous, as, for example, when a liquid is at or near thermal equilibrium with respect to its vapour, the principal cause of buoyancy is seldom the temperature field, even though it may be non-uniform. Under these circumstances, buoyancy is mainly attributable to the density difference between the phases, as may be illustrated by film condensation of a saturated, quiescent vapour on a vertical surface. Ignoring thermal buoyancy effects for simplicity, the functions F_X, where X is measured vertically downwards, reduce to

$$F_{X\mathrm{L}} = -\frac{\partial P_{\mathrm{I}}}{\partial X} + \rho_{\mathrm{L}} g$$

$$F_{XV} = -\frac{\partial P_V^o}{\partial X} + \rho_V g \tag{1.5}$$

for the liquid and for the vapour respectively. If the liquid film falls freely, and is thus unconstrained, $P_L^o = 0$ and therefore does not appear in the first of the above equations where P_I represents the interfacial pressure imposed on the liquid film by the vapour. From the second equation, hydrostatic equilibrium in the constrained vapour requires that

$$-\rho_V g = -\frac{\partial P_V^o}{\partial X} = -\frac{\partial P_I}{\partial X}$$

and hence

$$F_{XL} = (\rho_L - \rho_V) g$$

which permits the *Archimedean buoyancy* to be defined by $(\rho_L - \rho_V)g/\rho_L$. This acceleration, which vanishes when $\rho_L = \rho_V$, for example, at the critical point, is experienced by any liquid–vapour combination but will be used throughout this book in the context of condensation or evaporation, for which it is thermally induced. As with thermal buoyancy, the denser fluid is accelerated downwards; that is, in general, the less dense fluid tends to move in the opposite direction to the field acceleration vector.

In addition to density variations, there are two other important aspects of thermally induced buoyancy. The first of these is, of course, the body force field, which is unrestricted in the following pages, although the greatest emphasis will be placed on gravitational and centrifugal fields. The second originates in the relationship between the temperature field and the force field; more specifically, between the gradient vectors which represent them. Where the temperature gradient has a direction opposite to that of the force field vector, no thermal buoyancy[1] may occur; where the two vectors are normal to each other, convection will always occur, however small the temperature gradient. In general, the vigour, and indeed the very existence, of natural convection depends upon both the magnitude and the direction of the temperature gradient and force field vectors.

The very same buoyancy forces at work in the atmosphere and the oceans may be captured in a tube and used to great advantage in transferring heat from one place to another. The result is a thermosyphon, in which the phenomena mentioned above all take place in miniature. Despite these restrictions in scale, however, human ingenuity applied to the thermosyphon tube has created convective processes which are very efficient and very intense.

[1] That is, no 'positive' buoyancy; density variations which create 'negative' buoyancy act to stabilize the fluid. 'Positive' buoyancy implies a positive thermal expansion coefficient; obviously fluids with negative expansion coefficients exhibit the opposite behaviour.

As we shall see, heat flux densities which are thousands of times greater than those found in the atmosphere have been recorded. But intrinsic behaviour is the same, and this leads quite naturally to the definition of a thermosyphon as

> any fluid system in which the prevailing temperature and body force fields combine to produce a circulation the result of which is an enhanced thermal flux,

or, more simply,

> any fluid circulation produced by thermally induced buoyancy.

These descriptions emphasize the fact that the high heat transfer rate along the thermosyphon tube is attributable to circulation; specifically, to a *net* enthalpy flux down the temperature gradient. They also emphasize the cause of circulation, and thus point to the source of higher efficiencies: stronger force fields and greater density changes.

Either of the above definitions suffices to exclude non-circulatory convection; for example, near a heated plane surface. They also permit important subdivisions. In the *loop* thermosyphon, the hot and cold convective fluxes are physically separated; each flows unidirectionally in a separate tube, the tubes being connected in series to form the loop illustrated in Fig. 1.1(a). In the *tubular* thermosyphon, on the other hand, the two fluxes act in parallel within the same tube, as illustrated by the recirculating, or *refluent*, pattern shown in Fig. 1.1(b). Both types satisfy the requirement of circulation and both admit two further subdivisions which arise from closure, or its absence. For the tubular type, the subject of this book, we define the open system as one for which the tube is open at one end; the closed system, as its name suggests, is open at neither end.

It is now clear the the tubular thermosyphon owes its heat transfer ability to the buoyancy-induced circulation that follows from the imposition of a temperature difference between its ends. Typically, this ability is hundreds, and even thousands, of times better than the best metal conductor filling the same tube. The device thus acts as a 'superconducting rod'. The temperature difference may act in a simple manner as the result of conduction into, through, and from a homogeneous fluid fill which thereby experiences the thermal buoyancy forces associated with the internal distribution of temperature. Under these single-phase conditions, the enthalpy flux consists simply of sensible heat. Alternatively, the fluid may be made to evaporate at the hot end of the tube and condense at the cold end; the enthalpy change during vaporization then gives rise to a latent heat flux driven by Archimedean buoyancy. The evaporative mode usually yields a higher heat flux with a lower imposed temperature difference, and hence the heat transfer coefficient in a thermosyphon, as in nature, is usually higher if the fluid changes phase.

Whether the operating mode is simple (single-phase) or evaporative, the

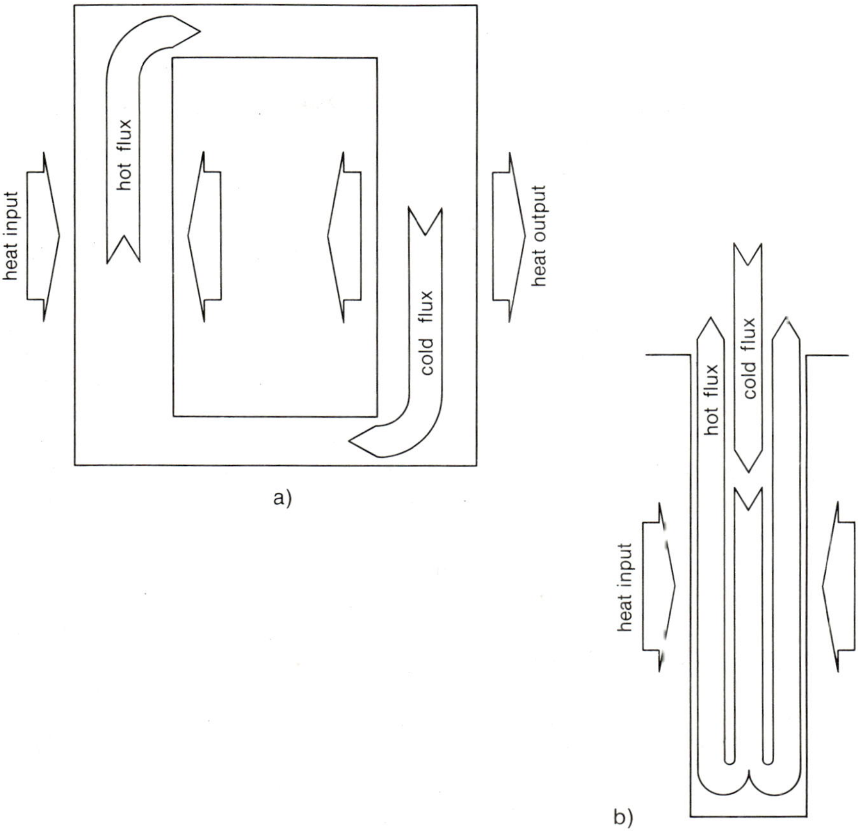

Fig. 1.1. Characteristic circulation patterns in simple thermosyphons: (a) simple, loop thermosyphon; (b) simple, tubular thermosyphon.

flow pattern in the tubular thermosyphon is always refluent, colder fluid moving up the temperature gradient while hotter fluid moves down it. The precise behaviour of these interacting streams varies enormously, as we shall see, but the basic requirement of refluence remains a characteristic feature, distinguishing it from the loop thermosyphon. A horizontal thermosyphon tube, for example, does not behave in the same manner as a vertical tube, but this reflects the difference between the buoyancy mechanisms acting over horizontal and vertical surfaces rather than a relaxation of the refluence characteristic. More generally, however, while the relative orientation of the temperature field and force field vectors does not destroy the counterflow behaviour in the tubular thermosyphon, it may introduce substantial alterations in the primary and secondary velocity fields.

Before closing this section it is worthwhile commenting briefly on auxiliary causes of circulation in thermosyphon tubes. These take a variety of forms, principal among which are: enhanced buoyancy, such as bubbling in liquids; concentration differences associated with osmosis; imposed pressure gradients introducing forced convection; and capillary pressure gradients caused by surface tension. With the exception of enhanced buoyancy, which is not thermally induced, none of these require either a body force field or a temperature gradient; none are intrinsic to thermosyphon operation. From time to time, however, they may be usefully employed as augmentation techniques. In recent years, the advent of the capillary-driven heat pipe has led some researchers, perhaps unaware of previous and parallel developments in thermosyphons, to describe the latter as a 'wickless heat pipe'. This is unfortunate, and is about as useful as describing a heat pipe as a 'gravity-free thermosyphon'. For the thermosyphon, the body force field is essential whereas the wick is not; the reverse is evidently true of the heat pipe. Likewise, other tubular devices which use additional flow-driving mechanisms may only be regarded as thermosyphons in so far as the fluid is dominated by thermally induced buoyancy according to the definitions given earlier.

1.2 Historical development: the Perkins tube

While the natural convection processes found in the atmosphere and the oceans have long been known to science, their application in engineering is comparatively recent. The first recorded example is attributed to an American, A. M. Perkins, in the early nineteenth century. Newly arrived in England, Perkins took out a patent for a system of hermetically sealed tubes by means of which he could transfer heat efficiently and cleanly from a furnace to wherever it was needed: in steam boilers, for sugar refining, the melting of alloys, etc. Figure 1.2 is an illustration of its original use in a steam boiler. Today, we would describe this as a single-phase closed-loop thermosyphon but in 1831, the date of the first patent, the terminology was unknown. At a time when high-pressure boilers were still experimental and their operation was plagued by scaling and fouling problems, this system represented excellence in design, the proof of which is to be found in its commercial success for over a century.

As the figure shows, the Perkins boiler physically separates the furnace from the evaporator, the thermal connection between them being provided by the thermosyphon. The expansion tube demonstrates a sound knowledge of thermophysics, even at this early date, by incorporating an air space which is about twice that needed for the liquid filling when it expands. The liquid was in fact water, and its thermal expansion coefficient was known approximately. Despite this allowance, the sealed air became compressed to pressures of several thousand pounds per square inch, thereby preventing the water in the thermosyphon tube from boiling, while still avoiding bursting. The hermetic

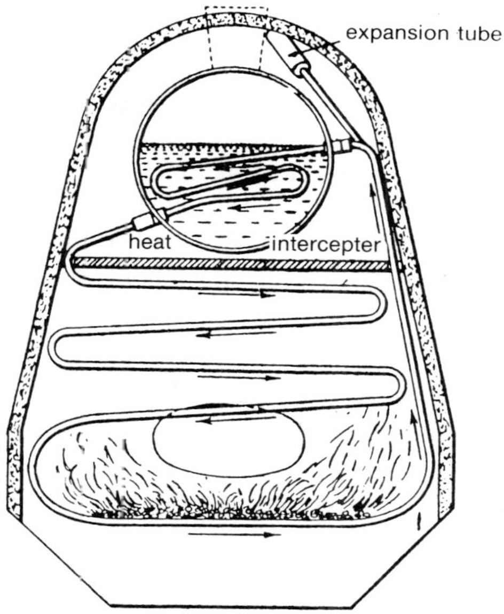

Fig. 1.2. Perkins' boiler (after King 1931).

seal also prevented accidental overheating and burnout which could occur in other systems as the result of gradual evaporation. The same concept was later used in a concentric tube boiler which was capable of raising superheated steam very quickly; so fast, in fact, that it was used to power a rapid-firing machine gun at the time of the American Civil War. More commonly, it was used in conjunction with marine steam engines where it provided 'a surprising economy of fuel . . . with lightness and compactness of form' (King 1931).

It is not known if Perkins ever used his single-phase heat transmission system in a straight tube, rather than a loop, but there are other examples of straight tubes being used with their contents boiling at one end and condensing at the other. In fact this system—the evaporative, tubular thermosyphon—eventually became known as the Perkins tube. In the context of steam generation, it consisted of a small quantity of fluid (water) sealed inside a tube, one end of which projected into the firebox while the other extended into the steam drum. With the firebox positioned below the steam drum, the water inside the tube boiled at the bottom and condensed at the top, thus transmitting the latent heat of evaporation as it circulated. This is the first known example of a tubular thermosyphon in commercial use.

Because the water was sealed within the tube, this system had all the advantages of the successful single-phase system; in particular, it facilitated the control of scaling and internal pressure. It also had the additional merit of replacing sensible heat with latent heat. This feature, and the amount of filling required to produce it efficiently, were carefully considered (King 1931). According to a Perkins[2] patent of 1836, each tube was 'to have a small quantity of water depending upon the degree of pressure required by the engine; and I recommend that the density of the steam in the tubes should be somewhat more than that intended to be produced in the boiler'. The pressure inside the tube was again controlled, this time by the relationship between the saturation pressure and temperature of the filling. Perkins goes on to specify for atmospheric boilers 'that the quantity of water to be applied in each tube is to be about 1:1800 part of the capacity of the tube; for a pressure of 2 atmospheres to be 2:1800 parts; for 3 atmospheres, 3:1800 parts, and so on'. As we shall see in Chapter 3, these are remarkably small fillings, and they demonstrate a clear intention not just to produce boiling and condensation in the same tube but to separate the two effects into two different regions with the space between them serving as a thermal conduit. This idea, with its attendant flexibility, is a further example of good design. Its earliest applications were in locomotive boilers and superheaters, which were noted for their cleanliness.

As experience with the device grew, improvements began to follow (Perkins[3] and Buck, in their patent of 1892). Among these is the treatment of filling technique in which the presence of incondensable gases (in this case air) is clearly seen as undesirable. It was therefore recommended that the air be purged by boiling off some of the filling before each tube was sealed. It was also recognized that since different liquids boil at different temperatures, the filling could be chosen to suit the operating temperature. By this means the range of the device could be extended not only upwards to even higher temperatures but downwards to include the freezing range below 0 °C. There was also a keen awareness of the effect of geometrical configuration on performance in general, and on safety in particular. Specifically, much attention was paid to the importance of installing or refitting the tubes at the correct angle and with the correct tube length exposed in the furnace; assembly and maintenance were thus facilitated.

The applications cited in the patent of Perkins and Buck (1892) are wide ranging. Apart from the original use in steam generation, and the evaporation of other liquids, they include domestic heating and the warming of greenhouses. In this moderate temperature range, the device was evidently seen in much the same light as the conventional 'radiator' supplied with either hot water or steam. With single glazing as the norm of the times, tubes were also recommended for the prevention of window fogging; this was accomplished simply by running the condensing section along the bottom of the

[2] Jacob Perkins, the father of A. M. Perkins.
[3] Ludlow Perkins, the son of A. M. Perkins.

window thus creating a warm air current on the inside of the pane. Similarly, they could be used in drying processes on a commercial scale. Claims were also made for their efficacy in the removal of heat from dairy products and from the cooling water of internal combustion engines, novel though they must have seemed at the time.

Perkins tubes were also proposed as the basis of heat exchangers or, more precisely, as the means by which heat transfer between two fluid streams could be facilitated. The particular application cited in the patent is the recovery of waste heat from a blast furnace by using the tuyere to preheat the incoming blast of air. This type of application may have caught the attention of Gay (1929) who also filed a patent in which the heat exchanger shown in Fig. 1.3 was proposed. This not only incorporates the extended surface area which accompanies the tube lengths themselves, but recognizes the secondary (though substantial) effect of fins attached to the outside of the tubes. In this, as in most other deployments seeking high heat transfer rates, the evaporative section is directly beneath the condensing section.

An example of inclined Perkins tubes is provided by the baking oven presented at the Paris Exhibition (1867). This was a portable bread oven designed for use by the British Army in the Victorian era. For this particular application, the rate of heat transfer is less important than fuel efficiency which was greatly enhanced, presumably because the separation of heat source and sink, so characteristic of thermosyphon usage, permitted the more efficient thermal design of both. In later years, and particularly in large establishments using gas or oil as fuel, the design took the form illustrated in Fig. 1.4 from which it can be seen that the tubes barely lie above the horizontal. Also noteworthy is the length of the evaporator in relation to that of the condenser. A typical ratio was about 0.02:1, with a connecting 'adiabatic' section which was sometimes longer than the evaporator. This is significantly different from the heat exchanger design of Fig. 1.3 for which the heat flux *density* is about the same for both sides. For the oven, the high-temperature, high-density heat transfer has been converted to a relatively low-temperature, low-density heat transfer. Because of the evaporation/condensation area ratio the heat flux is divergent, and the device acts as a thermal attenuator. More will be said about this feature in Section 1.4.

Two important scientific advances were also made during these years, probably during the development of the oven. The first of these was the growing awareness of the filling quantity as a design parameter. Perkins and Buck had noted that if the quantity was too small it may evaporate completely, leading to the condition of dryout and the likelihood of dangerous overheating. In later designs they proposed a filling of 32% of the tube volume because this quantity of water alone ensures that neither the liquid fraction nor the vapour fraction vanish if the critical pressure and temperature are approached (Moorhouse 1960). The second advance was a realization that the

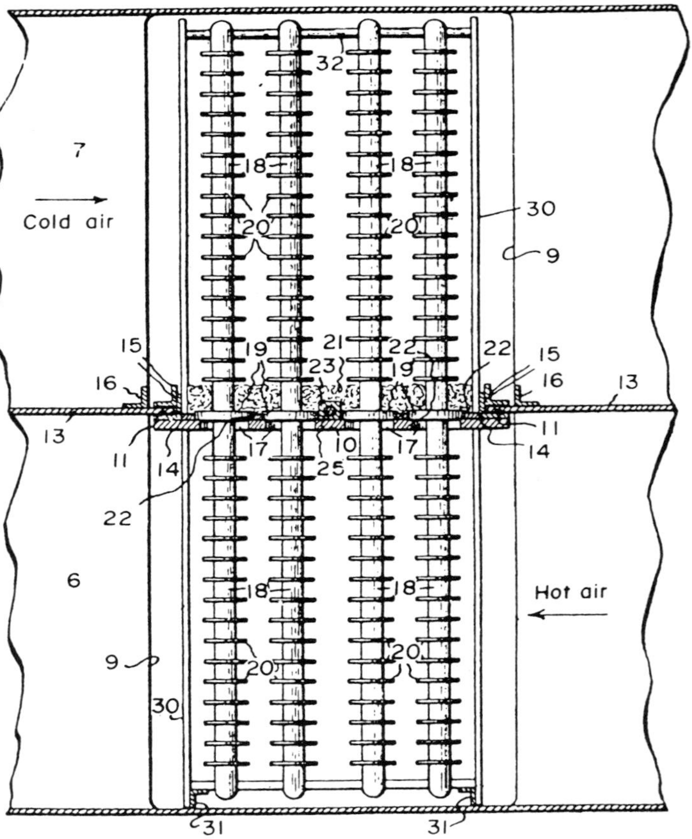

Fig. 1.3. Thermosyphon heat exchanger proposed by Gay (1929).

refluent nature of the tubular thermosyphon may give rise to significant flow opposition or internal impedance; this compromises the ability of the device to operate under high heat loads, which require large convective fluxes. In the absence of flow visualization techniques, such a realization reveals considerable insight with respect to the complex process of entrainment. Their solution was to take advantage of geometric flexibility by introducing U-bends where necessary, thus making the tube into a loop and thereby creating a one-dimensional flow.

A later application, this time for cooling, is illustrated in Fig. 1.5. As indicated, this device was designed to transfer heat from the cooling water of a car engine to the ambient air. It is a variation of the scheme proposed in Fig. 1.3, with two main differences. The hotter fluid is water circulating[4] over the

[4] It is not clear whether by forced or natural convection.

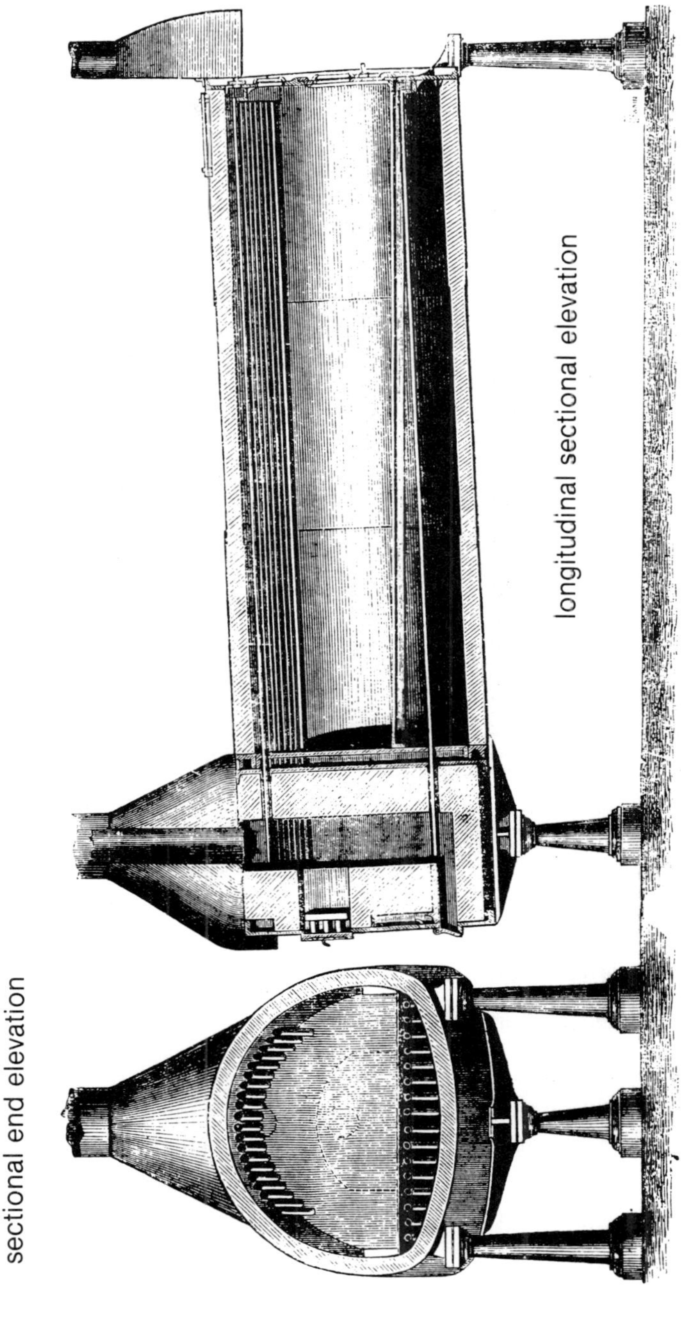

Fig. 1.4. Perkins' baking oven presented at the Paris Exhibition in 1867.

shorter evaporator sections of the thermosyphon tubes; heat transfer coefficients on the water side of the tubes were therefore likely to be greater than on the air side where augmentation was obtained partly by the fins shown and partly by the motion of the vehicle. The other important difference is the shallow angle which the tube banks make with the horizontal.

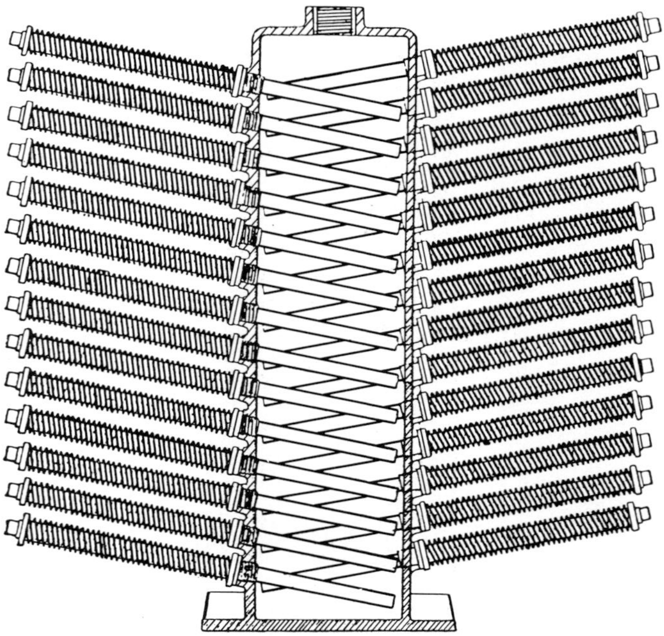

Fig. 1.5. The Critchley–Norris car radiator.

1.3 Modern development

Many of the needs cited in early thermosyphon patents are still with us today. Internal combustion engines still need cooling; rooms still need heating; bread still needs baking; and the demand for safe, clean, and economical steam seems to be as great as ever. In general terms, the transmission of heat from a source at one point to a sink at another continues to be a widespread problem which may often be solved with some type of thermosyphon. The examples described above are illustrative of the potential of the thermosyphon but they certainly do not constitute an exhaustive list. They are limited, for instance, to gravitational fields. Many more applications have since appeared and these too are worthy of discussion. However, before attempting a brief overall survey, particular attention must be given to two special applications each of which has acquired its own history of several decades: these are the cooling of gas turbine blades and the freezing of soil and water.

1.3.1 *Turbine blade cooling*

The idea of using a thermosyphon to cool gas turbine blades appears to have originated with Holzwarth (1938) at a time when maximum cycle temperatures were comparatively low. It was recognized that centrifugal accelerations in this machine are of the order of $10^4 g$ suggesting that density variations in a rotating fluid may be expected to give rise to substantial natural circulation velocities. Thus, if a turbine blade is first hollowed and then filled with a fluid, any heat transferred to the blade by the surrounding hot gas will be conducted into the fluid thereby introducing radial buoyancy forces. A schematic of the system tested by Schmidt (1951, 1962) is shown in Fig. 1.6. This is an open thermosyphon in which thermal buoyancy gives rise to a convective flux towards the blade root, that is towards the mouth of each of the cooling holes. The thermal energy is then removed either by removing the coolant, for secondary heat exchange externally, or by evaporation at the coolant's free surface, the vapour being withdrawn and replaced by coolant make-up.

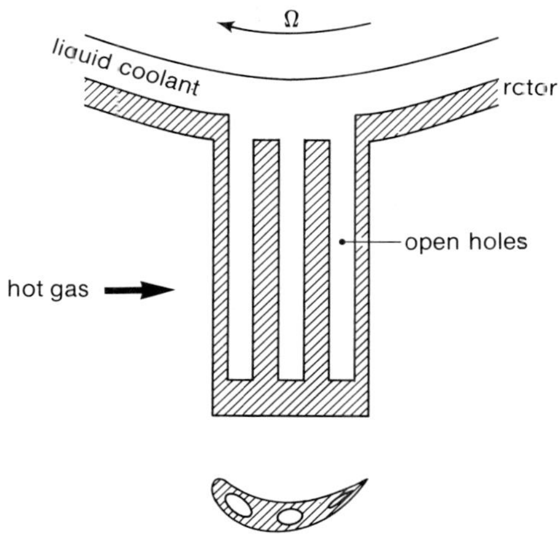

Fig. 1.6. Schematic of Schmidt's blade cooling system.

Schmidt chose to conduct his tests with a water filling under conditions close to the critical point, and was able to demonstrate that turbine inlet temperatures of at least 1200 °C are practical. This seminal piece of work spawned many later studies of the Holzwarth–Schmidt system[5] as the result of

[5] Japikse's (1973) paper contains an excellent historical review of this and other blade cooling systems.

which its merits and demerits gradually became clearer (Robinson 1950; Freche and Diaguila 1950; Friedrich 1962; Ivanov 1967). It has several disadvantages: large internal hydrostatic pressures; vibration from coolant instability around the root (mouth); and blockage attributed to oxidation. Taken together, these difficulties gradually turned the attention of designers towards the closed system depicted in Fig. 1.7 (Bayley 1955*a*; Cohen and Bayley 1955). This arrangement has the same advantages which sealing conferred on the Perkins tube; namely, control of internal scaling, and the separation of the primary and secondary coolants. Flexibility and reliability may thus be improved. Also, the internal pressure is controlled by the density and radial extent of the filling. Here again, greater flexibility is introduced; metallic liquids are denser than non-metallic liquids but they have a much greater thermal conductivity; and in the evaporative mode the filling need not extend over the full length of the cooling hole, as the Perkins tube had previously demonstrated.

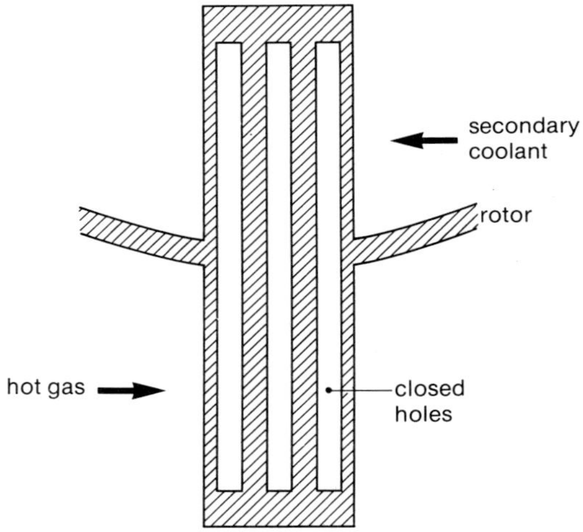

Fig. 1.7. Schematic of closed thermosyphon cooling system for gas turbine blades.

A closed, single-phase system similar to this has been installed in a marine gas turbine and successfully run for 400 hours during eight of which it was subjected to the same maximum turbine inlet temperature as the Holzwarth–Schmidt system but produced a much lower blade temperature (Brown, 1957; Ogale 1968). In this particular installation, the blade was completely filled with a sodium–potassium eutectic, and forced convection water cooling was applied at the root.

As with any other internal form of liquid cooling, the thermosyphon has the disadvantage of providing only limited ability to cool the slender trailing edge of the blade. Its most obvious advantage is the absence of any pumping equipment with its attendant power penalty. Equally important is the elimination of aerodynamic penalties attending the use of film or transpiration cooling in which compressor air is injected into the main gas stream. However, the thermosyphon is bound to introduce a thermodynamic penalty because heat is transferred out of the working gas. While this is not large, and is of course more than offset by the thermodynamic advantage of higher gas temperatures, it must be weighed carefully along with the power required to circulate the secondary coolant. In principle, the thermosyphon solution is very attractive if the heat transferred from the gas can be used in such a way as to recover some of the energy lost; part of it may then be used for secondary pumping.

Japikse (1973) has suggested that the recovery of lost work may be worth considering with the Holzwarth–Schmidt system; for example, vapour from the coolant itself, that is steam, could be used to drive a companion steam turbine. Majcen and Šarunac (1984), on the other hand, indicate that a similar benefit is obtainable from the closed thermosyphon when it is incorporated in a double gas turbine cycle. The arrangement they suggest is illustrated in Fig. 1.8 which reveals how two simultaneous gas cycles, one hot and one cold, may be connected thermally by heat transfer along blades which stretch between the two gas streams, the hotter being located further from the axis of rotation. A suitable blade design is shown in Fig. 1.9. The main gas stream, at a temperature of about 1250 °C, passes over the outermost portion of the twin aerofoil blade which is filled with a sodium–potassium alloy. The heat gained in this region of the closed thermosyphon is subsequently given up to the 'secondary coolant' flowing over the innermost portion. This 'coolant' is air being reheated; it subsequently expands and thereby helps offset the overall thermodynamic loss. A similar arrangement could be used as a regenerative heat exchanger located immediately downstream of the compressor in a conventional gas turbine cycle.

1.3.2 *Freezing of soil and water*

In many northern nations, growing demand for the basic amenities of community life, coupled with a burgeoning demand for fossil fuels, has led to various forms of construction in, and on, ground which would otherwise be frozen for all or part of the year. Thawing beneath a heated building, for example, or adjacent to a buried hot-oil pipeline, illustrates a general problem. The structural integrity of frozen soil is much greater than that of thawed soil, which is unable to support the same load; in addition, thaw settlement may occur, thereby causing displacement of the foundations to such an extent that the structure is rendered inoperative, if not dangerous. One solution to this type of problem is to maintain the soil in a frozen state all year round by

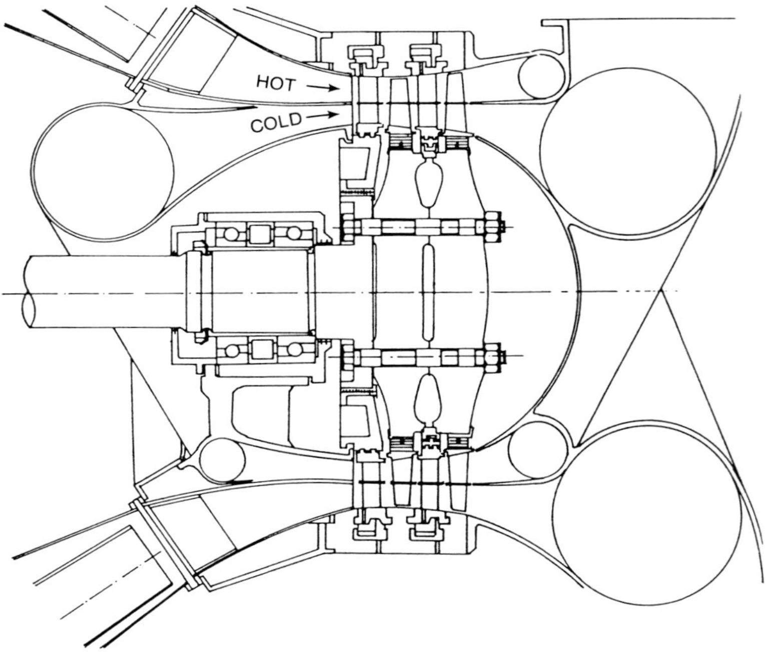

Fig. 1.8. Double-cycle gas turbine (after Majcen and Šarunac 1984).

boosting the natural loss of heat to the atmosphere during the winter period. This may be done with a tubular thermosyphon when its exposed upper portion is colder than its buried lower portion; the heat then removed both freezes and subcools the moist soil. During the summer, the temperature gradient is reversed, the system ceases to convect, and the consequent conduction downwards has only a limited ability to thaw back the soil. Acting thus as a one-way heat valve, the thermosyphon progressively freezes and subcools more soil each year.

This application of the gravitational thermosyphon is an excellent example of working with, rather than against, nature. On the North American continent, the first proposal for its use as a construction pile has been credited to Erwin Long (1963) whose design ingenuity was scarcely exceeded by the ingenuity he displayed in naming it: he called his device the Long thermopile! Similar devices proposed by others have been called cryopiles, cryosyphons, cryoanchors, etc. Here we will use the more general term *cryosyphon*.

Figures 1.10, 1.11, and 1.12 illustrate typical uses of the cryosyphon. In Fig. 1.10, the device is integrated into the piles which support a building. This is a typical situation in which the buried portion of the tube is designed as an

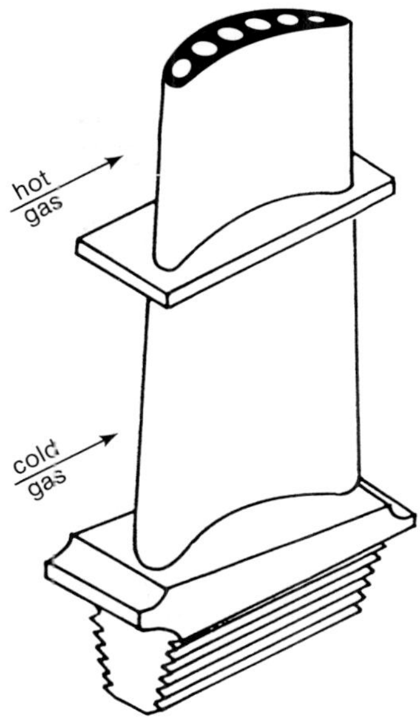

Fig. 1.9. Twin-aerofoil gas turbine blade (after Majcen and Šarunac 1984).

evaporator while the exposed portion, fitted with cooling fins, acts as a condenser; usually, the evaporator is much longer than the condenser, precisely opposite to the situations found in early applications of the Perkins tube. Figure 1.11 illustrates the many variations on this theme. Short condensers are also evident in Fig. 1.12 which shows the device being used in protecting an airstrip against summer thermal erosion. Also evident in Fig. 1.12 is a significant departure from the original linear tube design. The figure shows right-angled and cranked configurations, both being employed to create a substantial area of frozen ground near the surface. Long, almost horizontal, evaporators are commonly used to guard against permafrost degradation beneath heated buildings; a similar protection is often given to roads, which frequently have heat-absorbing surfaces. Construction is frequently mounted above a fill laid directly on top of the permafrost. Cryosyphon evaporators laid in the fill prevent any downward heat flux from reaching the permafrost beneath, thereby preserving its integrity.

Closely related to the freezing of soil is freezing in large bodies of water such as lakes or rivers. Typically, the purpose is very different and arises from the need to separate one submerged region from another. For example, construction on a river bed may have to take place inside a circular cofferdam or caisson, the walls of which can be made of solid ice produced by a ring of cryosyphons. Another example is shown schematically in Fig. 1.13. This is a

Fig. 1.10. Building piles fitted with cryosyphons. The vertical condensing sections protrude from the pile sides (courtesy E. Long).

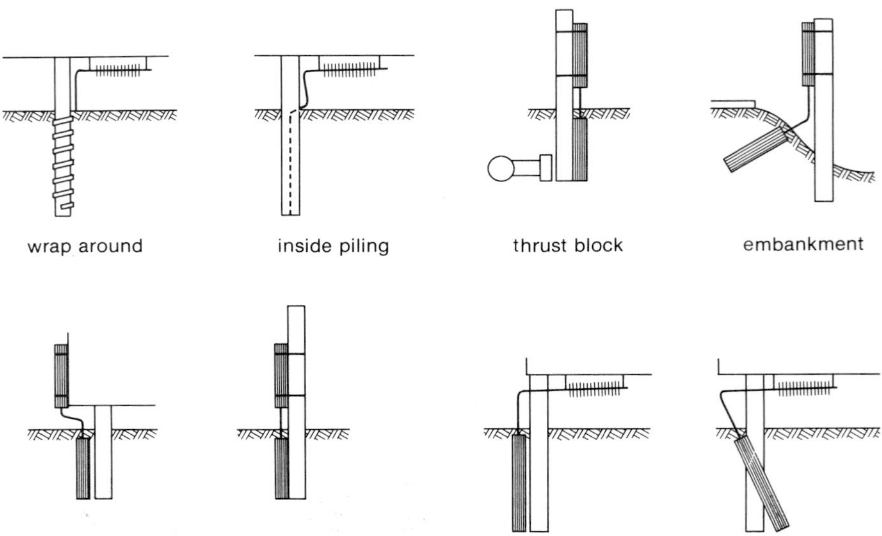

Fig. 1.11. Various schemes for freezing soil around a pile (after Waters 1974*b*).

Fig. 1.12. Cryosyphons installed beneath an airstrip. The evaporators are completely buried; only the condensers protrude (courtesy E. Long).

row of cryosyphons strung across a river in such a way that the ice growing on their submerged sections during the winter eventually forms a continuous wall thus blocking, or partially blocking, the flow of water. During the spring runoff period, this ice–tube structure may help control potential flooding downstream. By summer, the wall has melted and the river resumes normal hydraulic behaviour; ecological integrity is thus preserved (Lock 1968*a*,*b*).

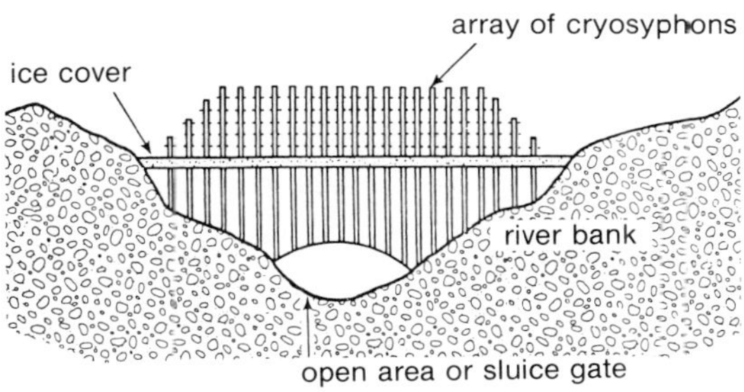

Fig. 1.13. Schematic representation of ice wall for control of spring runoff (after Lock 1986*a*,*b*).

1.3.3 *Current and future applications*

It is certain that the tubular thermosyphon will continue to provide a solution to blade cooling and geothermal freezing problems, but there are many other applications which, although more specialized, are also worthy of discussion. In general, the concept lends itself to a very wide range of situations in a variety of contexts: emergency cooling of nuclear reactors; capture and storage of solar energy; one-way heat walls (honeycombed with thermosyphons); cooling of electronic packages; geothermal and hydrothermal (e.g. OTEC) energy transmission over substantial variations in elevation; cooling of electric motors and machine bearings; anti-icing protection; waste heat recovery; and so on. Clearly it is impossible to describe every application in detail in a short space. This subsection will therefore be limited to a brief outline of selected, though representative, applications.

To begin with, it is worth noting that thermosyphons continue to play a role in heat exchanger design, provided that the hot fluid stream does not lie above the cold stream.[6] Given the ability of the thermosyphon to transfer heat effectively over some distance, the two fluid streams do not have to be adjacent to each other, but in practice they frequently are. Uses include waste heat recovery from boilers, furnaces, etc. and from the exhaust of large ventilated spaces (Larkin 1983*a*; Reay 1977; Barratt and Henderson 1979; Littwin and McCurley 1982; Vasiliev 1989; Ma and Hou 1989; Polášek 1989; Dobran 1989; Groll 1989). Figure 1.14 shows an example of a waste heat boiler consisting of 528 carbon steel/water thermosyphon tubes. Steam is generated by exhaust gases which enter at 650 °C and leave at 180 °C. In the same context, thermosyphon heat exchangers are widely used as preheaters and economizers. They are also used in the recovery of waste heat from car exhausts, in air-conditioning systems, and in agriculture. The range and versatility of these heat exchangers is well illustrated by the cassette design shown in Fig. 1.15.

An alternative form of heat exchanger (Niekawa *et al.* 1982) is shown diagrammatically in Fig. 1.16. In this scheme, the thermosyphons rotate about a horizontal axis parallel to their own, that is eccentrically. As indicated in Fig. 1.16(a), air and exhaust gas flowing in opposite directions enter the rings of thermosyphons more or less axially, but they are separated by a rotating plate through which the thermosyphon tubes penetrate. The gases discharge peripherally. In Fig. 1.16(b), the air side of the heat exchanger is replaced by a rotating evaporator, with the water being supplied along the shaft so that it is preheated before being sprayed on to the condensing sections of the thermosyphons. While the flue gas exhaust is essentially the same as in Fig. 1.16(a), the vapour discharges axially not peripherally. A photograph of such a waste heat boiler is given in Fig. 1.17, the vapour outlet being in the right foreground.

This application leads us naturally to the use of thermosyphons in meeting the cooling requirements of rotating machinery. Commonly, these stem from two major heating sources: the hot gases mentioned above and previously

[6] In fact, 'anti-gravity' thermosyphons would permit heat transfer even in this condition but no application has been cited.

Fig. 1.14. Waste heat boiler using vertical thermosyphons (from Ma and Hou 1989).

Fig. 1.15. Cassette heat exchanger showing trays of thermosyphon tubes which stretch across the two halves of a rectangular duct (from Groll 1989).

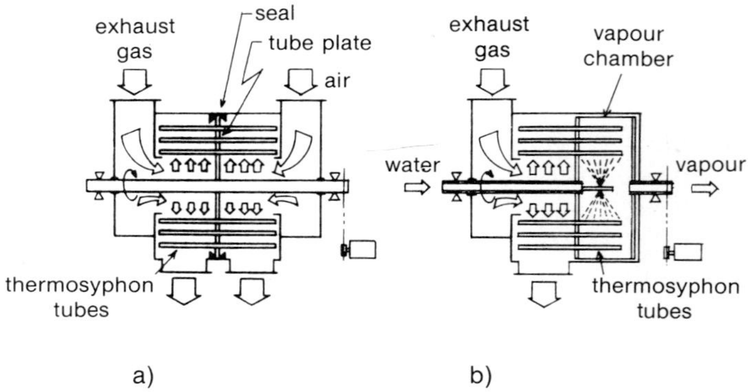

Fig. 1.16. Diagrammatic of rotating thermosyphon heat exchanger: (a) gas–gas type; (b) gas–vapour type (after Niekawa *et al.* 1982).

Fig. 1.17. Waste heat boiler using rotating thermosyphons (see previous figure) (courtesy Furukawa Electric Co. Ltd).

during reference to the gas turbine; and the Joulean dissipation that accompanies the operation of electrical motors and generators. In the second category, the heat released in the armature windings may be removed axially, either by using eccentric holes (which function much like the thermosyphon tubes shown in Fig. 1.16) or by means of a single concentric hole (Polášek 1973, 1989; Gray 1969; Nakayama *et al.* 1984*a*; Groll 1989; Ma and Hou 1989). The latter configuration is shown schematically in Fig. 1.18; in the evaporative form, tapering of the condenser facilitates the axial return of the condensate to the evaporator (Schneller *et al.* 1984; Vasiliev 1989). Other applications of axial thermosyphons are too numerous to describe in detail. They include: cooling of hydraulic turbine bearings (Nakayama *et al.* 1984*a*); cooling of high-temperature fan bearings (Schneller *et al.* 1984); cooling of lathe spindles (Vasiliev 1989); and the cooling of machine tools (Ma and Hou 1989).

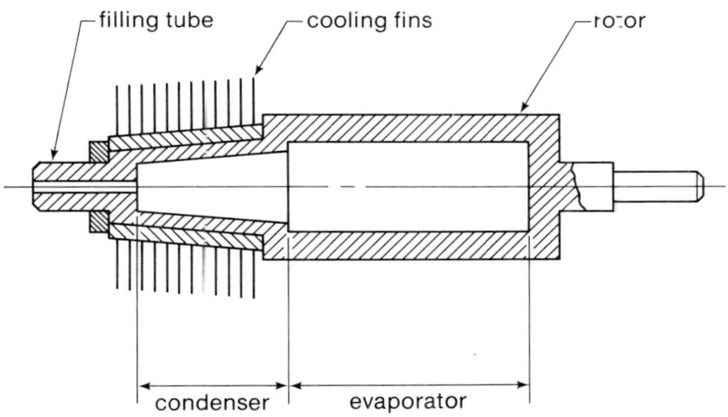

Fig. 1.18 Schematic of concentric thermosyphon in a rotor (after Schneller *et al.* 1984).

Returning to gravitational systems, and in particular to those designed to supply rather than withdraw heat, we find two categories dedicated to combating the formation of ice on engineering structures. In one category, the ice originates in a cloud of supercooled water droplets which freeze after impact on any surface moving relative to the cloud; this surface may be an aircraft wing, an overhead cable, a ship, etc. Anti-icing and de-icing[7] of marine structures, both floating and offshore, provide further illustrations of the thermosyphon's ability to transfer heat from a source to a sink (the ice) when these are not adjacent to each other. The source, which must not lie directly above the ice, may be an electric heater, an engine exhaust, etc. The heat is

[7] These terms usually mean prevention and removal of ice, respectively.

transmitted to those structural components which are most susceptible to icing and which, if not kept ice free, could compromise operations and, in an extreme situation, threaten the security of the entire structure and all the personnel on board (Lock 1972). Figure 1.19 illustrates a representative scheme. A more familiar situation arises from the need to keep railway points free of ice (Vasiliev 1989). Figure 1.20 illustrates the use of exceptionally long horizontal condensing sections to achieve this objective. A particularly interesting de-icing scheme for a buoy (Larkin and Dubuc 1976) is shown in Fig. 1.21. Since no external power is normally available in this application, thermal energy must enter the evaporator from the sea itself, the small temperature excess above the equilibrium freezing temperature of the water being sufficient to prevent, or at least mitigate and delay, ice accretion and, in any event, to speed the de-icing process.

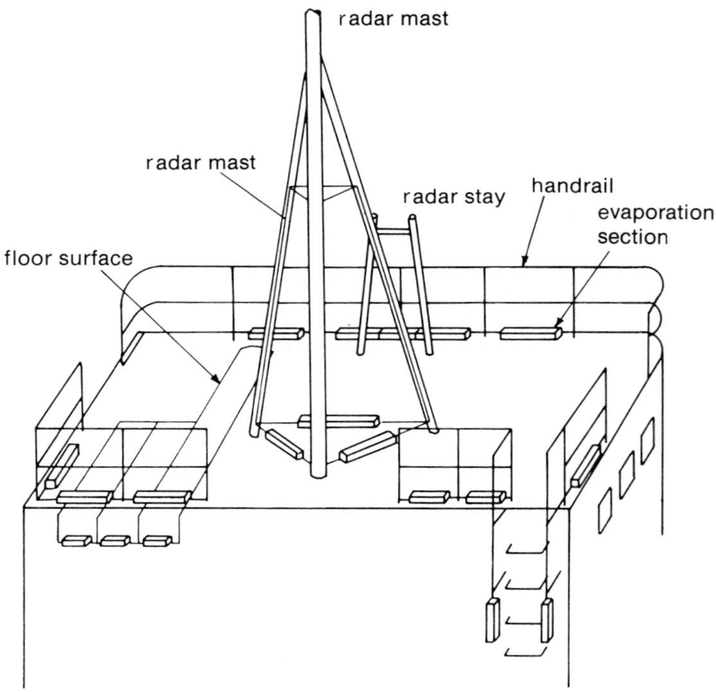

Fig. 1.19. Elements of a ship's superstructure which may be fitted with thermosyphon de-icing capability (after *Zosen* 1981).

The second type of ice which has an adverse and, in this instance, pervasive effect on human society is snow. Although this too originates in the atmosphere, its aggregate structure on the ground changes enormously as it undergoes metamorphosis from a light, porous medium which is easily moved

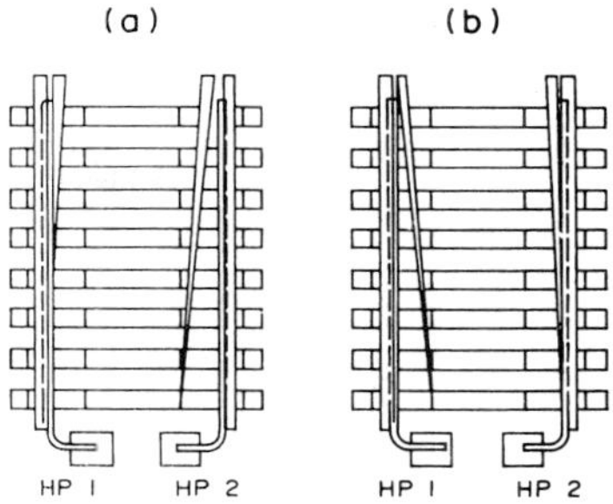

Fig. 1.20. Thermosyphons used in de-icing of railway points. Only the horizontal condensing sections are visible (after Vasiliev 1989).

and sheared, to one which is essentially solid ice. Melting and refreezing on a roof may thus lead to excessive structural loads, not to mention the threat posed by an overhang, which can break away without warning. On roads and access driveways, the presence of snow in various stages of metamorphosis may seriously limit the safe operation of vehicular traffic (Wilson *et al.* 1978; Tanaka *et al.* 1982). One solution is to tap thermal energy stored in the earth and thus provide heat immediately below the traffic surface through thermosyphons. Figure 1.22 illustrates the installation of thermosyphon condensing sections in a bridge deck with the evaporators being sunk into the neighbouring ground. This type of application usually requires the thermosyphon tube to be in a non-linear form. The elbow configuration seen in Fig. 1.22 (and earlier) is also evident in Fig. 1.23 which depicts a different purpose for which the elbow must be inverted.

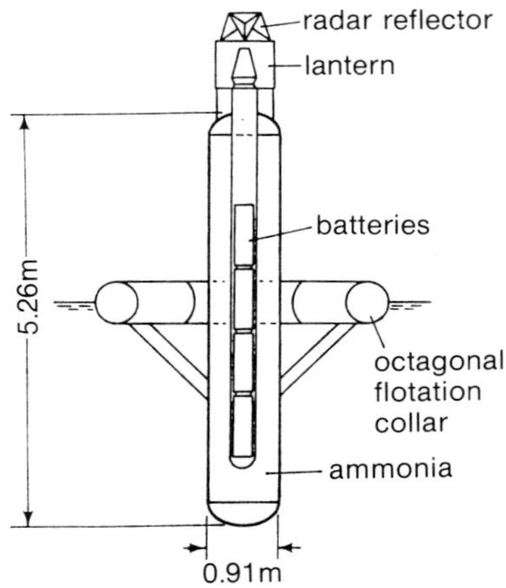

Fig. 1.21. Self-de-icing navigation buoy (after Larkin and Dubuc 1976).

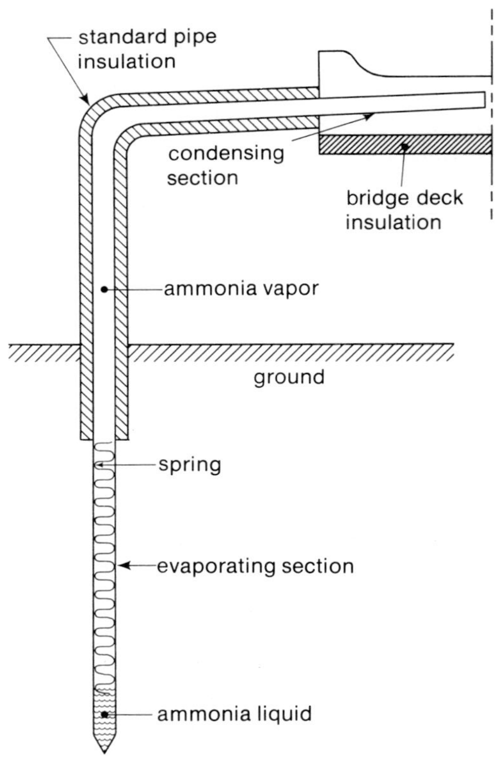

Fig. 1.22. De-icing thermosyphons installed in a bridge deck (after Wilson *et al*. 1978).

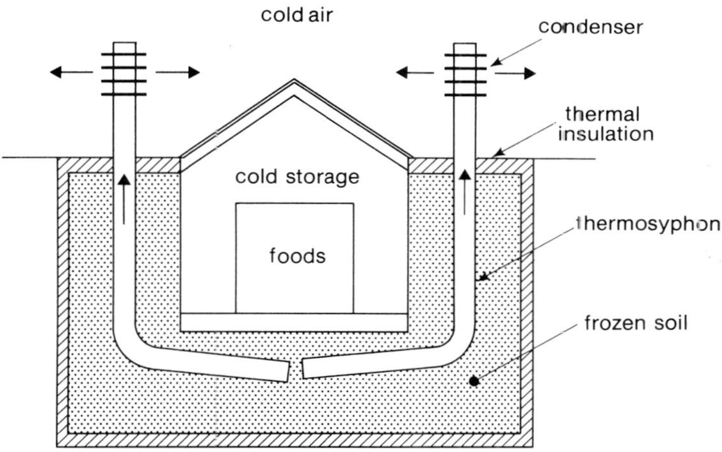

Fig. 1.23. The use of thermosyphons in the cold storage of foods (after Fukuda *et al.* 1990).

Finally, a brief word about the use of thermosyphons in the capture and storage of solar energy. Typically, they provide a means of heating water, the acquired sensible heat being stored elsewhere. Solar energy may be incident directly on the thermosyphon tubes, or may first be required to cross through a surrounding evacuated glass tube which reduces heat loss. Figure 1.24 illustrates a scheme in which the hot-water tank is simply located at the top (Ma and Hou 1989). An alternative storage scheme is depicted in Fig. 1.25. This is noteworthy on two accounts. Firstly, the energy storage consists of latent, not sensible, heat. Secondly, the non-linear tube arrangement seen earlier in elbow form is taken a step further.

1.4 Functional analysis

It is now apparent that the vast majority of thermosyphon applications are based on its ability to transmit substantial amounts of thermal energy over some distance separating a heat source from a heat sink. Even so, there are other functions which are worth mentioning, partly because they are important in their own right, but largely because they afford the opportunity to develop a broader perspective within which the tubular thermosyphon may be seen more clearly as a versatile thermal element. Consider the thermal resistors shown in Fig. 1.26 in which Fig. 1.26(a) is an initial representation: a heat source at temperature T_H^* is separated from a heat sink at temperature T_C^* by a thermal resistance R_{HC}^* considered to be too large for the situation in hand. The difficulty is removed by the introduction of a tubular thermosyphon (Fig. 1.26(b)) with a resistance $R_T \ll R_{HC}^*$, *provided also that the flanking*

Fig. 1.24. Thermosyphon solar collectors (from Ma and Hou 1989).

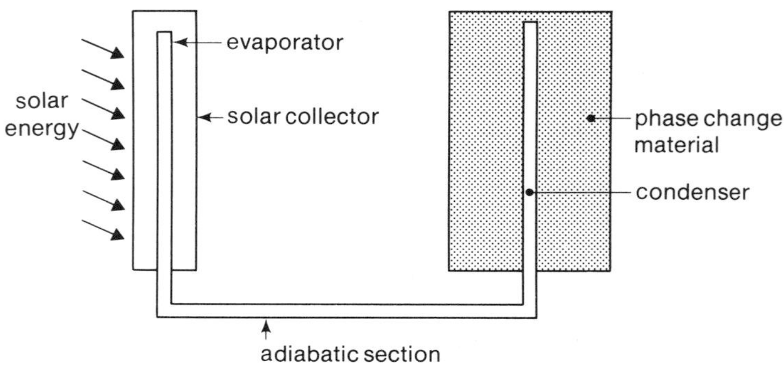

Fig. 1.25. The use of thermosyphons in the storage of solar energy (after Xia *et al.* 1990).

resistances at both ends of the thermosyphon are comparable with R_T. Obviously, the smaller R_H^* and R_C^* the closer the tube wall temperatures T_H and T_C will approach T_H^* and T_C^*, respectively, and the greater will be the heat flux transmitted by the thermosyphon. Within the thermosyphon itself, the overall resistance may be divided between the heated and cooled ends, as suggested in

Fig. 1.26(c). The intermediate temperature T_I is not equal to the mean temperature T_m unless $R_H = R_C$.

The heat flux transmitted under steady conditions is thus given by

$$\dot{Q} = \frac{T_H - T_C}{R_T} = \dot{Q}(T_H, T_C) \tag{1.6}$$

since R_T is a function of both T_H and T_C for any given thermosyphon. From this description in terms of the primitive variables T_H, T_C it follows that

$$d\dot{Q} = \left(\frac{\partial \dot{Q}}{\partial T_H}\right)_{T_C} dT_H + \left(\frac{\partial \dot{Q}}{\partial T_C}\right)_{T_H} dT_C$$

or, using eqn (1.6),

$$d\dot{Q} = \frac{1}{R_T}\left[1 - \dot{Q}\left(\frac{\partial R_T}{\partial T_H}\right)_{T_C}\right] dT_H - \frac{1}{R_T}\left[1 + \dot{Q}\left(\frac{\partial R_T}{\partial T_C}\right)_{T_H}\right] dT_C. \tag{1.7}$$

Alternatively, the description may be given in terms of the 'natural' variables[8]

$$\theta = T_H - T_C$$

and

$$T_I = T_I(T_H, T_C)$$

so that eqn (1.6) must be restated as

$$\dot{Q} = \frac{\theta}{\mathfrak{R}_T} \tag{1.8}$$

where $\mathfrak{R}_T(\theta, T_I) = R_T(T_H, T_C)$. Hence

$$d\dot{Q} = \frac{1}{\mathfrak{R}_T}\left[1 - \dot{Q}\left(\frac{\partial \mathfrak{R}_T}{\partial \theta}\right)_{T_I}\right] d\theta - \frac{\dot{Q}}{\mathfrak{R}_T}\left(\frac{\partial \mathfrak{R}_T}{\partial T_I}\right)_{\theta} dT_I. \tag{1.9}$$

Equations (1.6)–(1.9) cover any application of the tubular thermosyphon under steady conditions.

For a high heat flux, $\dot{Q}$, it is clear that R_T, R_H^*, and R_C^* should all be as small as possible. The heat flux density $\dot{q}$, on the other hand, is influenced by the area A of the surface through which the heat passes; that is, $\dot{q} = \dot{Q}/A$. It is therefore possible either to intensify or attenuate the heat flux *density* by simply varying the surface areas of the thermosyphon. Thus, for example, thermal energy from a small high-temperature source may be dispersed to heat a large room if

[8] Any convenient pair of variables $\theta(T_H, T_C)$, $T_I(T_H, T_C)$ will suffice; θ determines the heat flux directly, while $T_I = T_m$, the mean temperature, provides a useful reference value for fluid properties. Under evaporative conditions, $T_I = T_{sat}$ may also be convenient.

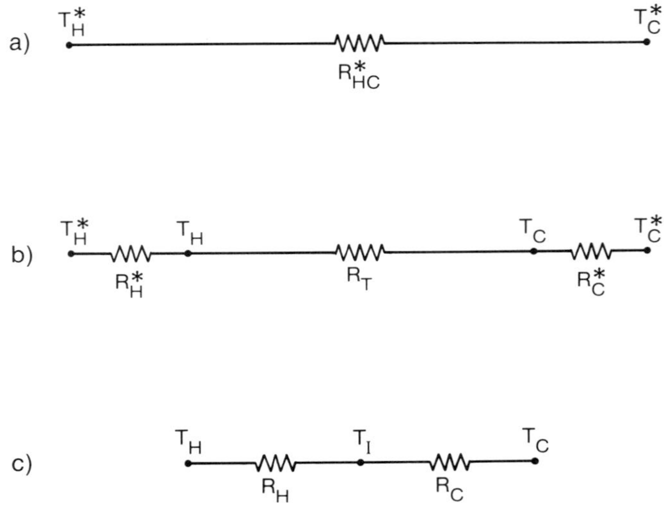

Fig. 1.26. Arrangement of resistive elements between a heat source and sink.

the area of the heated section of the thermosyphon is small but that of the cooled section is large; the heat flux has been made to diverge. Similarly, the energy from a large waste heat source may be intensified in a small boiler heated by the small, cooled section of the thermosyphon; the heat flux has converged.

With a given configuration, the device not only creates high heat fluxes but has an inherent ability to control them through a switching feature. This feature takes two forms: the unique ability to limit heat transfer to one direction, and the more general ability to reduce the flux to a negligible value. The first of these has given rise to the term 'thermal diode'[9] applied, for example, to a vertical thermosyphon tube which is very effective in transmitting heat upwards when the heat source lies beneath the sink. It was noted earlier that if the positions of the source and sink are reversed, the fluid inside the tube becomes stably stratified and hence downward heat transfer is limited to the much less effective mechanism of conduction. For many practical purposes, the thermosyphon then operates as a one-way heat valve with the downward value of R_T being much greater than the upward value.

Under typical operating conditions, an increase in the temperature difference θ results in an increased heat flux, but there are circumstances when it is desired to reverse this tendency so as to shut off the heat flow. It is clear from eqn (1.8) that this may be accomplished if $\Re_T = \infty$, or is at least very large, but it is equally evident from eqns (1.8) and (1.9) that a reduction in $\Re_T$,

[9] Using the same analogy, a thermal triode may be devised. See, for example, Section 5.2.

which commonly accompanies an increase in θ, does not facilitate a reduction in $\dot{Q}$. The coefficient of the differential in the first term on the right-hand side of eqn (1.9) is normally positive, but it may be offset by the coefficient of the second term provided that $(\partial\mathfrak{R}_T/\partial T_I)_\theta > 0$; however, this second coefficient is usually negative. In an evaporative thermosyphon, where T_I may be taken as the interfacial (saturation) temperature, an increase in T_H may be used to create an increase in T_I. This moves the equilibrium state of the (constant volume) liquid–vapour mixture contained in the tube towards the dry saturated condition; the liquid fraction is thus reduced, as illustrated by Fig. 1.27. As we shall see in Chapter 3, a reduction in the liquid fraction of the filling increases the thermal resistance $\mathfrak{R}_T$ rapidly once the fraction falls beneath a critical value. Thus, by making a judicious choice of filling charge it is possible to allow normal operation of the thermosyphon up to a specified source temperature. At, or above, this temperature the emergence of a positive derivative $(\partial\mathfrak{R}_T/\partial T_I)_\theta$ and the consequent increase in $\mathfrak{R}_T$ make it possible to shut down the thermosyphon even though T_H continues to rise.

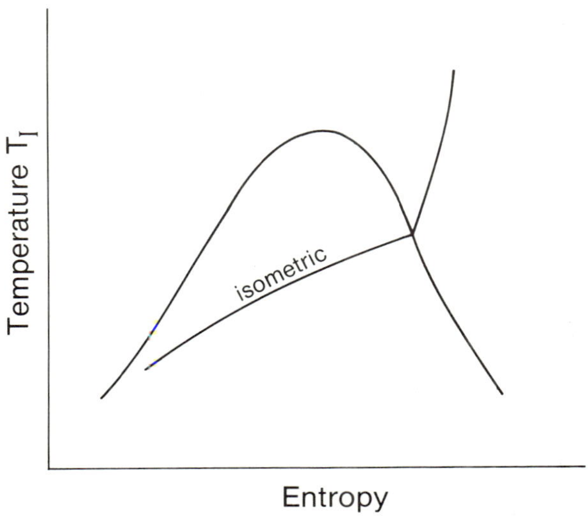

Fig. 1.27. Isometric curve on the T_I–S diagram

This type of switching behaviour may be used to prevent overheating of a finite sink, the temperature of which generally tends to rise with increases in source temperature (Larkin 1984). Shutdown then provides a form of temperature control for the sink which may, for example, contain temperature-sensitive instruments. More commonly, it is the source temperature which requires regulation; for example, when the thermosyphon is employed to maintain the temperature of some body (source) at a steady value by regulating the flow of heat from it. Under ideal circumstances, this requires that two conditions be satisfied: firstly, a sympathetic inverse relation exists between T_H and R_T; secondly, that regulation be independent of the sink temperature T_C. It is at once evident from eqn (1.7) that the second requirement cannot be satisfied automatically unless the coefficient of the second differential on the right-hand side is zero. Since this is unlikely to be true, the only way to remove the effect of sink temperature is to maintain T_C at a constant value.

With this understanding, eqn (1.7) may be reduced to

$$\mathrm{d}T_H = R_T\,\mathrm{d}\dot{Q}\Big/\left[1-\dot{Q}\left(\frac{\partial R_T}{\partial T_H}\right)_{T_C}\right]$$

from which the fluctuation in source temperature with heat flux may be calculated if the resistance of the thermosyphon is known; the variable resistance (conductance) thus provides the control. For the fluctuation to be small, R_T must be as small as possible while the slope $-(\partial R_T/\partial T_H)_{T_C}$ must be as large as possible. Typically, an increase in T_H reduces R_T but also reduces the slope. Should their combined natural effect not help limit the temperature fluctuation to an acceptable range, alternative measures must be found. In the closed, evaporative thermosyphon, for example, a more sympathetic $R_T \sim T_H$ relation may be obtained through the introduction of an incondensable gas. Under steady conditions, this tends to collect at the cold end of the tube where, in excluding the condensing vapour, it adds to R_C and hence to R_T. An increase in T_H, however, raises the saturation temperature T_I and along with it the saturation vapour pressure. This vapour pressure increase has the effect of compressing the gas into a smaller space thereby increasing the effective length of the condenser and reducing R_T; the slope $-(\partial R_T/\partial T_H)_{T_C}$ is thus increased.

For simplicity, the above discussion has been limited to steady conditions which are assumed extensively throughout this book. In the majority of circumstances, the thermosyphon will be designed to operate steadily; if changes do occur, they will be slow enough to produce quasi-steady behaviour, the only additional complication arising when the $\dot{Q} \sim \theta$ relation is not a single-valued function, and hysteresis occurs. More rapid transients, such as occur naturally during startup, shutdown, or sudden changes in T_H or T_C, are beyond the scope of this text, although it might be mentioned that the addition of thermal capacitors to Fig. 1.26 would accommodate transients, at least in general terms. A similar comment applies to unstable behaviour, although this is more troublesome and is not well understood. Hysteresis, for example, usually refers to two separate but stable paths, any particular state being determined from the initial conditions. Instability, on the other hand, results in a *spontaneous* change to another state from which the system may or may not return, depending upon whether or not it is a stable state. As we shall see in later chapters, there are various situations in which the state of the tubular thermosyphon is not stable, thus leading to a sudden change of flow régime. Under certain circumstances, the new régime is maintained and steady operation ensues; in others the new régime dissolves back into the old and the process repeats itself in a predictable though oscillatory matter. Complex transients such as these are beyond an introductory treatment and will be mentioned later only in passing.

1.5 Scope of the text

The aim of this book is to bring to graduate students and practising engineers an understanding of the tubular thermosyphon, and to introduce them to the wide range of circumstances in which it offers a solution to the general problem of transferring heat efficiently from a source to a sink. Emphasis is placed on the *modus operandi* of the device because this provides the basis for understanding the thermal performance which makes it so attractive in its many forms. While this occasionally leads to a considerable amount of detail, particularly in the treatment of basic types, it stops well short of being an exhaustive treatment. Thus, within a comprehensive introduction to the subject the reader will still encounter limits imposed by technical and spatial constraints.

In general, the approach lies in the statement and interpretation of the physics of the thermosyphon. A deceptively simple device, it nevertheless houses a fluid which transmits thermal energy from one end of the tube to the other by circulating in a seemingly infinite set of flow patterns. An important step towards understanding its behaviour is therefore an appreciation of not only the general cause of circulation, thermally induced buoyancy, but its particular manifestations: in various body force fields, various fluids, with or without phase change, etc. Accompanying the physical behaviour is, of course, mathematical description. This will be undertaken wherever practical by using the differential equations which govern continuity, motion, and conservation of energy in a continuous fluid. The treatment of these equations is generally limited to steady conditions and to fluids with properties, other than density, which are constant. While this restricts the range of behaviour to be considered, and excludes flow instabilities in particular, it none the less permits a thorough and wide-ranging discussion of basic circulation patterns and the heat transfer rates which accompany them.

Physics and mathematics are interwoven in the construction of physico-mathematical models. Ideally, these models suggest the forms of solutions to the governing differential equations but thermosyphons are so complex that simple solutions are rarely available. Mathematical modelling will therefore be used for two purposes: to analyse the primary flow pattern which carries the thermal energy; and to follow the development of this pattern as it changes in response to changing circumstances. The discussion, which generally incorporates both primary and secondary flow, leads naturally to the description of régime behaviour and its effect on thermal performance. Heat transfer characteristics are thus revealed in terms of variations in the internal flow patterns.

The book is organized into four main chapters. In Chapter 2, thermal buoyancy is discussed in detail, beginning with a review of natural convection in cavities, both open and closed. Various types of simple (single-phase) thermosyphons are then considered using theoretical analysis complemented by experimental data. Chapter 3 shifts attention to thermosyphons within which boiling and condensation are taking place. After a review of these processes as they occur within the confines of a tube, the discussion evolves into a detailed study of thermosyphon performance under the influence of Archimedean buoyancy.

Chapter 4 is devoted to tubular thermosyphons operating in a centrifugal field. Two configurations are considered in detail: tubes with radial axes, and tubes with their axes parallel to the axis of rotation. The latter further subdivides into concentric and eccentric configurations. Discussion of both the simple and evaporative forms is provided. The final chapter gives an introduction to related heat transfer systems which, although incorporating different operating principles, may be used in augmenting the performance of the tubular thermosyphon. These include the heat pipe, the aerosyphon, and the bayonet tube. Also included is a brief discussion of anti-gravity thermosyphons.

2
THE SIMPLE, TUBULAR THERMOSYPHON

Newton's law of cooling, in which the rate of cooling of a body is taken to be proportional to its temperature measured relative to that of the environment, is the first quantitative statement ever made about natural convection. Implicit within this law is the requirement that the heat transfer coefficient at the body surface be constant. Today, not only is it known that the heat transfer coefficient h is a function of the temperature difference θ, but it is also understood that the natural convective system established around the body is governed directly by this temperature difference. It is a commonplace observation that $h \propto \theta^n$ where n is a small positive fraction. Over a small temperature difference range this power law relation may be approximated by a constant.

In the late nineteenth century, Lorenz (1881) took the local heat transfer coefficient in natural convection to be independent of position on the body surface. It was later discovered (Prandtl 1904; Langmuir 1912; Schmidt and Beckmann 1930) that this spatial restriction was also incorrect. And again it was noted that it was often a good averaging approximation.

Since those early investigations, and especially during the second half of this century, natural convection has received a great deal of attention from which it has gradually become apparent that h is generally a complex function of many variables. Since the buoyant motion derives from a body force field, it is obvious that the strength and spatial distribution of this field must play an important role. Equally, it is clear that the diffusion of heat, which establishes the temperature field, is coupled to the diffusion of momentum, which gives rise to the velocity field; that is, fluid properties in general, and diffusivity in particular, must also influence h. Moreover, it is to be expected that the heat transfer coefficient will be tied to the temperature distribution over the bounding solid surfaces and to the geometry of those surfaces.

In the first section of this chapter, the non-dimensional heat transfer coefficient will be sought, explicitly and implicitly, as a function of the above variables. Coverage will be restricted to single-phase natural convection within enclosures, the objective being not to provide an exhaustive review of the subject but simply to introduce the reader to those features which are characteristic of internal, thermally buoyant flows. In this way it should be possible to provide the insight necessary for the understanding of gravitational thermosyphons filled with homogeneous fluids. The treatment will be both physical and mathematical, beginning with an analysis of the equations governing steady, two-dimensional, laminar flow. The discussion then divides into two parts, one covering cavities heated from the side and the other covering cavities heated from beneath. The section ends with a brief introduction to instability, transition, and turbulence in cavity flows. Subsequent sections in this chapter use these general observations as a basis for interpreting the behaviour of tubular thermosyphons: open or closed, vertical or inclined, linear or non-linear. It will be discovered that behaviour is generally complicated and highly variable, and that the adjective 'simple' in the title of this chapter must be taken not in an absolute sense, meaning elementary and easily understood; it is meant only in a relative sense, offering a comparison with devices to be discussed later.

2.1 Natural convection in cavities

2.1.1 *Governing equations for steady, laminar flow*

The equations of continuity, motion, and energy governing steady[1], laminar convection are given, respectively, by:

$$\nabla \cdot \mathbf{V} = 0$$

$$\rho(\mathbf{V} \cdot \nabla)\mathbf{V} = -\nabla P + \rho \mathbf{G} + \mu \nabla^2 \mathbf{V}$$

$$\rho c_p (\mathbf{V} \cdot \nabla) T = k \nabla^2 T. \tag{2.1}$$

When the fluid is at rest with a uniform reference temperature T°, the equation of motion reduces to

$$0 = -\nabla P^\circ + \rho^\circ \mathbf{G} \tag{2.2}$$

thereby prescribing the hydrostatic pressure distribution P^0 for a fluid with uniform density ρ°. If any departure from this pressure is defined by

$$P' = P - P^0$$

it follows that

$$-\nabla P + \rho \mathbf{G} = -\nabla P' + (\rho - \rho^\circ)\mathbf{G}$$
$$= -\nabla P' - \rho^\circ \beta (T - T^\circ)\mathbf{G}$$

where $\beta = -(1/\rho)(\partial \rho / \partial T)_p$ is the thermal expansion coefficient of the fluid. Hence, if the density is dependent on temperature alone, the equation of motion may be written approximately as

$$(\mathbf{V} \cdot \nabla)\mathbf{V} = -\frac{1}{\rho} \nabla P' - \beta \theta \mathbf{G} + \nu \nabla^2 \mathbf{V} \tag{2.3}$$

[1] Apart from the obvious absence of time, these equations omit any reference to bulk viscosity, viscous dissipation, and compression work. They also imply that fluid properties are constant.

where $\theta = T - T^{\circ}$. This incorporates the Boussinesq approximation; namely, that density is constant, except in concert with the body force field. The second term on the right-hand side represents the physical effect of thermal buoyancy which acts oppositely[2] to the field vector **G**.

Taking the curl of each term in eqn (2.3) eliminates the pressure, and the result, together with the continuity and energy equations, may be expressed in the two-dimensional coordinate system described in Fig. 2.1. Thus

$$\frac{\partial U}{\partial X} + \frac{\partial V}{\partial Y} = 0 \tag{2.4a}$$

$$\frac{\partial}{\partial X}\left(U\frac{\partial V}{\partial X} + V\frac{\partial V}{\partial Y}\right) - \frac{\partial}{\partial Y}\left(U\frac{\partial U}{\partial X} + V\frac{\partial U}{\partial Y}\right)$$
$$= \beta g\frac{\partial \theta}{\partial X} - \nu\frac{\partial}{\partial X}\left(\frac{\partial^2 V}{\partial X^2} + \frac{\partial^2 V}{\partial Y^2}\right) - \nu\frac{\partial}{\partial Y}\left(\frac{\partial^2 U}{\partial X^2} + \frac{\partial^2 U}{\partial Y^2}\right) \tag{2.4b}$$

$$U\frac{\partial \theta}{\partial X} + V\frac{\partial \theta}{\partial Y} = \kappa\left(\frac{\partial^2 \theta}{\partial X^2} + \frac{\partial^2 \theta}{\partial Y^2}\right) \tag{2.4c}$$

where $\kappa = k/\rho c_{\text{p}}$. These equations possess a very wide set of solutions, the precise details depending upon the thermal and geometrical conditions imposed on the cavity. It is difficult to generalize but useful insight may be gained by considering a few particular situations in which temperature variations along the horizontal walls are much greater than those along the vertical walls.

2.1.2 *Primary characteristics of the vertical cavity*

Consider first the situation shown in Fig. 2.2(a), often described as heating (and cooling) from the side. The horizontal walls are adiabatic (but not isothermal) and the circulation shown is principally the result of *direct* thermal buoyancy in the Y direction. When the temperature difference scale $\theta^{\text{c}} = T_{\text{H}} - T_{\text{C}}$ is large enough, this circulation consists essentially of a single pair of boundary layers driven by buoyancy forces operating in the vicinity of the vertical walls; the system is illustrated in Fig. 2.3(a). If these forces are balanced principally by the viscous forces acting along the vertical walls, then from eqn (2.4b)

$$\beta g\frac{\partial \theta}{\partial X} = O\left(\nu\frac{\partial^3 V}{\partial X^3}\right).$$

[2] In a heated fluid if $\beta > 0$. Obviously the reverse is true if the fluid is cooled or if $\beta < 0$, the latter condition being rare.

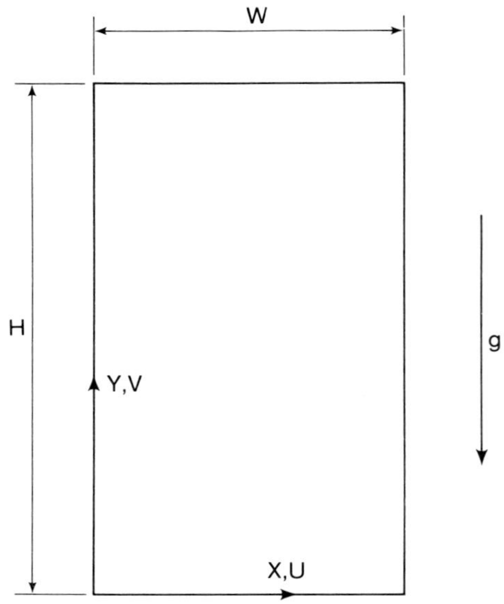

Fig. 2.1. Coordinate system for a rectangular cavity.

Thus,

$$\frac{\beta g \theta^c (X^c)^2}{\nu V^c} = O(1) \tag{2.5}$$

where V^c is the vertical velocity scale and X^c is the width scale of the boundary layer, neither of which are known *a priori*. At the same time, heat conducted laterally into the boundary layer is largely carried away as an upward convective flux, and hence from eqn (2.4c)

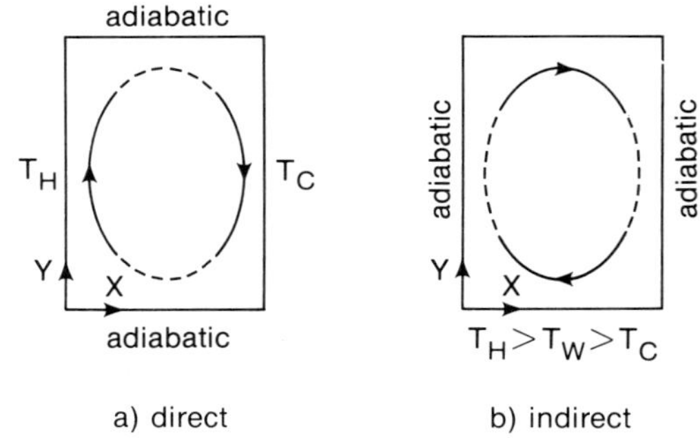

Fig. 2.2. Direct (a) and indirect (b) buoyancy effects.

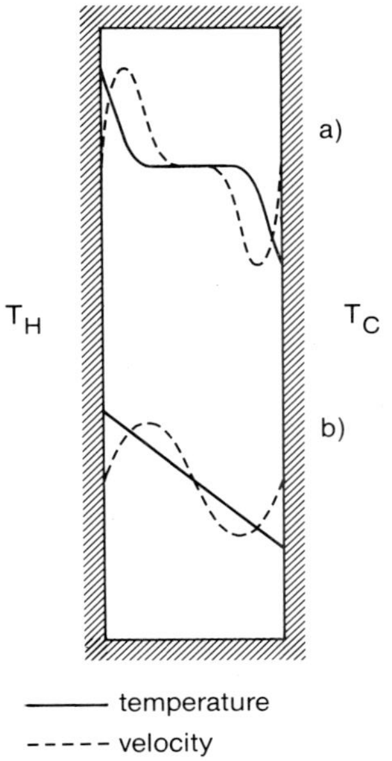

Fig. 2.3. Temperature and velocity profiles in a rectangular cavity: (a) boundary layer régime; (b) conduction régime.

$$V\frac{\partial\theta}{\partial Y}=O\left(\kappa\frac{\partial^2\theta}{\partial X^2}\right).$$

Thus

$$\frac{V^c(X^c)^2}{\kappa Y^c}=O(1) \tag{2.6}$$

in which $Y^c=\mathrm{O(H)}$, the cavity height. Equations (2.5) and (2.6) enable the two unknown scales X^c and V^c to be found. Thus,

$$X^c=O(H/Ra_{\mathrm{H}}^{1/4}) \tag{2.7}$$

$$V^c=O\left(\frac{\kappa Ra_{\mathrm{H}}^{1/2}}{H}\right) \tag{2.8}$$

in which $Ra_{\mathrm{H}}=\beta g\theta^c H^3/\nu\kappa$ is a Rayleigh number. The horizontal velocity scale U^c may be determined from the continuity equation by virtue of the fact that

$$\frac{\partial U}{\partial X}=O\left(\frac{\partial V}{\partial Y}\right)$$

and hence

$$U^{\mathrm{c}}=0\left(\frac{V^{\mathrm{c}}X^{\mathrm{c}}}{Y^{\mathrm{c}}}\right)=O\left(\frac{\kappa Ra_{\mathrm{H}}^{1/4}}{H}\right). \tag{2.9}$$

The above scales now enable eqns (2.4) to be written in normalized form by defining

$$x=\frac{X}{X^{\mathrm{c}}},\quad y=\frac{Y}{Y^{\mathrm{c}}},\quad u=\frac{U}{U^{\mathrm{c}}},\quad v=\frac{V}{V^{\mathrm{c}}},\quad \text{and}\quad \phi=\frac{\theta}{\theta^{\mathrm{c}}} \tag{2.10}$$

which are non-dimensional variables having unit orders of magnitude. Substituting in eqns (2.4) we obtain

$$\frac{\partial u}{\partial x}+\frac{\partial v}{\partial y}=0 \tag{2.11a}$$

$$\frac{1}{Pr}\frac{\partial}{\partial x}\left(u\frac{\partial v}{\partial x}+v\frac{\partial v}{\partial y}\right)-\frac{1}{PrRa_{\mathrm{H}}^{1/2}}\frac{\partial}{\partial y}\left(u\frac{\partial u}{\partial x}+v\frac{\partial u}{\partial y}\right)$$

$$=\frac{\partial\phi}{\partial x}+\frac{\partial}{\partial x}\left(\frac{\partial^2 v}{\partial x^2}+\frac{1}{Ra_{\mathrm{H}}^{1/2}}\frac{\partial^2 v}{\partial y^2}\right)-\frac{\partial}{\partial y}\left(\frac{1}{Ra_{\mathrm{H}}^{1/2}}\frac{\partial^2 u}{\partial x^2}+\frac{1}{Ra_{\mathrm{H}}}\frac{\partial^2 u}{\partial y^2}\right) \tag{2.11b}$$

$$u\frac{\partial\phi}{\partial x}+v\frac{\partial\phi}{\partial y}=\frac{\partial^2\phi}{\partial x^2}+\frac{1}{Ra_{\mathrm{H}}^{1/2}}\frac{\partial^2\phi}{\partial y^2} \tag{2.11c}$$

in which $Pr=\nu/\kappa$ is the Prandtl number. These equations describe laminar, boundary layer natural convection, and reveal that the solutions will, in general, depend upon the magnitudes of two parameters: Pr and Ra_{H}. For very viscous liquids, $Pr \gg 1$ and the inertial terms are thereby suppressed. Likewise, the condition $Ra_{\mathrm{H}}^{1/2} \gg 1$ also introduces simplifications; in particular, longitudinal viscous and conduction effects may be ignored.

The heat transfer rate for laminar, boundary layer flow is readily obtained by recalling the physical requirement which led to eqn (2.6); namely, that lateral conduction creates longitudinal convection. Heat transfer between the two vertical walls may thus be represented by the vertical convective flux. Accordingly, the average heat flux density at the hot wall may be estimated from

$$\bar{\dot{q}}H=O(\rho V^{\mathrm{c}}X^{\mathrm{c}}c_{\mathrm{p}}\theta^{\mathrm{c}})$$

from which it follows that

$$\bar{\dot{q}}=O\left(\frac{k\theta^{\mathrm{c}}}{H}Ra_{\mathrm{H}}^{1/4}\right),$$

and hence the average Nusselt number is given by

$$\bar{N}u_{\mathrm{H}} = \frac{\bar{\dot{q}}H}{\theta^{\mathrm{c}}k} = O(Ra_{\mathrm{H}}^{1/4}). \tag{2.12}$$

As the Rayleigh number increases, the heat transfer rate increases while the boundary layer width X^{c} decreases until the flow becomes unstable and eventually becomes turbulent. The boundary layer régime is also limited by Rayleigh numbers which are small enough for X^{c} to become comparable with the cavity width W, as illustrated in Fig. 2.3(b). When this occurs, it is necessary to take $X^{\mathrm{c}} = O(W)$; eqn (2.6) no longer applies and eqn (2.5) then gives the vertical velocity scale directly as

$$V^{\mathrm{c}} = \frac{\beta g \theta^{\mathrm{c}} W^2}{\nu} = \frac{\kappa}{W} Ra_{\mathrm{W}}, \tag{2.13}$$

still implying that the direct buoyancy forces are balanced principally by viscous forces. These scale changes alter the normalized forms of eqns (2.4) which now transform to

$$\frac{\partial u}{\partial x} + \frac{\partial v}{\partial y} = 0 \tag{2.14a}$$

$$\frac{W}{H}\frac{Ra_{\mathrm{W}}}{Pr}\frac{\partial}{\partial x}\left(u\frac{\partial v}{\partial x} + v\frac{\partial v}{\partial y}\right) - \left(\frac{W}{H}\right)^3 \frac{Ra_{\mathrm{W}}}{Pr}\frac{\partial}{\partial y}\left(u\frac{\partial u}{\partial x} + v\frac{\partial u}{\partial y}\right)$$
$$= \frac{\partial \phi}{\partial x} + \frac{\partial}{\partial x}\left[\frac{\partial^2 v}{\partial x^2} + \left(\frac{W}{H}\right)^2 \frac{\partial^2 v}{\partial y^2}\right] - \frac{\partial}{\partial y}\left[\left(\frac{W}{H}\right)^2 \frac{\partial^2 u}{\partial x^2} + \left(\frac{W}{H}\right)^4 \frac{\partial^2 u}{\partial y^2}\right] \tag{2.14b}$$

$$Ra_{\mathrm{W}} \frac{W}{H}\left(u\frac{\partial \phi}{\partial x} + v\frac{\partial \phi}{\partial y}\right) = \frac{\partial^2 \phi}{\partial x^2} + \left(\frac{W}{H}\right)^2 \frac{\partial^2 \phi}{\partial y^2} \tag{2.14c}$$

which yields solutions that are markedly different from those of eqns (2.11). The Prandtl number plays the same role as before, but a new parameter, the aspect ratio H/W, appears. Most important is the implication associated with a low Rayleigh number $Ra_{\mathrm{W}} = \beta g \theta^{\mathrm{c}} W^3/\nu\kappa$ or, more precisely, with a low value of the product $Ra_{\mathrm{W}} W/H$. Clearly as $Ra_{\mathrm{w}} W/H \rightarrow 0$, the convective terms in the energy equation are suppressed and its solution, then uncoupled from the equation of motion, corresponds to pure conduction heat transfer, as suggested in Fig. 2.3(b).

2.1.3 *Primary characteristics of the horizontal cavity*

Turning now to the situation shown in Fig. 2.2(b), we face a very different problem. Although the general circulation may appear to be similar, it is driven by an entirely different mechanism[3]. The absence of conduction through the vertical walls typically reduces their effect to a secondary role. The primary driving force is now exerted horizontally by virtue of the horizontal temperature gradients which the walls impose on the fluid. This *indirect* thermal buoyancy force was evident in eqn (2.4b). Again considering conditions under which buoyancy effects are balanced by viscous effects, eqn (2.4b) reveals that

$$\beta g \frac{\partial \theta}{\partial X} = O\left(\nu \frac{\partial^3 U}{\partial Y^3}\right)$$

and hence

$$\frac{\beta g \theta^{c} (Y^{c})^3}{\nu U^{c} X^{c}} = O(1) \tag{2.15}$$

in which $X^{c} = O(W)$. For a boundary layer flow, the balance between longitudinal convective and lateral conductive effects now demands that

$$U \frac{\partial \theta}{\partial X} = O\left(\kappa \frac{\partial^2 \theta}{\partial Y^2}\right).$$

Thus

$$\frac{U^{c} (Y^{c})^2}{\kappa X^{c}} = O(1). \tag{2.16}$$

From eqns (2.15) and (2.16) it follows that the scales for the horizontal velocity and boundary layer width are

$$Y^{c} = O(W / Ra_{W}^{1/5}) \tag{2.17}$$

and

$$U^{c} = O\left(\frac{\kappa}{W} Ra_{W}^{2/5}\right) \tag{2.18}$$

analogous to eqns (2.7) and (2.8). The vertical velocity scale is obtainable from the continuity equation. Substitution of these scales into the normalized variables of eqns (2.10) once again enables eqns (2.4) to be transformed into an appropriate normalized set. The result is analogous to the boundary layer equations (2.11) with the implications being the same, bearing in mind that Ra_{H} is replaced by Ra_{W} and the fractional exponents are different. The heat transfer rate may also be found in the same manner. Thus

[3] The same mechanism that drives land and sea breezes.

$$\bar{\dot{q}}W = O(\rho U^{c} Y^{c} c_{p} \theta^{c})$$

and hence

$$\bar{\dot{q}} = O\left(\frac{k\theta^{c}}{W} Ra_{W}^{1/5}\right)$$

from which it follows that

$$\overline{Nu}_{W} = \frac{\bar{\dot{q}}W}{\theta^{c}k} = O(Ra_{W}^{1/5}), \tag{2.19}$$

analogous to eqn (2.12). When the boundary layer width given by eqn (2.17) is comparable with the cavity height H, the same arguments employed earlier may be used to show that as $Ra_{H}(H/W)^{2} \to 0$, heat transfer is limited to pure conduction[4].

It should now be apparent that the determination of scales and relative orders of magnitude helps uncover valuable information on the behaviour of buoyant cavity flows. It is interesting, for example, to compare the thermal and spatial dependencies of the heat transfer coefficient with those assumed by the pioneers mentioned earlier. More important perhaps, the analysis reveals the changing dependency of the Nusselt number on the Rayleigh number, from conduction as a lower limit to boundary layer convection as an upper (laminar) limit. This type of physico-mathematical modelling provides powerful insights which aid in the understanding of thermosyphon behaviour, but it is important to recognize that it is not without limitations. These will now be discussed.

To begin with, the boundary layer width appropriate to the equation of motion is not generally the same as that appropriate to the energy equation; the former, the viscous width, depends upon the momentum diffusivity ν while the latter, the thermal width, depends upon the thermal diffusivity κ. The ratio of these two fluid properties, the Prandtl number, indicates the relative lateral extent of the two boundary layers. The suppression of the inertial terms for $Pr \gg 1$ is thus only acceptable in non-metallic liquids; when $Pr \ll 1$, the buoyancy forces are balanced by the inertial forces, and viscous forces then play a less important role[5]. The second difficulty is associated with the tacit assumption that the coefficient of each normalized term in the governing differential equations specifies the relative importance of that term. This is based on the fact that the normalized quantities are, by definition, of order 1. However, unit orders of magnitude in the variables do not necessarily produce unit order in their derivatives. The ordering of terms is therefore limited to

[4] Since the vertical walls have been taken as adiabatic, this implies conduction into and out of the fluid along both the top and bottom walls.

[5] Since they represent the highest-order term in the equation of motion, this introduces mathematical difficulties.

regions in which the derivatives are well behaved and, in particular, exhibit no singularities. A third note of caution concerns the completeness of the statement presented by the governing equations. The boundary and initial conditions were not mentioned above because they added no further information, but in general they do. Normalization of these conditions may reveal additional parameters or, perhaps more significantly, may determine some of the scales and orders of magnitude; the temperature scale, for example, is frequently imposed by the boundary conditions. The incorporation of boundary conditions in the normalization process becomes particularly important when the adjacent region is also part of the overall system being considered. Combined conduction–convection problems such as the double-pane window shown in Fig. 2.4 illustrate this.

2.1.4 *Modifications to the primary flow*

Probably the greatest limitation of scaling and ordering is encountered when the flow throughout the cavity is not uniform; that is, when it has subdivided into smaller regions each with its own scales. In the context of thermosyphons, the subdivisions are created by secondary and tertiary effects which superimpose themselves upon the primary motion caused by direct or indirect buoyancy acting along the tube axis. The use of the primary flow alone to estimate the longitudinal convective flux continues to be valid but with a reduced accuracy. In general, secondary flows are not simply independent additions; typically, they are coupled to the primary flow with which their scales must be consistent. In the cavity flow discussed above, for instance, no reference was made to the core of fluid in the central region of the cavity. This is a subregion which changes according to the thermal and geometrical constraints.

For both circular and rectangular section cavities heated from the side it has been observed that the core may be stagnant or it may be stratified, with the hotter fluid being on top (Martini and Churchill 1960; Eckert and Carlson 1961; Elder 1965*a*; Brooks and Ostrach 1970; Linthorst *et al.* 1981*b*). For $H/W \simeq 5$, increases in Ra_W above the conduction régime gradually produce in the stratified core a pair of secondary rolls (near $Ra_W = 1.5 \times 10^5$) with the same rotation as the primary flow. These are associated with a change in sign of the horizontal temperature gradient at the edges of the core, and are illustrated in Fig. 2.5. Further increases in Rayleigh number create a tertiary flow (near $Ra_W = 2.5 \times 10^5$) rotating in the opposite direction. This is also seen in Fig. 2.5. On the other hand, if $H/W > 10$, a single primary loop exists until $Ra_W \simeq 10^4$ but breaks down before 10^5 while the system is still in the conduction régime (Elder 1965*a*; Vest and Arpaci 1969). Longitudinal waves appear and eventually establish a series of transverse cells, as illustrated in Fig. 2.6. This is in marked contrast to the secondary transition in the boundary layer régime where the stratified core acts to inhibit any such tendency.

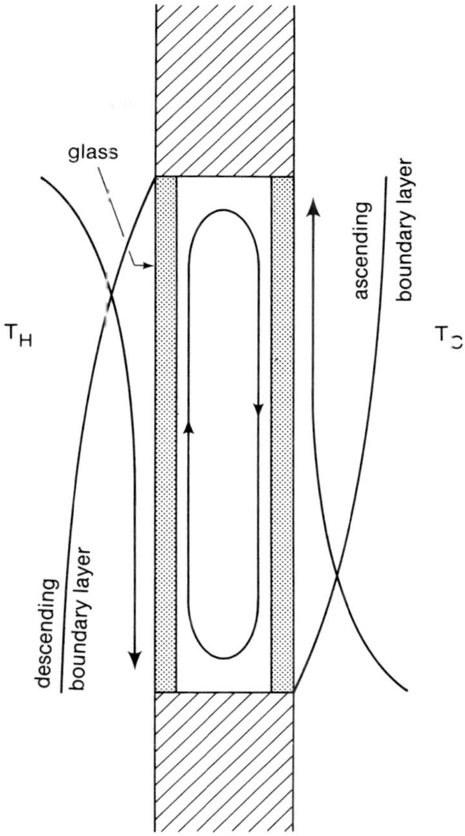

Fig. 2.4. Natural convection inside and outside a double-pane window.

As might be expected, aspect ratios less than 1.0 (i.e. horizontal cavities) produce an analogous set of phenomena but the situation is complicated further by the choice of thermal boundary conditions. For example, if the horizontal walls are adiabatic and the vertical walls are isothermals, it is the latter which tend to drive the flow (Bejan and Tien 1978; Bejan *et al.* 1981; Simpkins and Chen 1986; Blythe *et al.* 1983). Under these conditions, the flow pattern consists of three regions: a central stratified 'core' and two end regions in which behaviour varies from a sluggish, but simple, return to a vigorous boundary layer. On the other hand, if the horizonal wall temperature varies linearly, that is the wall is not adiabatic, behaviour is more closely related to that in the vertical cavity. The work of Ostrach *et al.* (1980), for example, has revealed secondary end cells for $H/W > 0.2$.

The orientation of the cavity with respect to either the force field or the

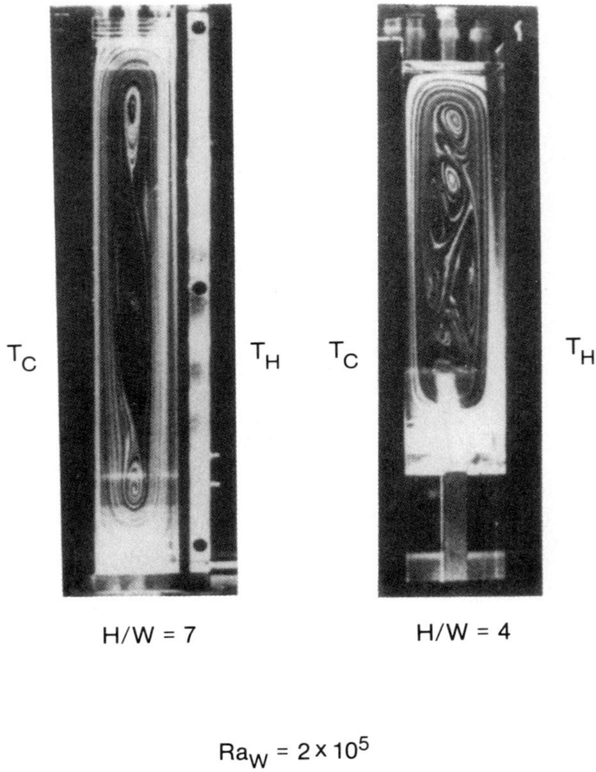

Fig. 2.5. Secondary and tertiary flow in a rectangular cavity (after Linthorst *et al.* 1981*b*).

prevailing temperature field also has a significant effect on the flow and heat transfer characteristics. This is readily explored by considering the effect of inclination in a gravitational field. Consider, for example, the rotation of the cavity shown in Fig. 2.7, with the heated side tilted backwards at an angle α to the vertical. The initial effect is to modify the secondary loops, as indicated in Fig. 2.8(a and b). A slightly greater tilt tends to suppress the secondary motion entirely, but the details of this progression cannot be fully described in two-dimensional terms. Figure 2.8(c and d), for example, shows two primary circulation patterns observed in different vertical planes displaced from each other along the axis of circulation, that is in a direction normal to the page. This behaviour heralds the appearance of an entirely different flow pattern associated with heating from below, a situation produced either by rotating the cavity or the prevailing temperature gradient through 90°.[6] Below a

[6] However, they must rotate in opposite direction.

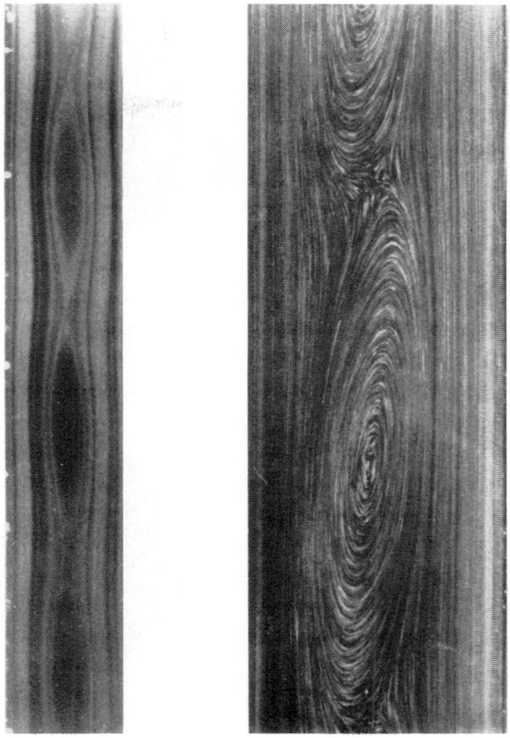

H/W = 33 $Ra_W = 6.7 \times 10^3$

H/W = 20 $Ra_W = 4.5 \times 10^5$

Fig. 2.6. Secondary flow in a high aspect ratio cavity (after Vest and Arpaci 1969).

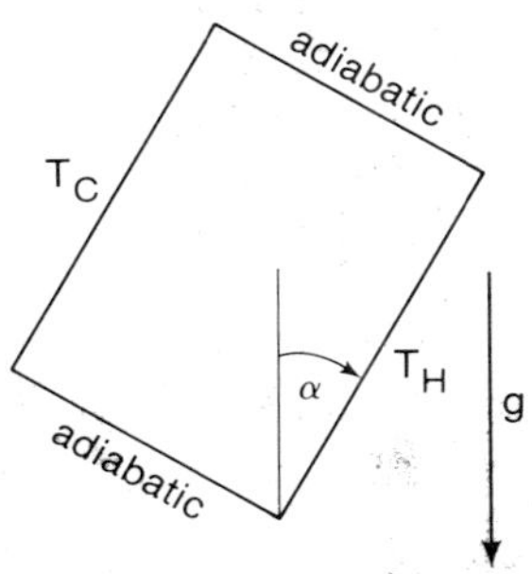

Fig. 2.7. Tilting of a rectangular cavity.

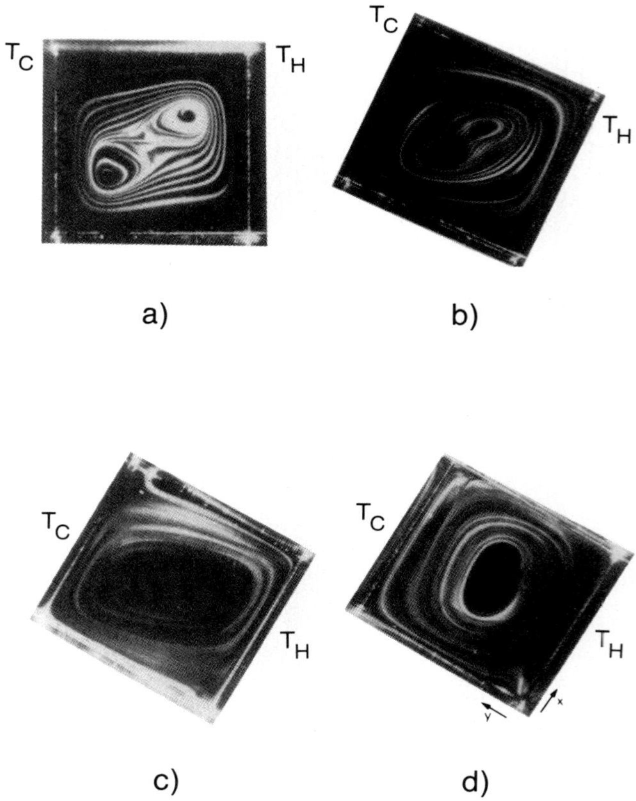

Fig. 2.8. Effect of tilt on flow in a square cavity (after Linthorst *et al.* 1981*b*).

critical Rayleigh number (based on the gap between the isothermal walls) the fluid is stationary; heat is transferred upwards by conduction alone. Above this critical value, a cellular structure appears. For a rectangular cavity, the cells or rolls tend to align themselves parallel to the shorter side, thus giving rise to the patterned structure evident in Fig. 2.9. These developments have been observed by many authors (Linthorst *et al.* 1981b; Symons and Peck 1984; Yang *et al.* 1987, 1988; Lock and Han 1990).

Another important aspect of thermally buoyant cavity flow which has a bearing on thermosyphon behaviour is the presence of openings in the walls. Thus far, it has been assumed that the cavity was completely closed, but the entry or exit of fluid is a common feature, and one which may profoundly affect both the flow pattern and the heat rate. In general terms, the removal of one of the cavity walls may be regarded simply as a change of boundary conditions,

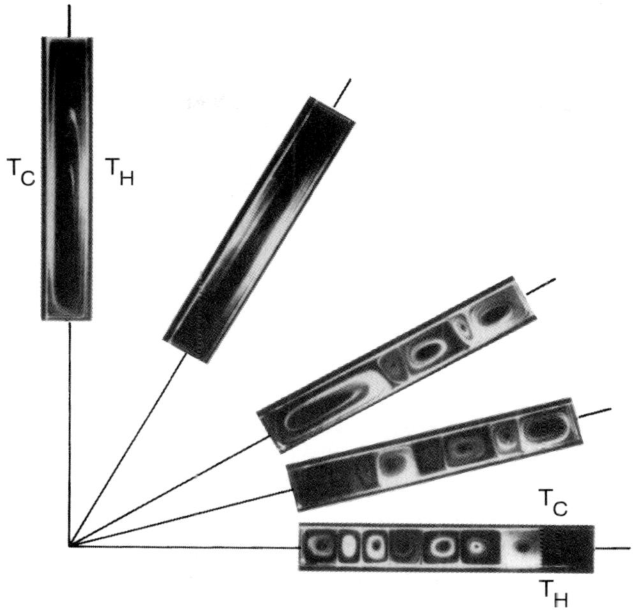

Fig. 2.9. Effect of tilt on flow in a rectangular cavity (after Symons and Peck 1984).

but on closer examination it becomes clear that these conditions are difficult to specify without considering the space beyond the cavity mouth. When the top wall of an isothermal cavity is absent, for example, the external flow pattern is intimately connected to the flow within, as suggested in Fig. 2.10. If the cavity is shallow, its heated base creates a buoyant jet above. This is illustrated in Fig. 2.10(a) which also shows incipient secondary circulation produced near the short vertical walls. As the cavity is deepened, the second circulation grows in strength. Eventually, this internal motion dominates, as indicated in Fig. 2.10(c). At the same time, flow in the vicinity of the mouth becomes increasingly unstable because the upward internal flow originating near the side walls has to break through the flow entering the core. More will be said about this in the next section.

When one of the side walls is removed instead, the flow pattern is perhaps less sensitive to cavity length (horizontally) but is still coupled to external conditions, the coupling increasing with Rayleigh number. This is illustrated in Fig. 2.11 which shows the theoretical streamlines calculated for a square, two-dimensional cavity (Chan and Tien 1985). For $Ra = 10^3$, Fig. 2.11(a), the streamlines which fully incorporate the external flow system, the upper set, are not too much different from those obtained by assuming a smooth entry[7] on

[7] Suppressing horizontal gradients at the mouth.

(a)

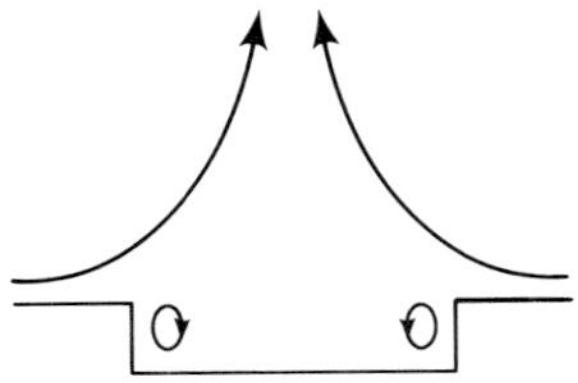

(b)

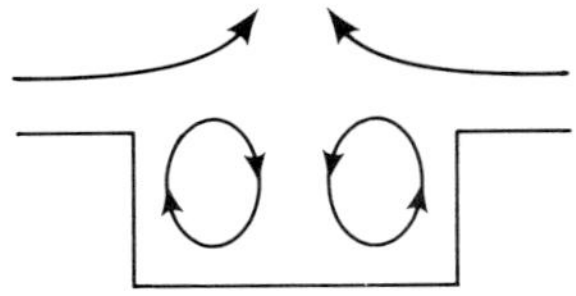

(c)

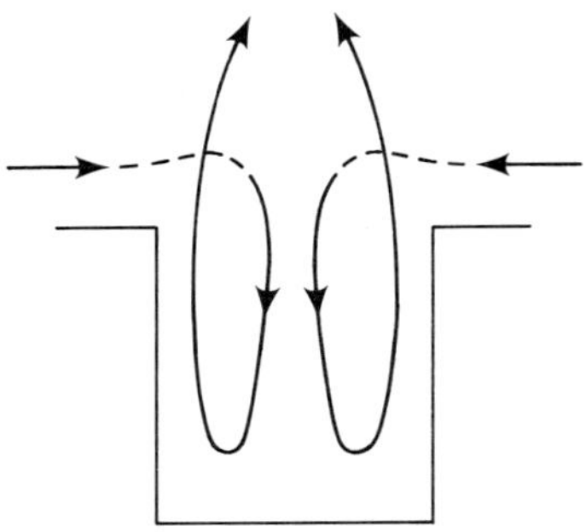

Fig. 2.10. Effect of depth on natural convection in an open cavity.

the plane of the missing face. However, for $Ra = 10^9$, Fig. 2.11(b), the external flow is seen to become very important. In particular, it signals separation of the flow entering over the lower wall. This separation has been observed experimentally by Chen *et al.* (1983) who studied its behaviour when the cavity was tilted. Figure 2.12 shows some of their visual observations from which it is evident that inclination not only affects separation but will also influence any turbulence, its structure, and indeed its very existence.

2.1.5 *Instability and turbulence in a cavity*

This brings us naturally to the question of turbulence in thermally buoyant cavity flows. Not surprisingly, this is closely linked to flow stability which, being particularly complex in cavity flows, makes the treatment of turbulence the more difficult. In Fig. 2.13, Hart (1971) indicates the range and evolution

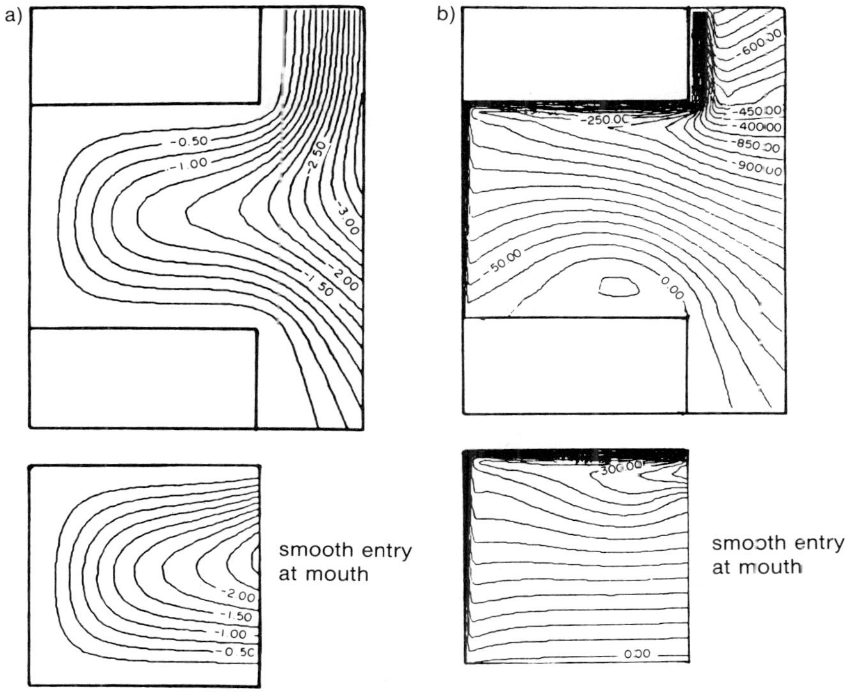

Fig. 2.11. Streamlines in an open horizontal cavity (after Chan and Tien 1985): $Ra_D = 10^3$; (b) $Ra_D = 10^9$.

of flow patterns as they occur with increasing Rayleigh number in a slender, laterally heated cavity; the long side is vertical when the inclination if zero, and positive angles indicate heating from beneath, as defined in Fig. 2.7. It is evident that the Bérnard–Rayleigh type of instability which occurs at large positive inclinations leads to turbulent flow only after a protracted transition in which unsteady behaviour is important. On the other hand, it is the growth of a Tollmien–Schlichting type of instability which is characteristic of near-vertical cavities. Longitudinal rolls form the base flow under the more stable conditions associated with negative inclinations. In general, the structure of the low Rayleigh number base régime, and the evolution of subsequent instabilities, thus map out various routes to turbulence for cavities in general, and thermosyphons in particular (Hart 1985). The situation may be complicated further by the existence of bifurcations in the base solutions (Winters and Brown 1985; Yang 1988; Lock and Han 1990).

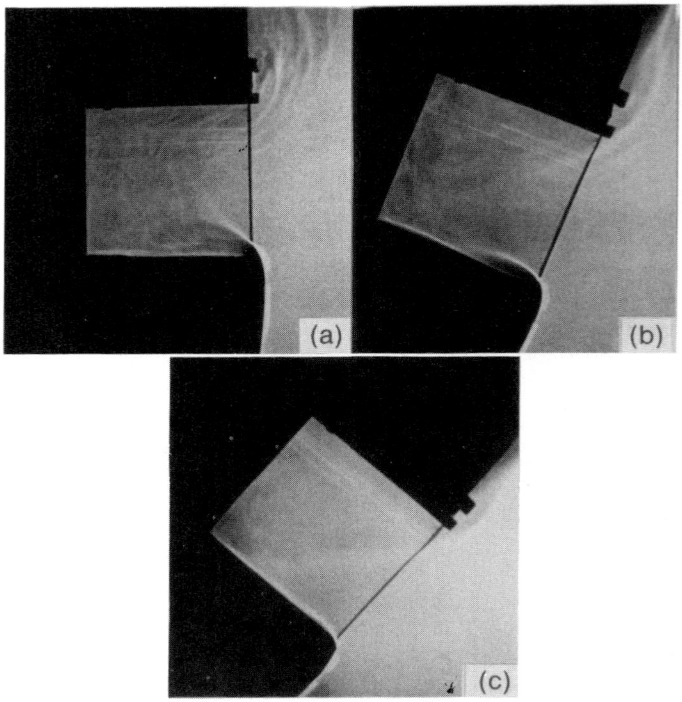

$\mathrm{Ra} = 3.84 \times 10^6$

Fig. 2.12. Visual observations on flow in an open, inclined cavity (after Chen *et al.* 1983).

Turbulent natural convection is often distinguishable from turbulent forced convection when it exists in conditions which are not too far removed from the transitional zone. In the boundary layer adjacent to a heated vertical surface, for example, the scale of turbulence may be of the order of the boundary layer thickness and the intensity may be of the order of 50% (Lock and Trotter 1968). In these conditions, intermittent fluctuations are observed very close to the wall, thus throwing doubt on the concept of a steady, laminar sublayer. Despite this, Prandtl's extension of the mixing length model does provide an overall description which can be reconciled with empirical heat transfer data (Eckert and Jackson 1951; Bayley 1955*b*).

In attempting to describe turbulent natural convection in a thermosyphon tube, Lighthill (1953) was guided by the characteristics of unidirectional pipe turbulence, thus reflecting the influence of wall shear stress; this is not inconsistent with the existence of a local maximum shear stress in the interior

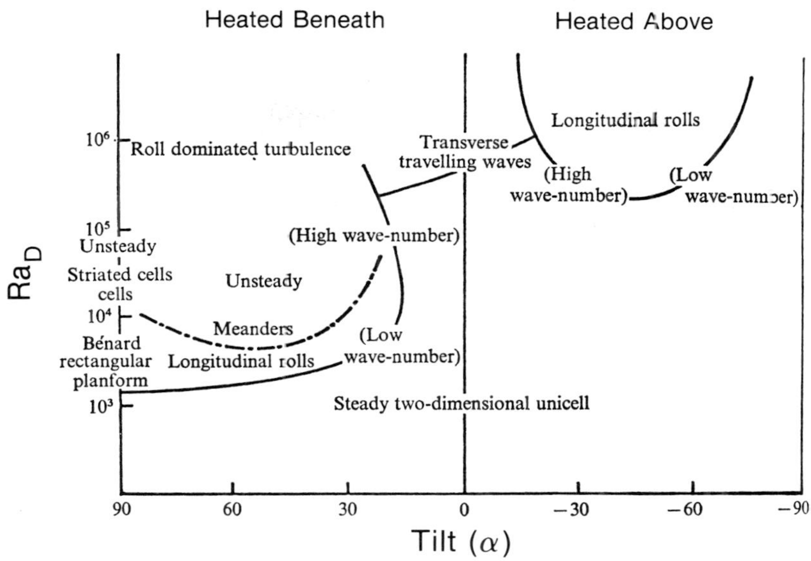

Fig. 2.13. Flow and stability régimes for a slender, laterally heated cavity of width D (after Hart 1971).

of a cavity flow, which is usually refluent. He also posulated that the scale of turbulence would be of the order of the turbulent boundary layer thickness when this is less than the cavity width. This is also consistent with observations on turbulent boundary layers and pipe flows, and implies that a boundary layer type of flow will generate the higher heat transfer rate. If so, the cavity width will exert an additional influence by limiting at least the scale of turbulence; narrower cavities may thus transfer less heat at the same (turbulent) Rayleigh number. Later work by Thomas and colleagues (1971, 1979) explores the use of surface renewal theory in which the penetration of eddies to the wall is explicitly recognized. More recently, Giel and Schmidt (1990) have compared experimental data for a vertical cavity with their low Reynolds number predictions obtained by adapting a pair of two-dimensional k–ε models. While the comparison revealed only qualitative agreement, it also demonstrated that such formulations are capable of spanning three major régimes: laminar, transitional, and fully turbulent. Eventually, it may be possible thus to model three-dimensional flows, at least under steady boundary conditions.

In concluding this section, it is worthwhile making a few summary remarks on thermally buoyant cavity flows. Perhaps their most obvious feature is their complexity which is matched by,and largely created by, their variability. It has been suggested above that the easiest way to describe the influence of the various thermal, hydrodynamic, and geometrical parameters is through the use of scaling and ordering techniques; this strategy is well suited to the estimation of convective fluxes dominated by the primary flow but has its limitations in cavities occupied by a variety of subregional flows. Orientation of the cavity and the surrounding temperature field relative to the force field are key factors in behaviour, as reflected in the division between cavities heated from the side and those heated from below. While solutions to the former are frequently stable and unique, the same cannot be said of the latter which frequently exhibit transition through more than one régime before turbulence appears. Turbulence itself is not well understood in the refluent flow patterns characteristic of cavity flows.

In such a short space, it has only been possible to introduce the characteristic features of natural convection in cavities, and to point out the imperfect state of our knowledge. This knowledge should also include features which have not been covered above: transient behaviour, three-dimensionality, non-gravitational fields, non-Newtonian fluids, thermosolutal coupling, etc. All of these complexities form part of the body of knowledge which must be completed if the behaviour of the tubular thermosyphon is ever to be fully understood.

2.2 The open, vertical tube

2.2.1 *Primary characteristics*

The simple, open tubular thermosyphon is a generic type to which many other forms claim relation. In its prototypical form it consists of no more than a heated, vertical tube closed at the bottom and open at the top where it communicates with a reservoir of cooler fluid. As indicated in Fig. 2.14, hotter fluid next to the wall rises in an annular space which creates and defines a central core space where cooler fluid descends at the same rate; the net mass flux along the tube is zero. A purely refluent circulation is thus established by thermal buoyance forces generated in the presence of the gravitational field. The precise details of the flow, and of the heat transfer it produces, depend principally upon the difference in temperature between the wall and the reservoir fluid, the length and diameter of the tube, and the fluid properties.

Early experiments with this system date back to the early 1950s when Foyle (1961) and Foster (1953) undertook limited studies from which the importance of the principal variables began to emerge, but it was not until the appearance of the seminal paper by Lighthill (1953) that theoretical and experimental investigations alike could be placed in a logical and general framework. Shortly after, the experiments of Martin (1954, 1955) confirmed this framework and added substantially to our understanding. Most of the

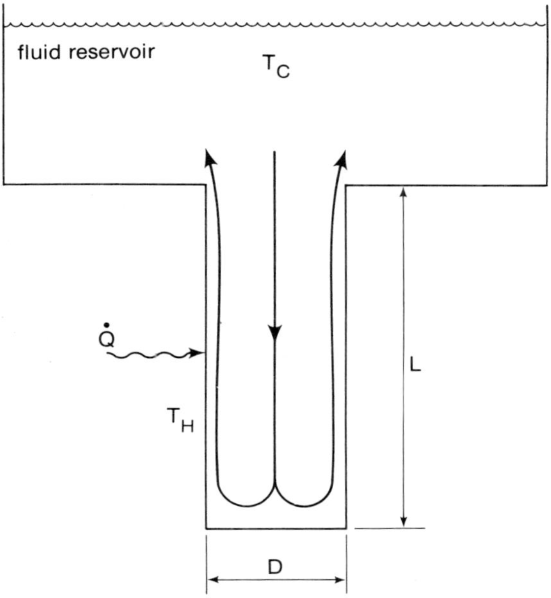

Fig. 2.14. The vertical, open thermosyphon.

characteristic features of the open, tubular thermosyphon were identified during this brief period.

It was noted, for example, that laminar flow extended over at least two régimes. For large temperature differences, a thin boundary layer surrounds a large and sluggish core which has very little effect on heat transfer; behaviour closely resembles that of a flat plate wrapped into the shape of the thermosyphon wall. On the other hand, as the temperature difference decreases, a thickening boundary layer is increasingly opposed by a smaller, but faster, core; this is appropriately known as the impeded régime. Ultimately, as $T_H \rightarrow T_C$, the temperature field becomes increasingly independent of the weakened motion and the system enters the conduction régime.

The equations of continuity, motion, and energy for steady, laminar flow in a circular[8] tube are given in cylindrical polar coordinates X, R by

$$\frac{\partial U}{\partial X}+\frac{\partial V}{\partial R}+\frac{V}{R}=0 \tag{2.20a}$$

[8] For simplicity, axisymmetry will be assumed. A discussion of the effect of cross-sectional geometry is given in Section 2.4.

$$U\frac{\partial U}{\partial X}+V\frac{\partial U}{\partial R}=-\frac{1}{\rho}\frac{\partial P'}{\partial X}+\beta g\theta+\nu\left(\frac{\partial^2 U}{\partial X^2}+\frac{1}{R}\frac{\partial U}{\partial R}+\frac{\partial^2 U}{\partial R^2}\right) \tag{2.20b}$$

$$U\frac{\partial V}{\partial X}+V\frac{\partial V}{\partial R}=-\frac{1}{\rho}\frac{\partial P'}{\partial R}+\nu\left(\frac{\partial^2 V}{\partial X^2}-\frac{V}{R^2}+\frac{1}{R}\frac{\partial V}{\partial R}+\frac{\partial^2 V}{\partial R^2}\right) \tag{2.20c}$$

$$U\frac{\partial \theta}{\partial X}+V\frac{\partial \theta}{\partial R}=\kappa\left(\frac{\partial^2 \theta}{\partial X^2}+\frac{1}{R}\frac{\partial \theta}{\partial R}+\frac{\partial^2 \theta}{\partial R^2}\right) \tag{2.20d}$$

where $\theta=T-T_{\mathrm{C}}$ and X is measured vertically upwards from the closed end, as indicated in Fig. 2.15. In the boundary layer régime, the appropriate velocity and length scales are determined as follows. Balancing the buoyancy and principal viscous forces in the longitudinal momentum equation gives

$$\frac{\beta g \nu^{\mathrm{c}}(R^{\mathrm{c}})^2}{\nu U^{\mathrm{c}}}=O(1), \tag{2.21}$$

while balancing the lateral conduction and longitudinal convection terms in the energy equation gives

$$\frac{U^{\mathrm{c}}(R^{\mathrm{c}})^2}{\kappa X^{\mathrm{c}}}=O(1). \tag{2.22}$$

Since $X^{\mathrm{c}}=O(L)$, the tube length, eqns (2.21) and (2.22), yield

$$R^{\mathrm{c}}=O\left(\frac{L}{Ra_{\mathrm{L}}^{1/4}}\right) \tag{2.23}$$

and

$$U^{\mathrm{c}}=O\left(\frac{\kappa}{L}Ra_{\mathrm{L}}^{1/2}\right) \tag{2.24}$$

in which $Ra_{\mathrm{L}}=\beta g\theta^{\mathrm{c}}L^3/\nu\kappa$ and $\theta^{\mathrm{c}}=T_{\mathrm{H}}-T_{\mathrm{C}}$. It is immediately obvious that these scales are essentially the same as those obtained previously for a rectangular cavity under boundary layer conditions. In fact they apply to any vertical surface on which a laminar, natural convection boundary layer exists. Following the argument of the previous section, the heat transfer rate into such a boundary layer may be expressed by

$$\bar{N}u_{\mathrm{L}}\propto Ra_{\mathrm{L}}^{1/4}. \tag{2.25}$$

This well-established relation only describes the heat transfer behaviour of a vertical open thermosyphon so long as the laminar boundary layer thickness is

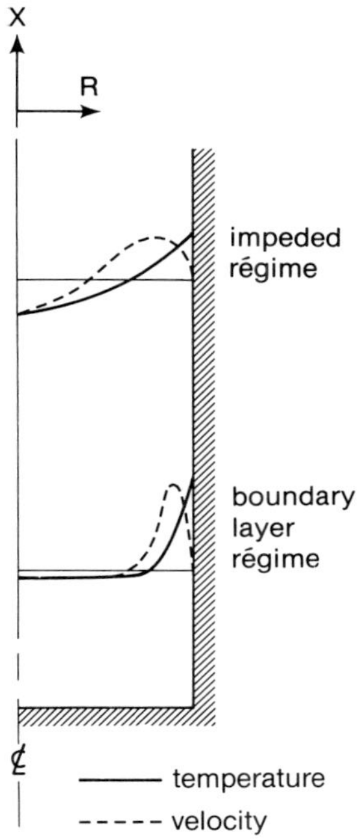

Fig. 2.15. Velocity and temperature profiles in a vertical, open thermosyphon.

much less than the tube radius, that is $R^c \ll D/2$ or $Ra_L^{1/4} \gg 2L/D$, which may also be written as $t_D^{1/4} \gg 2$, where $t_D = Ra_D\ D/L$. Equation (2.25) thus becomes increasingly inaccurate when t_D falls below $O(10^5)$ and the flow enters the impeded régime. In a thermosyphon, this régime differs from that in a slender, laterally heated cavity where two opposing boundary layers interfere directly with each other and leave no intermediate region in which the temperature and velocity gradients are necessarily small. However, as Fig. 2.15 illustrates, the region surrounding the thermosyphon tube axis always contains fluid in which velocity and temperature gradients are small in comparison with elsewhere[9]. In particular, the velocity gradient in the core interior is much lower than at the core boundary near which an inflection point harbours the potential for a Tollmien–Schlichting instability. The transition between the boundary layer and impeded régimes is sometimes associated with flow instability and oscillation but the precise mechanism has yet to be identified.

Scaling in the impeded régime demands the relaxation of the boundary layer energy balance which led to eqn (2.22). Hence from eqn (2.21) directly,

[9] The location of the velocity maximum provides the singular exception.

$$U^{c} = O\left(\frac{\kappa}{4D} Ra_D\right) \tag{2.26}$$

in which R^c has been replaced by $D/2$. The heat transfer rate associated with these conditions may be estimated from the longitudinal convective flux. Thus

$$\bar{\dot{q}}\pi DL = O\left(\rho U^{c} c_p \theta^{c} \frac{\pi D^2}{8}\right)$$

from which we obtain

$$\bar{\dot{q}} = O\left(\frac{k\theta^{c}}{32L} Ra_D\right)$$

and hence

$$\bar{Nu}_L \propto \frac{Ra_D}{32} \tag{2.27}$$

which contrasts sharply with eqn (2.25). In fact both results may be put in a more succinct form by writing them in terms of the diameter-based Nusselt number. Thus

$$\bar{Nu}_D = C_1 t_D^{1/4} \tag{2.28}$$

in the boundary layer régime, while

$$\bar{Nu}_D = C_2 t_D \tag{2.29}$$

in the impeded régime. Experiments reveal that $C_2 \ll C_1 = O(1)$.

As noted above, the boundary layer régime does not extend below $t_D = O(10^5)$. The upper limit of the impeded régime is more difficult to determine *a priori* but is found to be just below this point. It is therefore possible to estimate C_2 from C_1 because at the knee

$$C_1 t_D^{1/4} \simeq C_2 t_D$$

and hence

$$C_2 \simeq \frac{C_1}{t_D^{3/4}} = O\left(\frac{2}{10^4}\right). \tag{2.30}$$

The lower limit of the impeded régime is equally difficult to predict, except in the sense that it must somehow lead to a transition between this and a conduction régime. When the thermosyphon wall temperature *decreases linearly* from a maximum base value of T_H to T_C (the fluid temperature at the mouth, the conduction limit is simply specified by $\bar{Nu}_D = D/L$, provided it is understood that the total heat transfer rate is given, in general, by

$$\dot{Q} = \bar{q}_w \pi DL + \bar{q}_B \pi D^2/4$$

where $\bar{q}_w$ is the average wall flux density and $\bar{q}_B$ is the average base flux density[10]. The conduction and impeded régime relations taken together suggest that the transition between them is determined by a threshold value of Ra_D not t_L.

When the tube wall is *isothermal*, at a temperature T_H, the lower limit of the impeded régime is even more difficult to define. The isothermal wall solutions of Lighthill (1953), for example, reveal the emergence of a stagnant region at the bottom of the tube for t_D less than the value which yields a similarity[11] solution. At this point, any further reduction in t_D simply shortens the effective heated length of the tube until, eventually, it is of the same order as the diameter; the boundary layer approximations are then questionable and the weakened flow undergoes the fundamental change suggested earlier in Fig. 2.10.

Solutions to eqns (2.20) have been obtained by a number of authors. The earliest solutions were obtained by Lighthill (1953) followed later by Japikse and Winter (1971), both of whom employed the Von Karman–Pohlhausen integral method after the equations had been reduced to parabolic form. Solutions of the full elliptic forms have since been obtained numerically by Gosman *et al.* (1971) and Simpson (1986). These solutions and representative experimental data[12] from Martin (1955) are combined in Fig. 2.16 to demonstrate the good agreement which has been obtained between theory and experiment for non-metallic liquids. The two principal laminar régimes are clearly evident and the effect of L/D is satisfactorily absorbed[13] in t_D. Detailed comparison of the theoretical velocity and temperature profiles reveals excellent agreement in the impeded régime where the annulus and core have about the same width. In the boundary layer régime, however, Lighthill's model of an isothermal core is confirmed but his assumption of plug flow is not, presumably because when $Pr \gg 1$ the diffusion of momentum cannot be restricted to the thermal boundary layer. The data shown in Fig. 2.16 accurately fit the forms predicted in eqns (2.28) and (2.29) to give $C_1 = 0.6$ and $C_2 = 1.75 \times 10^{-4}$. The lower limit of the boundary régime occurs at $t_a = 10^{3.8}$ (corresponding to $t_D = 10^5$) while the knee is just below at $t_a = 10^{3.5}$. It is thus evident that primary flow analysis based on natural scales is a useful tool for predicting the longitudinal convective flux in a thermosyphon containing viscous liquids. Heat transfer coefficients observed in a 5 cm diameter tube are found to be in excess of 100 W/m^2 K in the boundary layer régime.

[10] Thus $\dot{Q} \simeq \bar{q}_B \pi D^2/4$ in the conduction régime but is otherwise given approximately by $\bar{q}_w \pi DL$ because $\bar{q}_B < \bar{q}_w$ and $D \ll 4L$ in a typical thermosyphon.

[11] Not to be confused with similarity in boundary layer flow.

[12] Note that these authors have chosen to present their results using tube radius a, not diameter: $t_L = 16t_a$.

[13] A paper by Yamagishi *et al.* (1989) suggests that a separate L/D effect may exist.

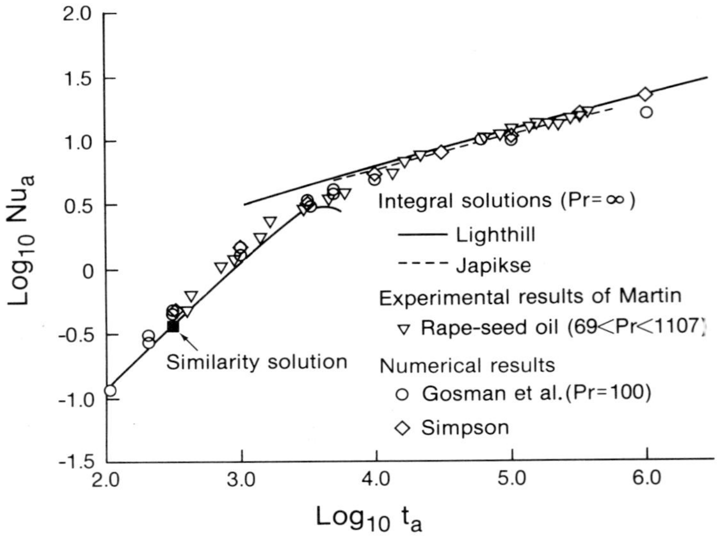

Fig. 2.16. Heat transfer behaviour of the vertical, open thermosyphon: laminar flow (after Simpson 1986).

What is not evident from Fig. 2.16 is the possibility of unsteady behaviour in the vicinity of the inter-régime transition. This was first observed by Martin (1955) using glycerine, and is indicated clearly in Fig. 2.17; curiously, this behaviour has not been reported in other fluids. The data are seen to be in excellent agreement with Lighthill's predictions for $Pr = \infty$, at least in the mean. The oscillations evidently have a considerable magnitude which decays slowly as the flow is better established in either régime. It is significant that in a fluid as viscous as glycerine the oscillations are noticeable well into the boundary layer régime. As mentioned earlier, the origin of these oscillations has not yet been determined but it is certain that they do not mark the laminar–turbulent transition, about which more will be said later. It is not unreasonable to suppose that they are determined by conditions at the tube mouth if, and when, the momentum boundary layer fills the tube, but this explanation may be facile. In fact, stability of the flow in the vicinity of the mouth raises a much more general problem.

Entry effects in the vertical, open thermosyphon have been studied carefully by Martin and Lockwood (1963) using flow visualization techniques. In general, they observed that the hot annulus of fluid exiting had a tendency to neck down to form a plume while the cold entering fluid behaved in a manner which depended upon Ra_D. In the impeded régime, the entering fluid created a series of egg-shaped holes in the annular discharge through which it passed, apparently with little or no mixing. This situation was suggested in Fig.

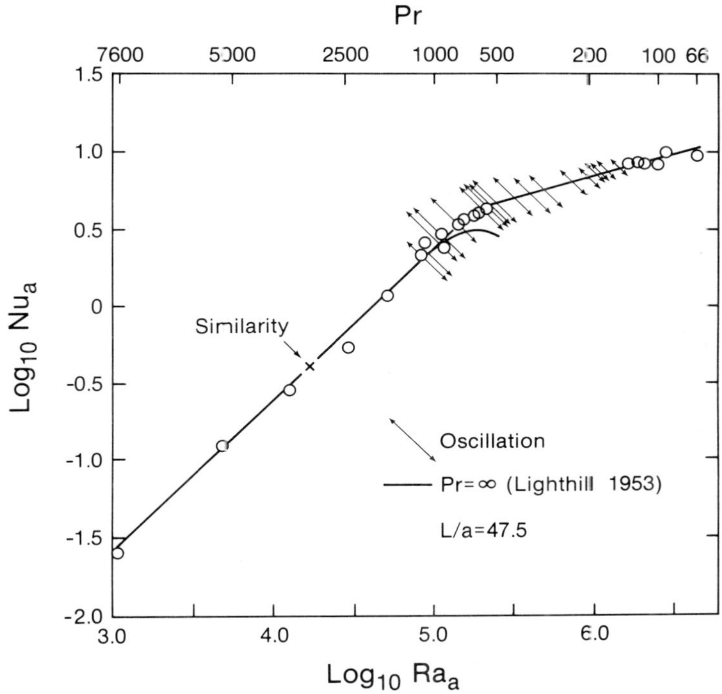

Fig. 2.17. Flow oscillations in a glycerine-filled, vertical thermosyphon (after Martin 1955).

2.10(c), and is depicted schematically in Fig. 2.18(a). In the boundary layer régime, the parting of the hot 'curtain' was effected by the asymmetric juxtaposition depicted in Fig. 2.18(b). Around some parts of the tube mouth, the exiting hot fluid simply entrained any external fluid nearby, while elsewhere around the mouth the entering fluid caused the hot fluid to recirculate, thus giving rise to local mixing. This type of recirculation has also been noted in a heated, open channel (Sparrow *et al.* 1984) but for Rayleigh numbers (based on channel width) of about 3.5×10^4.

2.2.2 *The origin and effect of turbulence*

Although it is to be expected that mouth conditions would influence oscillatory behaviour in the vicinity of the impeded–boundary layer transition, they are not believed to be the cause of it. Nor are they considered to be the cause of the instability leading to the laminar–turbulent transition. The criterion for this is not yet clear, and evidently depends upon the régime in which the transition occurs. Japikse and Winter (1971) have argued that t_D must reach a critical value, thus implying that $(Ra_D)_{cr} \propto L/D$. The data they

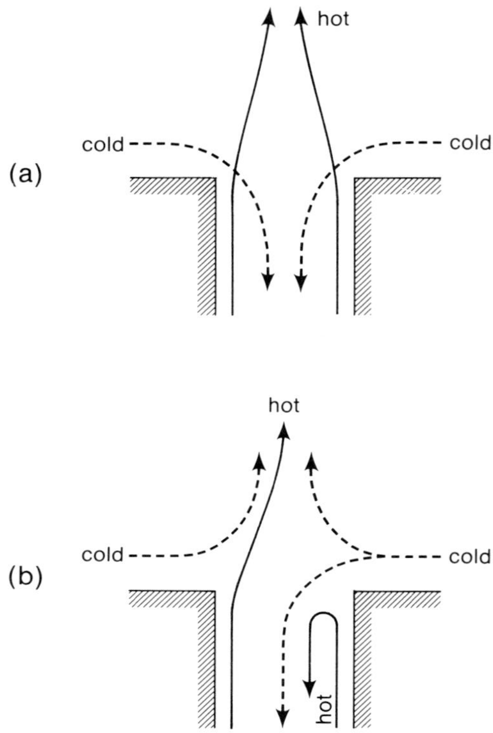

Fig. 2.18. Laminar entry flows for the vertical, open thermosyphon.

present in support of this, however, follow a correlation of the form $(Ra_D)_{cr}Pr^{1/3} = \text{constant}$; and it is clear from Martin's data for air that $(Ra_D) \simeq 10^{5.1}$ marks a transition in the impeded régime, and is independent of length–diameter ratio. But the alternative suggestion that the Rayleigh number alone provides the criterion is not supported by Martin's data for water or ethylene glycol. The latter fluid undergoes transition in the boundary layer régime, the change causing an increase in heat transfer, while the transition for water and air both occur in the impeded régime, and both create a chasm in the $\bar{N}u_D$–t_D curve before restoring the heat transfer rate to its value immediately before transition.

In the boundary layer régime, it may be argued that transition is reflected in a critical Reynolds number, based on the boundary layer thickness. Since this is defined by

$$Re = \frac{U^c R^c}{\nu} = \frac{Ra_L^{1/4}}{Pr},$$

the implication is that

$$(Ra_{\mathrm{L}})_{\mathrm{cr}} \propto (Pe)^{4}_{\mathrm{cr}}. \tag{2.31}$$

For air moving over a flat plate, $(Ra_{\mathrm{L}})_{\mathrm{cr}} = O(10^9)$ and hence $(Re)_{\mathrm{cr}} \simeq 250$. However, this implies that $(Ra_{\mathrm{L}})_{\mathrm{cr}}$ for water ($Pr \simeq 8$) is 16×10^{12}, much higher than that observed by Martin (1955). Likewise, if it is argued that the diameter-based Reynolds number defined by

$$Re_D = \frac{U^{\mathrm{c}} D}{\nu} = \frac{1}{4} \frac{Ra_D}{Pr}$$

reaches a critical value in the impeded régime then, according to Martin's air data, $(Ra_D)_{\mathrm{cr}} = 10^{5.1}$ and hence $(Re_D)_{\mathrm{cr}} \simeq 10^{4.6}$ which is about an order of magnitude higher than the value for the classical Reynolds transition, even though the flow is refluent. It thus appears that the laminar–turbulent transition is best represented by

$$(Ra_D)_{\mathrm{cr}} = N f(Pr) \tag{2.32}$$

where N is a large number which depends upon the régime in which the transition occurs, and $f(Pr)$ is a weak monotonic function[14].

In shallow tubes, the transition to turbulence is likely to occur in the boundary layer régime. A turbulent boundary layer then develops in the presence of a laminar core, thus creating an increased heat transfer rate which may be estimated from the mixing length models employed by Eckert and Jackson (1951) and Bayley (1955*b*). For longer tubes, the transition is more likely to occur in the impeded régime with the result that the core and annular flows will intermix. Lighthill (1953) attempted to describe this situation for a fluid with $Pr = 1$ by adopting an eddy diffusivity model based on the observations of Reichardt (1951). Although the latter referred to pipe flow, Lighthill postulated that the eddy diffusivity distribution outside of the laminar sublayer (and a transitional layer) would have a similar form in a thermosyphon tube when the turbulence 'seeks to fill the whole tube'. He also hypothesized that transition to turbulence would depend upon the local Reynolds number. His findings are presented in Fig. 2.19 from which it is immediately evident that the predicted transition to turbulent flow in the impeded régime, or slightly above it, creates a dramatic drop in heat transfer. This drop, together with an equally abrupt rise, has been observed by Martin (1955) using air, but his data were not extensive enough to test the predicted re-establishment of a boundary layer régime at higher Rayleigh numbers. It is difficult to check Lighthill's proposed transition criterion but the previous discussion places it in some doubt.

For viscous liquids, Martin (1955) found that the reduction in heat transfer

[14] Increasing according to Martin and decreasing according to Japikse and Winter.

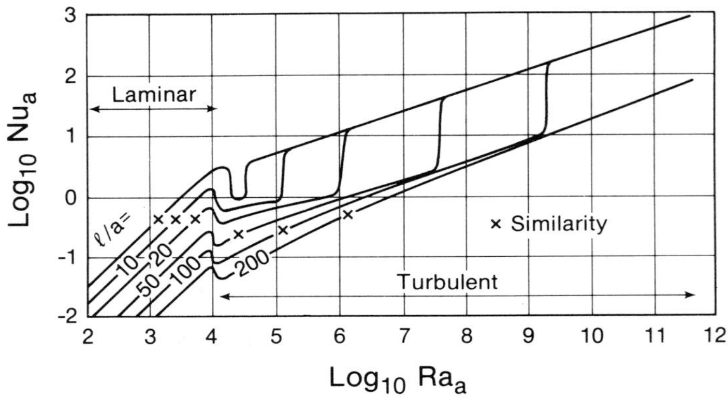

Fig. 2.19. Predicted laminar and turbulent heat transfer characteristics in a vertical, open thermosyphon with $Pr=1$ (after Lighthill 1953).

rate in the fully mixed turbulent régime is not as great as suggested by Lighthill, although Hasegawa *et al.* (1963) report closer agreeement. Figure 2.20 provides some comparisons. Martin found no evidence of a turbulent boundary layer in these liquids. Even so, the data in Figs 2.19 and 2.20 do draw attention to the effect of turbulence in the core, and suggest how it increases with increases in length–diameter ratio. It is also interesting to note the the slopes of the fully mixed data, theoretical and experimental, are not constant; Rayleigh number and length–diameter ratio both have an effect. Some later experiments with water in small-bore tubes (Lock and Liu 1989; Lock and Kirchner 1991) tend to confirm these observations, specifically revealing how the presence of fully mixed turbulence may reduce the heat transfer rate for a given Rayleigh number.

2.2.3 *The role of fluid properties*

The Prandtl number is not the only fluid property of relevance in natural convection but it does provide the single most important measure of property influence on thermosyphon behaviour. For most liquids and gases, the balance of the viscous forces with the buoyancy forces provides a suitable description of the physics of flow; the laminar heat transfer relation thus take the general form

$$\bar{N}u_D = F(t_D, Pr). \tag{2.33}$$

Theoretical and experimental work both indicate that F is a weak function of Prandtl number provided $Pr \geq O(1)$. For liquid metals, on the other hand, this model is inappropriate and must be replaced by one in which the buoyancy

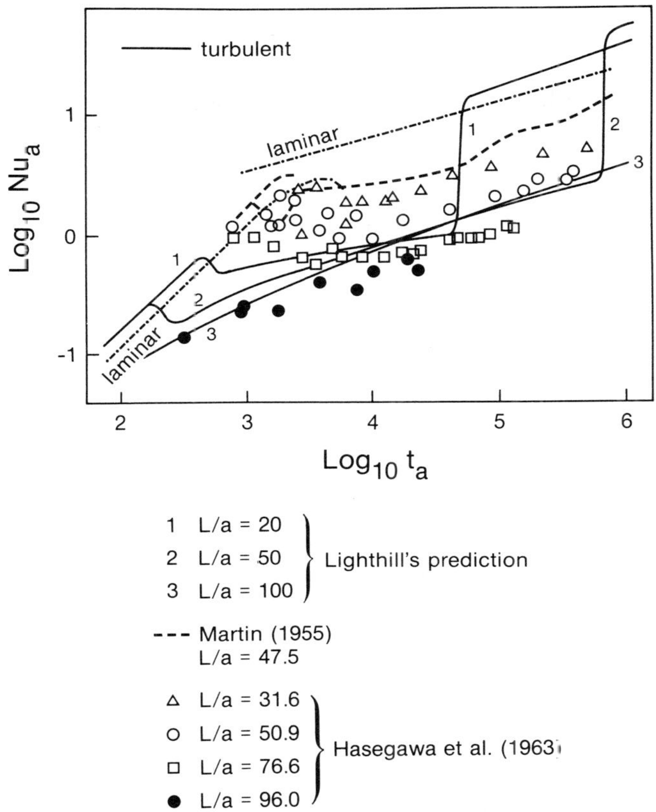

Fig. 2.20. Turbulent heat transfer in a vertical, open thermosyphon (after Japikse 1973).

forces are balanced by the inertial forces. This implies that eqn (2.21) must be replaced by

$$\frac{\beta g \theta^{c} X^{c}}{(U^{c})^{2}} = O(1) \tag{2.34}$$

which, together with eqn (2.22), yields the corresponding width and velocity scale in the boundary layer régime as

$$R^{c} = O\left(\frac{L}{Pr^{1/4} Ra_{L}^{1/4}}\right) \tag{2.35}$$

and

$$U^{c} = O\left(\frac{\kappa}{L} Pr^{1/2} Ra_{\mathrm{L}}^{1/2}\right). \tag{2.36}$$

Comparison of these equations with eqns (2.23) and (2.24) reveals the stronger influence of the Prandtl number in a liquid metal boundary layer[15]; in particular, the layer thickness is increased.

In the impeded régime, the annulus and core would each fill about half of the tube. The boundary layer régime is thus restricted to

$$R^{c} \ll \frac{D}{2}$$

or, for liquid metals,

$$(Prt_D)^{1/4} \gg 2$$

which implies that $Prt_D = O(10^5)$ marks the lower reaches of the régime in which, following the use of eqns (2.35) and (2.36), the average heat flux density is given by

$$\bar{q} \propto \frac{k\theta^{c}}{L} (PrRa_{\mathrm{L}})^{1/4},$$

and hence

$$\overline{Nu}_D = C_1 (Prt_D)^{1/4} \tag{2.37}$$

as compared with eqn (2.28) for non-metals. In the impeded régime, the use of eqn (2.34) implies that

$$\overline{Nu}_D = C_2 (Prt_D)^{1/2} \tag{2.38}$$

as compared with eqn (2.29).

The existence of a laminar–impeded régime in liquid metals is doubtful, although evidence is mixed. Theoretically, the régime cannot exist beyond $t_D \simeq 10^5/Pr$ which is much higher than the value for non-metallic fluids; at the same time, the Nusselt numbers are substantially lower. These facts suggest that the impeded régime may make only a brief appearance, if at all, in a gradual transition from the boundary layer régime to the conduction régime. Leslie and Martin (1959) concluded that similarity (impeded) solutions were impossible if $Pr < 0.4$, whereas the data of Bayley and Czekanski (1963) indicate a definite knee not much below $Prt_D = 10^5$. However, the steeper data of the latter authors, which are shown in Fig. 2.21, indicate that $\overline{Nu} \propto t_D$, thus contradicting the square root relationship given by eqn (2.38). The existence of the laminar–boundary layer régime is confirmed in Bayley *et al.*'s earlier data

[15] Again it should be noted that no distinction has been made between the thermal and momentum boundary layers. Obviously, the former would be thicker in a liquid metal.

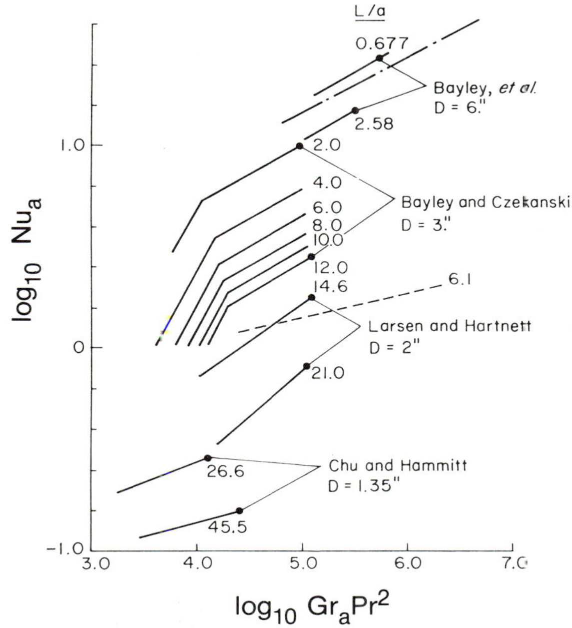

Fig. 2.21. Heat transfer behaviour of metal-filled vertical, open thermosyphons (after Japikse 1973).

(1961) which were obtained with smaller length–diameter ratios and thus closely resemble flat-plate data. It is thus reasonable to expect the laminar heat transfer data of liquid metals to correlate with Prt_D, but it is important to recognize that transition to turbulence may occur for values of this parameter which are below 10^5.

Figure 2.21 is Japikse's (1973) summary of liquid metal heat transfer data in the vertical, open thermosyphon. It reveals that tube geometry exerts a very great influence but its precise role is not clear. The use of Prt_D rather than $PrRa_D$ may offer an improvement in correlating the data but is evidently unable to absorb geometrical effects fully. For all but the shallowest tubes, radial heat conduction penetrates into the core where it acts to raise the fluid temperature with increasing depth; that is, the core is adversely stratified and may, depending upon boundary conditions at the bottom, become unstable. This prospect, coupled with Martin's observation that transition to turbulence occurred in the impeded régime for fluids with $Pr \leq 10$, suggests that turbulent flow will be common in metal-filled thermosyphons. This may explain the apparent dependency on tube diameter indicated in Fig. 2.21. The scale of turbulence cannot be greater than the tube diameter; with fully mixed turbulence, therefore, the scale, and presumably the intensity, of turbulence could be expected to decline with a reduced tube diameter, as noted earlier.

In ending this brief discussion of fluid property effects in the open thermosyphon, it is worth mentioning that temperature and pressure dependency may also be important. The effect of temperature dependence is most often seen in non-metallic liquids which have temperature-sensitive viscosities, thus presenting a problem in deciding on an appropriate choice of reference temperature for both ν and Pr. The average properties of the integral forms of the boundary layer equations provide a useful guideline; namely, that since integration takes place from the wall to the interior (where temperature gradients are substantially less), the properties are best evaluated at the wall. Pressure dependency is usually less important, except near the critical point. The coefficient of expansion

$$\beta = \frac{1}{v}\left(\frac{\partial v}{\partial T}\right)_{\mathrm{P}} = -\left(\frac{\partial P}{\partial T}\right)_{\mathrm{V}} \Big/ v\left(\frac{\partial P}{\partial v}\right)_{\mathrm{T}} \rightarrow \infty$$

at the critical point[16]; this makes the Boussinesq approximation $\beta\theta^{c} \ll 1$ invalid. The variation of β with temperature is also important in the narrow temperature range surrounding a density maximum, that is where $\beta = 0$, but this has little practical significance in typical thermosyphons.

2.3 The closed, vertical tube

The application of the Holzwarth–Schmidt open thermosyphon system to gas turbine blade cooling creates a number of operating difficulties: high hydrostatic pressure; oxidation leading to blockage; uncontrolled boiling; and restriction to inexpensive, chemically suitable, coolants. These limitations led Cohen, among others, to regard the closed system as a better solution (Cohen and Bayley 1955; Andvig 1956). Hahnemann (1951) made an early attempt to develop an empirical correlation for a gravitational version of the system, but the first systematic study appears to be that of Bayley and Lock (1965) who added a cooled upper section to Martin's apparatus. Japikse (1973) and others later expanded this work with both fundamental and empirical contributions.

2.3.1 *The effect of closure*

At first glance, one would not expect the closing of an open thermosyphon tube to alter its behaviour significantly but the brief discussion of openings in Section 2.1 suggests otherwise. Opening or closing one end of the tube is more than a mere modification of the end boundary conditions. The governing

[16] Where $-(\partial P/\partial T)_{\mathrm{V}}/v \neq 0$, but $(\partial P/\partial V)_{\mathrm{T}} = 0$.

equations, of course, remain the same; and the properties of their solutions, as reflected in the various flow régimes, may be expected to reappear. But the means by which thermal energy leaves (and enters) the tube changes fundamentally with closure; from convection through the mouth to conduction through the tube sides and the end itself. It is often useful to think of the closed system as two open systems joined mouth to mouth.

The heat transfer data of Bayley and Lock (1965) for glycerine clearly reveal the existence of the laminar–impeded régime in the closed tube thermosyphon. Visual confirmation has been provided by Schenk *et al.* (1969), and is shown in Fig. 2.22 which not only suggests impeded flow in the lower heated section, but impeded flow together with an almost stagnant end region in the upper cooled section. This immediately raises an important question. Can the flow régimes in the heated and cooled sections be different and, if so, under what conditions? In answering the first question in the affirmative, Japikse (1969) noted that temperature dependence of the fluid properties may be one cause, and cited as evidence some glycerine 'boundary layer' data (Bayley and Lock 1965) which lay well beneath data for fluids with much lower Prandtl numbers. This apparant contradiction of well-established open thermosyphon behaviour was resolved by the fact that the glycerine in the heated section was in the boundary layer régime while that in the cooled section, having a much greater viscosity, was in the impeded régime. A similar argument may explain the reduced circulation near the upper end of the tube in Fig. 2.22. The cooled section of a closed thermosyphon may thus exhibit régime behaviour which is displaced down[17] the t_D scale relative to the heated section, and it is therefore important to treat the two sections separately even though they are intimately connected in the behaviour of the system overall. In particular, fluid properties in each section must be evaluated at the appropriate section wall temperature (Japikse and Winter 1971); this is a logical extension of the referential use of the wall temperature suggested in the previous section.

Boundary layer régimes are evident in the heat transfer data displayed in Fig. 2.23. Transition to turbulent flow, as detected by irregular and rapid fluctuations[18] in the wall temperatures, is marked by the arrows. To the left of the arrow, the data show a slope of slightly less than the theoretical value of 0.25, presumably because of the temperature-dependent Prandtl number. Typically, the transition is marked by a slight decrease in heat transfer followed by a curve of steeper slope. This is illustrated in the two upper sets of data between the point of transition and $t_D \simeq 10^{7.5}$. Similar behaviour is evident in the water data shown in Fig. 2.24.

Evidence of the fully mixed turbulent régime in short tubes is equivocal,

[17] Assuming typical temperature-dependent characteristics for the fluid, for example, the viscosity becoming less at higher temperature.

[18] These are markedly different from the almost periodic, lower-frequency fluctuations characteristic of transition between the impeded and boundary layer régimes.

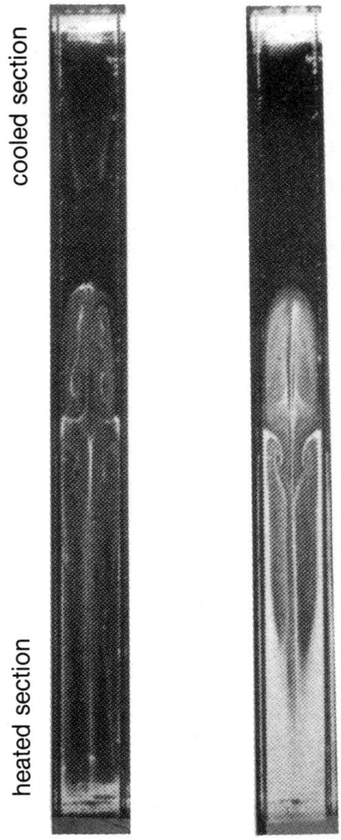

Fig. 2.22. Flow visualization of the laminar-impeded régime in a vertical, closed thermosyphon (after Schenk *et al.* 1969).

despite the fact that the upper t_D range is almost the same as that covered by Martin (1955) who presented strong evidence for its existence in the open thermosyphon. Whether or not the data above $t_D = 10^{7.5}$ in Fig. 2.23 show the characteristic deterioration in heat transfer associated with complete mixing is not conclusive. However, what does become clear, especially from Fig. 2.24, is that the performance of the turbulent closed system improves more rapidly with t_D than does its open counterpart. It thus appears possible that the closed system may outperform the open system for very large values of t_D. Further support for this trend is seen in Fig. 2.25 which displays data obtained with methanol (Lock and Simpson 1989). Extrapolating to $Nu_D = 10^2$ corresponds to a heat transfer coefficient of 600 W/m^2 K with a water filling. These data are also interesting on two other accounts. First, the heat transfer rate in the lower

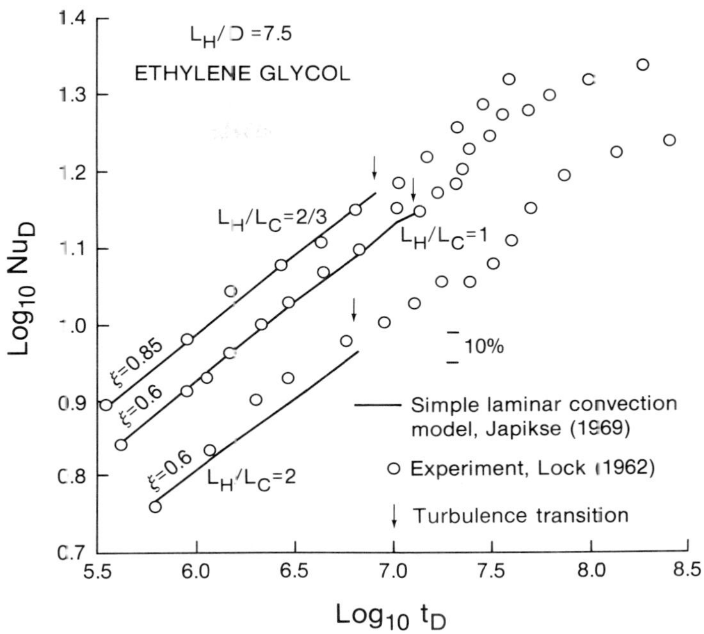

Fig. 2.23. Boundary layer heat transfer régimes in the vertical, closed thermosyphon: the effect of L_H/L_C (after Japikse 1973).

range of each data cluster shows a tendency to fall off[19], thus raising the question of possible chasmic behaviour associated with the transition from laminar–impeded flow to fully mixed turbulence or, more likely, a transition to impeded flow. Secondly, the empirical correlation draws attention to the geometrical parameter $(D/L_H)^3(L_C/L_H)$ which reappears in non-linear thermosyphons and in the aerosyphon discussed in Section 5.3.

One further example of open behaviour reflected in the closed system may be found in the upper reaches of the laminar–impeded régime. Substantial flow oscillations in the vicinity of transition to laminar, boundary layer flow in the open system were noted in the previous section (see Fig. 2.17). These were also detected in the closed system (Bayley and Lock 1965), thus suggesting that they do not originate externally from disturbances which enter the system through the mouth. They were, however, found to be smaller in amplitude in the closed system, indicating that some damping results from confinement deep within the tube. A common feature is also found in the air data: those of Bayley and Lock display the same chasmic transition to turbulence observed

[19] This is clearly seen when $L_H/D=10$ and $L_H/L_C=1$, but is also evident in other data sets, including some not shown here.

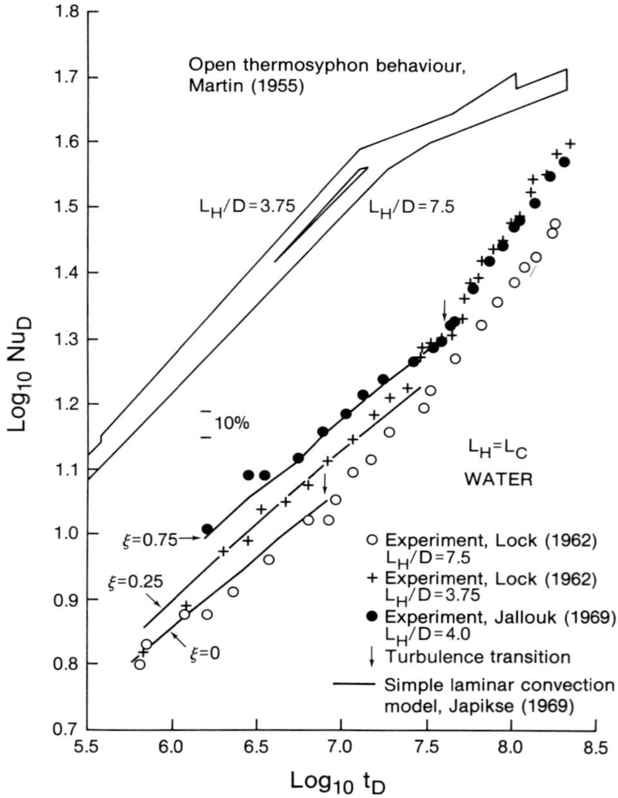

Fig. 2.24. Boundary layer régimes in the vertical, closed thermosyphon: the effect of L_H/D (after Japikse 1973).

by Martin; the Rayleigh number at the trough of the chasm is essentially the same in both systems. However, there is an important difference. In the closed thermosyphon, the dramatic rise in heat transfer beyond the chasm restored the Nusselt number to values normally associated with boundary layer flow; typically, these are comparable with laminar, boundary layer liquid data extrapolated backwards.

2.3.2 *Coupling mechanisms and models*

A model of the closed thermosyphon thus emerges. It is depicted in Fig. 2.26 as a pair of open tubes coupled at their mouths, and therefore accommodates open-system behaviour automatically. Differences between open and closed behaviour are thus ascribed to the mechanism by which coupling takes place. As described more fully later, the coupling region extends about one tube

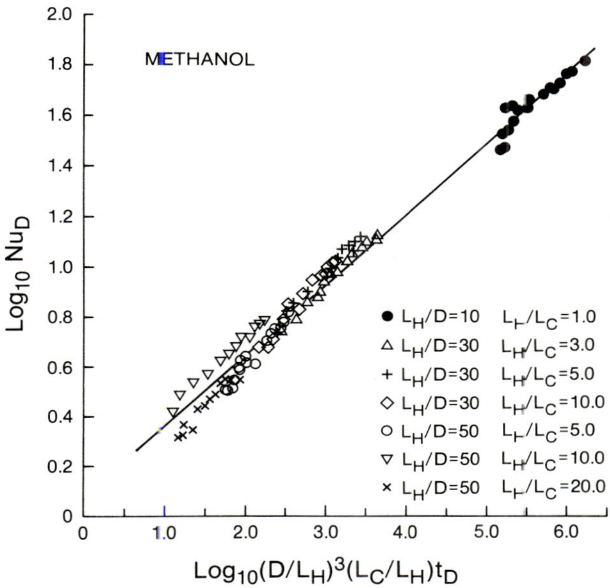

Fig. 2.25. Turbulent heat transfer in the vertical, closed thermosyphon (after Lock and Simpson 1989).

width on either side of the plane nominally separating the heated and cooled sections. It receives an annulus of hot fluid beneath and a corresponding annulus of cold fluid from above; at the same time, it discharges two central fluid cores in opposite directions. The means by which this balance between incoming and outgoing streams is achieved is unknown *a priori*.

Bayley and Lock (1965) suggested that the entire coupling process could be viewed as a radially focusing 'jet' formed at the confluence of the annular streams. Such a 'jet' is subject to adverse vertical temperature gradients which influence its stability and thereby affect its course and structure. Three hypothetical limits are indicated in Fig. 2.26. For small temperature differences $(T_H - T_C)$ across very viscous fluids, the 'jet' would be stable; consequently, each annulus would double back to become its own core, and heat would then be transferred across the coupling region by conduction. This is represented in Fig. 2.26(a). Larger temperature differences in the same viscous fluids could eventually destabilize the 'jet' causing it to break into alternating fingers: cold fingers moving downwards while hot fingers move upwards, much like the interdigitative pattern one may form with the hands. This is shown schematically in Fig. 2.26(b) which suggests that the cool fingers collectively become the downward moving core in the heated section, while a corresponding process applies to the formation of the core above. Should the

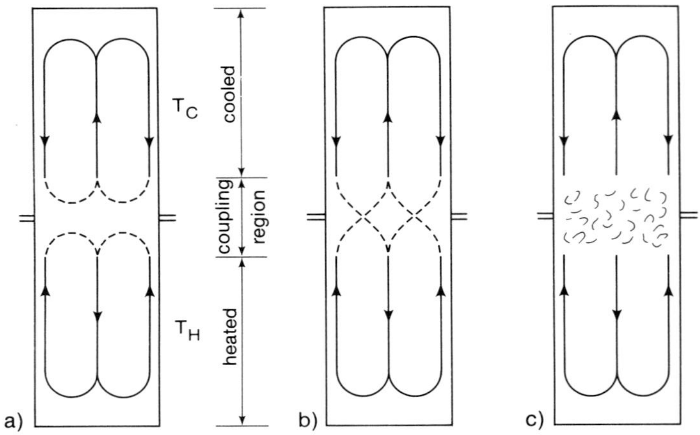

Fig. 2.26. Models of the vertical, closed thermosyphon: (a) conduction; (b) convection; (c) mixing.

temperature become too great, or the fluid have a low viscosity, or both, the convective process may deteriorate into vigorous mixing, as depicted in Fig. 2.26(c). This basic model of coupling, not as a single process but as one which alters with Prandtl number and Rayleigh number, has since been verified visually (Japikse *et al.* 1971; Schenk *et al.* 1969). The study of Japikse (1969) is particularly noteworthy and will be taken up more fully later.

The practical merit of the above description is that it suggests a way by which open thermosyphon data may be used to predict the performance of the closed thermosyphon (Bayley and Lock 1965; Japikse *et al.* 1971; Japikse 1973). In essence, this consists of satisfying two requirements: that the total heat conducted through the wall into the heated section must be equal to that conducted out through the wall of the cooled section; and that the mass flux in each of the cores must be some specified function of the mass fluxes in the annuli. The first of these is simply a particular example of the first law of thermodynamics. The second is a statement of continuity which carries the essential information on *how* the material is transferred from the annuli to the cores.

The general situation is modelled in Fig. 2.27. Of the ascending annulus mass flow $\dot{m}_{HA}$, a fraction x_H crosses into the cold core. At the same time, a fraction $1-x_C$ of the descending annulus mass flow $\dot{m}_{CA}$ also enters the cold core. Applying the continuity requirement (zero net mass flux) to the cooled section,

$$\dot{m}_{CA}=x_H\dot{m}_{HA}+(1-x_C)\dot{m}_{CA} \quad \text{or} \quad \frac{x_C}{x_H}=\frac{\dot{m}_{HA}}{\dot{m}_{CA}}. \tag{2.39}$$

The convective flux transmitted up through the system may be written in a number of ways. For example, in the cooled section just above the coupling region

$$\dot{Q}=[x_{\mathrm{H}}\dot{m}_{\mathrm{HA}}+(1-x_{\mathrm{C}})\dot{m}_{\mathrm{CA}}]c_{\mathrm{p}}\bar{T}_{\mathrm{CC}}-\dot{m}_{\mathrm{CA}}c_{\mathrm{p}}\bar{T}_{\mathrm{CA}}$$

where $\bar{T}_{\mathrm{CC}}$ and $\bar{T}_{\mathrm{CA}}$ represent the mean fluid temperatures in the cold core and cold annulus, respectively. By virtue of continuity, this simplifies to

$$\dot{Q}=\dot{m}_{\mathrm{CA}}c_{\mathrm{p}}(\bar{T}_{\mathrm{CC}}-\bar{T}_{\mathrm{CA}}), \tag{2.40}$$

corresponding to which

$$\dot{Q}=\dot{m}_{\mathrm{HA}}c_{\mathrm{p}}(\bar{T}_{\mathrm{HA}}-\bar{T}_{\mathrm{HC}}) \tag{2.41}$$

in the heated section just below the coupling region. The problem thus becomes the determination of the temperatures $\bar{T}$ for open thermosyphons which are coupled together in some manner prescribed through x_{C} and x_{H}.

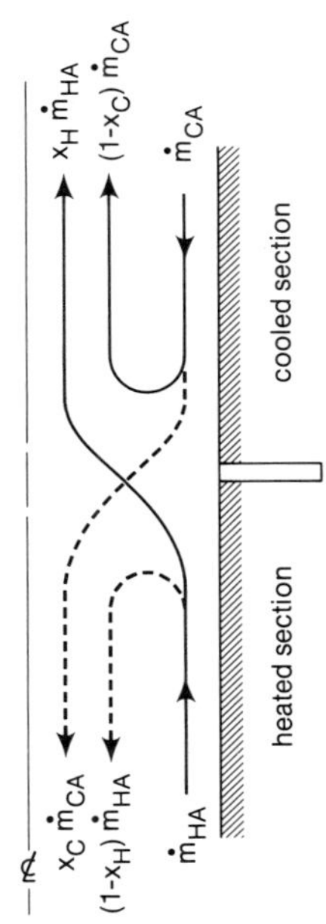

Fig. 2.27. Coupling circulation model.

When the coupling is conductive, $x_{\mathrm{C}}=x_{\mathrm{H}}=0$; the convective flux in the heated section core as it leaves the coupling region is then equal to the convective flux of its own ascending annulus at entry to the coupling region minus the heat conducted into the cooled section across the interfacial plane separating the two sections. When the coupling is convective, $x_{\mathrm{C}}=x_{\mathrm{H}}=1$; the ascending core leaving the coupling region has the same convective flux as the ascending annulus entering beneath[20]. For a mixing mechanism, x_C and x_H are indeterminate but the core temperatures are then both equal to T_m, the mixed mean temperature of the entering fluid annuli.

To solve the coupling problem it is necessary to specify the ratio x_C/x_H, at the very least. However, this is unknown and is not easily related to the physics of coupling. Two alternative simplifications have therefore been proposed, each expressing a direct relationship between the convective fluxes. Japikse (1973) employed the form

$$A_{CC} = \xi A_{HA} \tag{2.42}$$

where $A_{CC} = \dot{m}_{CC} c_p (T_{CC} - T_m)$ and $A_{HA} = \dot{m}_{HA} c_p (\bar{T}_{HA} - T_m)$ are the cold core and hot annulus convective fluxes, respectively. The quantity ξ thus represents the fraction of A_{HA} which reappears in the cold core. This definition has the disadvantage of ignoring the role of the cold annulus but has the merit of simplicity. Equation (2.42) forms the basis of the theoretical 'predictions' shown earlier in Figs 2.23 and 2.24 under laminar, boundary layer conditions for which the values of ξ chosen to fit the experimental data provide valuable insight into the effect of viscosity and geometry on the coupling mechanism. Figure 2.23 suggests that the coupling is largely convective in short tubes filled with highly viscous liquids and does not alter appreciably with the heated–cooled length ratio. Figure 2.24, on the other hand, indicates that ξ falls off with decreasing viscosity and as the tube is lengthened. Typically, $0 < \xi < 1$, although negative values are possible (Japikse 1973), presumably because of a contribution from both annular flows. The comparisons suggest that $\xi \simeq 1$ in short tubes if $Pr \gg 1$, but $\xi \to 0$ as L_H/D increases and/or Pr decreases.

The second simplified coupling model (Simpson 1986) uses the convective fluxes in both annuli according to the relation

$$A_{CC} = A_{HA} - 0.5K(A_{HA} - A_{CA}) \tag{2.43}$$

with a corresponding equation for A_{HC}. This has the ready interpretation that $K=0$ for convection while $K=1$ for mixing; furthermore, $K=2$ corresponds to refluent mixing[21]. The model is useful in revealing the deterioration in heat transfer performance which occurs as K increases; that is, as convection→ mixing→refluent mixing. This is displayed in Fig. 2.28 which is founded on numerical solutions to the full elliptic equations governing laminar flow and heat transfer in a vertical, open thermosyphon (Simpson 1986). As indicated, the loss of convective breakthrough in both the impeded and boundary layer

[20] Ignoring any secondary conduction between streams crossing through each other.

[21] Strictly, $K=2$ implies no heat transfer across the interfacial plane; thus $K \lesssim 2$ corresponds to non-convective heat transfer with a low effective conductivity.

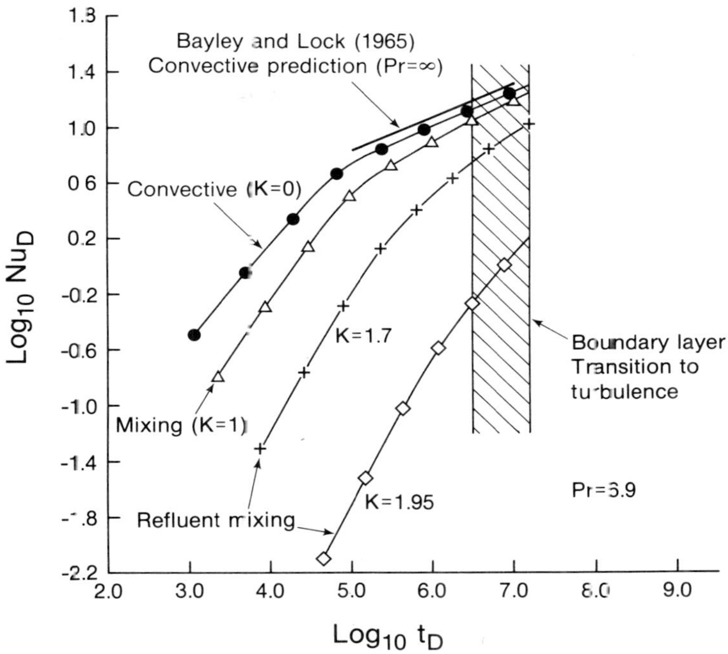

Fig. 2.28. Numerical study of the effect of coupling mechanism on laminar heat transfer in the vertical, closed thermosyphon (after Simpson 1986).

régimes introduces a significant penalty which increases substantially once the annuli begin to discharge an appreciable fraction of their contents back into their own cores. It will be recalled that under the turbulent conditions reflected in Fig. 2.25 the closed thermosyphon may become superior to an open system operating between the same temperatures. The explanation may lie in the coupling mechanism.

2.3.3 *Observations on the coupling region*

The general circulation model of coupling depicted in Fig. 2.27 masks a great many details which, as noted above, vary widely with conditions and constraints. To obtain a better understanding of this variability we conclude this section with a review of observations made on the coupling region itself. To begin with, it is worth recalling that small temperature differences in high Prandtl number fluids are likely to produce a completely refluent flow, but without secondary mixing; that is, heat will be transferred across the interfacial plane by conduction alone. The work of Mallinson *et al.* (1981) on a short, square thermosyphon provides clear evidence of this mechanism, as illustrated in Fig. 2.29 with $Ra_D = 10^3$. This shows the refluent circulation in each section; heat transfer calculations revealed very close agreement with a purely conductive solution. The same authors also obtained numerical solutions for fluids with $Pr > 10^2$ and $10^4 \leq Ra_D \leq 10^5$. Their data, which they corroborated with visualization studies, are in accord with the earlier work of Japikse (1969) who first recorded the details of convective coupling using a circular section thermosyphon.

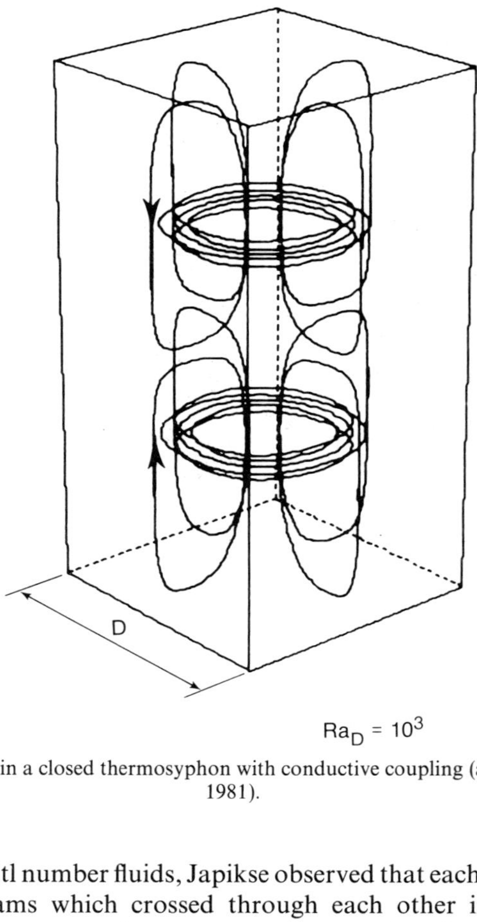

Fig. 2.29. Circulation in a closed thermosyphon with conductive coupling (after Mallinson *et al.* 1981).

For high Prandtl number fluids, Japikse observed that each annulus divided into several streams which crossed through each other in interdigitative fashion. In this process, each stream becomes partly refluent, even though much of it continues in the original axial direction to contribute to the core in the opposite section. The overall process is thus partly convective and partly refluent, thereby lending support to the circulation model described in Fig. 2.27. This observed flow pattern has features in common with the entry patterns of the open thermosyphon described previously in Fig. 2.18; in particular, the alternations of refluence (inflow) and convection (outflow) are strikingly similar. It appears to be the Rayleigh number which controls these details of the coupling process because it determines both the inertial forces (through the velocity scales in the tube) and the buoyancy forces in the vicinity of the coupling region. Below a critical value of Ra_D, conduction is found; immediately above the critical value, a complex, open cellular flow exists; well above the critical value, the cellular flow becomes turbulent.

Japikse noted that lowering the Prandtl number or raising the Rayleigh number had a similar effect; the interdigitating streams increased in number and their behaviour became increasingly capricious. Conversely, lowering Ra_D or increasing Pr decreases the number of streams and stabilizes their behaviour. Coupling behaviour is particularly interesting in the impeded régime, for $t_D < 10^5$ and $Pr > 90$. Under these conditions, Japikse noticed as many as 10 streams moving downwards (and, of course, 10 moving up) and as few as two. These findings guided the study of Mallinson *et al.*, who confirmed the existence of two, four, and six streams, all organized by the square geometry[22] they employed.

Other numerical studies of the laminar–impeded régime (Crespo *et al.* 1987; Lock and Zhao 1990*a*) have revealed a two-stream convective process. These avoided a step change in wall temperature in the coupling region by using a continuously varying profile, typically linear, between the heated and cooled ends. The paper of Lock and Zhao, which deals with a square section thermosyphon, makes an interesting comparison with that of Mallinson *et al.* It discusses solutions for water ($Pr \simeq 10$) in the laminar–impeded régime, each obtained from converged solutions previously obtained with the system tilted slightly. These not only exhibited bifilamental convective coupling but revealed it to be skewed relative to the original plane of tilt; that is, with tilting parallel to the side planes the bifilamental structure in the coupling region was symmetric about a diagonal, and vice versa. Apparently, this is an example of bifurcation resulting from the ambiguity found in systems heated from below. The same study explored the effect of the wall temperature profile by applying a simple power (n) law distribution; at one extreme ($n \gg 1$) the curve approximates the two-isothermal condition, while at the other ($n \ll 1$) it approximates a central isothermal with rapidly varying temperatures in the end regions. The latter condition, which corresponds roughly to adiabatic long walls, again produced bifilamental coupling but the end flows were completely reversed, in accord with visual observations (Müller *et al.* 1984). The steep end gradients evidently created a Rayleigh instability with the resulting central 'plume' moving away from the end along the tube axis.

Under turbulent conditions, the details of coupling are still not clear. It has

[22] A discussion of the role of tube cross-sectional geometry is given in Section 2.4.

been suggested by Japikse (1973) that while the existence of a turbulent boundary layer eventually leads to a turbulent core, the degree of intermixing is actually less than expected, and the fully mixed régime may not be established. The turbulent heat transfer data in Figs 2.24 and 2.25, which show a slope in the range 0.3–0.4, certainly suggest a turbulent boundary layer type of flow, in marked distinction to the behaviour of the open system. The absence or, at the very least, the delay of the fully mixed condition in the closed system may perhaps be attributed to the coupling mechanism, and this introduces the paradox that the closed system may perform better than the open system.

The explanation may be found in the open thermosyphon mouth or, more precisely, in the region just outside of the mouth where the bulk of the reservoir fluid is supposedly quiescent. Theoretically, cold fluid should flow under laminar conditions down into the core of an open thermosyphon but the discussion of entry conditions given in the previous section strongly suggests that under turbulent internal conditions this expectation is unlikely to be realized. The turbulent open system data for the higher values of t_D attained in Figs 2.20 and 2.24 indicate that adverse mixing is extensive. Such mixing in an open thermosyphon produces a performance deterioration which brings to mind the large perturbations associated with the laminar–turbulent transition in the impeded régime, for example, in air. It appears that the absence of flow control in the mouth may permit a permanent set of large perturbations, presumably initiated by a shifting, and perhaps separated, entry stream (see Fig. 2.18). If so, the remedy is to add a device such as that shown in Fig. 2.30. This everts the flow at the mouth so that the hot discharge from the turbulent annulus forms a central and stable plume while the cold entry flow is drawn from the quiescent surroundings and directed into the core. Such a measure would eliminate the effects of mouth instabilities and presumably generate heat transfer rates even higher than those observed in the closed system. A similar improvement might be expected from the insertion of a smaller-bore concentric tube extending well above the below the mouth. With the core and annulus thus separated artificially, the thermosyphon becomes a particular form of the bayonet tube, a device which is fully discussed in Section 5.3.

2.4 Tilted and bent tubes

Tilting a thermosyphon tube relative to the vertical produces two complementary effects: the component of the gravitational acceleration acting longitudinally decreases from a maximum of g while the component acting laterally increases from a minimum of zero. The changing relative magnitudes of these components may be expected to create changes in both the flow pattern and the heat transfer rate; and this will be true for both the open and the closed system. Orientation in a gravitational field is an important feature in its own

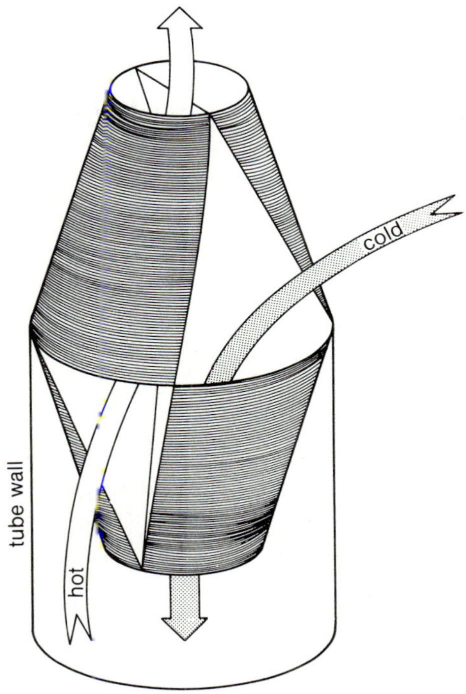

Fig. 2.30. Flow evertor.

right, but it also provides a valuable, and simple, way of simulating the superposition of body force fields. For example, Eckert and Jackson (1951) and Alcock (1951) suggested independently that the role of the Coriolis acceleration in a radial, rotating thermosyphon could be explored through the tilting of a gravitational system. This section, in discussing the effect of tilting, is thus addressing the wider question of thermosyphon performance when the body force field has components which act both longitudinally and laterally.

2.4.1 *Characteristics of the horizontal thermosyphon*

In one sense, the other extreme to the vertical thermosyphon is the horizonal thermosyphon. This is depicted in closed form in Fig. 2.31 from which it is immediately obvious that the circulation pattern in the horizontal position is very different from that in the vertical position. In both heated and cooled sections the refluent pattern of the primary flow is no longer a core–annulus structure but a bifilamental structure. Moreover, there is a complete absence of a coupling mechanism other than the convective exchange attributable to the primary flow itself. It is thus to be expected that the performance of the

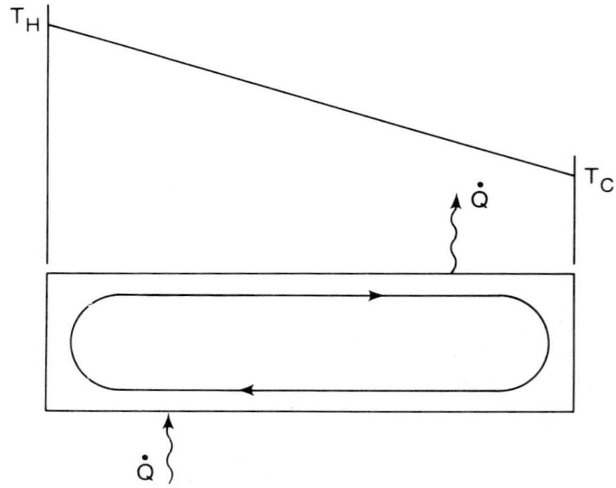

Fig. 2.31. The horizontal, closed thermosyphon.

horizontal thermosyphon would be markedly different from that of its vertical counterpart.

The principal cause of this difference is the thermal buoyancy which, as noted in Section 2.1, now acts indirectly through horizontal temperature gradients; at the same time, however, the buoyancy also acts directly over the vertical side and end surfaces. Both of these mechanisms operate simultaneously, with their relative importance depending upon the thermal boundary conditions. Adiabatic ends reduce the direct effect while adiabatic long walls tend to reduce the indirect effect. For a thermosyphon, the latter restriction seldom applies and thus our attention is more naturally focused on indirect buoyancy effects, especially over the upper and lower surfaces where most of the heat is transferred.

The equations governing steady, laminar flow in the horizontal thermosyphon have been presented previously (see eqns (2.1)). Using the rectangular coordinate system defined in Fig. 2.32 for a square section tube, the appropriate velocity and displacement scales may be determined as follows. If indirect (longitudinal) buoyancy is balanced by viscous forces

$$\frac{\beta g \theta^{c}(Z^{c})^{3}}{\nu X^{c} U^{c}} = O(1); \tag{2.44}$$

and if lateral conduction is balanced by longitudinal convection

$$\frac{U^{c}(Z^{c})^{2}}{\kappa X^{c}} = O(1). \tag{2.45}$$

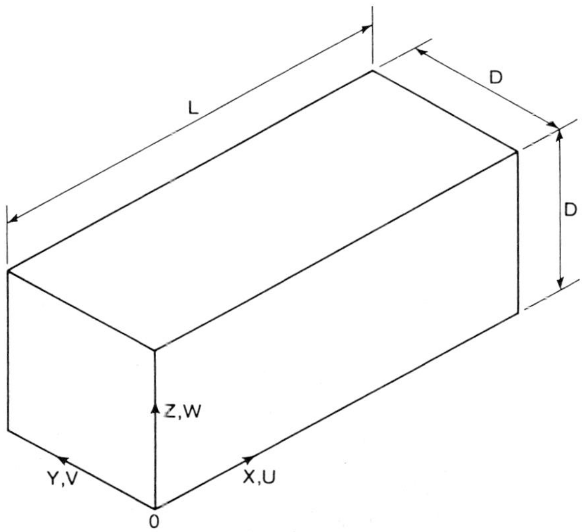

Fig. 2.32. Coordinate system for a square section thermosyphon.

From these it follows that the longitudinal velocity scale is given by

$$U^c = \frac{\kappa}{L} Ra_L^{2/5} \tag{2.46}$$

while the lateral (vertical) displacement scale is given by

$$Z^c = \frac{L}{Ra_L^{1/5}}. \tag{2.47}$$

The heat transfer rate may thus be estimated from the horizontal primary flow by

$$\dot{q}LD = O(\rho c_p \theta^c U^c Z^c D)$$

so that

$$\bar{N}u_D = C_1 \left[\left(\frac{D}{L}\right)^2 Ra_D\right]^{1/5}. \tag{2.48}$$

This relation ceases to apply when $Z^c = O(D/2)$, that is $[(D/L)^2 Ra_D]^{1/5} = O(2)$. Below this value of $(D/L)^2 Ra_D$ lies an impeded régime in which eqn (2.44) yields directly

$$U^c = O\left(\frac{\beta g \theta^c D^3}{8\nu L}\right) \tag{2.49}$$

from which it follows that the primary heat transfer rate may then be written in the form

$$\bar{N}u_D = C_2\left(\frac{D}{L}\right)^2 Ra_D. \tag{2.50}$$

For very low values of $(D/L)^2 Ra_D$, the system enters a conduction region.

It is evident from eqns (2.48) and (2.50) that once fluid motion becomes important the diameter-based Nusselt number is a function solely of $(D/L)^2 Ra_D$, corresponding to the variable $(D/L)Ra_D$ which is appropriate to a vertical thermosyphon. Hence the general relation is

$$\bar{N}u_D = F\left(Pr, \left(\frac{D}{L}\right)^2 Ra_D\right). \tag{2.51}$$

The form of this relation for laminar flow in a short, air-filled thermosyphon is indicated in Fig. 2.33 (Lock and Han 1989*a*). Also shown (dashed) is the primary heat transfer relation based on eqns (2.48) and (2.50); these intersect near $(D/L)^2 Ra_D = 3\times10^6$ or, with $L=5D$, at $Ra_D = 10^7$, revealing that $C_2 = O(10^{-5})$. The difference between the two curves is evidently caused by secondary motion, principally through side wall conduction. This secondary effect is not easily calculated in a three-dimensional flow but the lateral temperature difference θ_z^c causing it may be estimated by noting that for impeded flow the heat conduction rate may be represented by

$$DL\frac{k\theta_z^c}{D/4} = Lk\theta^c C_2\left(\frac{D}{L}\right)^2 Ra_D$$

and hence

$$\frac{\theta_z^c}{\theta^c} = \frac{C_2}{4}\left(\frac{D}{L}\right)^2 Ra_D. \tag{2.52}$$

Bearing in mind the magnitude of C_2, this suggests that $\theta_z^c \to \theta^c$ at the upper end of the impeded régime where the shrinking thickness of the boundary flow reduces the relative importance of secondary heat transfer. At the other end of the impeded régime, $\theta_z^c \ll \theta^c$ and secondary effects die out, although it should be noted that they grow more quickly than the primary effect, presumably because the onset of convection over the vertical sides occurs at a lower value of Ra_D. Individual contributions to the total heat transfer rate *to* this system are shown in Fig. 2.34. It is evident that the bottom and (both) side walls each contribute almost half of the total; the end walls play only a minor role, as confirmed by the adiabatic end wall data incorporated in Fig. 2.33.

The secondary flow consists of two pairs of vortices, one for each filament. Each pair is caused by conduction through the side walls and begins in the end

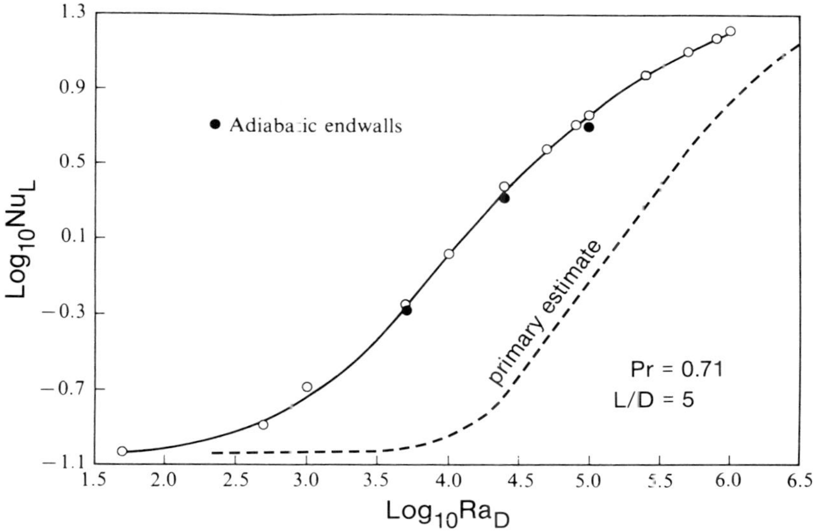

Fig. 2.33. Laminar heat transfer relation for a horizontal, closed thermosyphon (after Lock and Han 1989a).

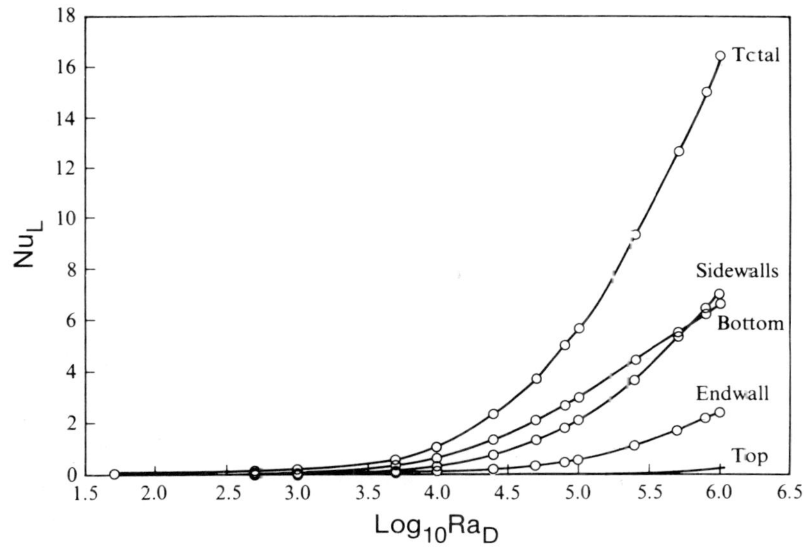

Fig. 2.34. Components of the heat transfer rate to a square, horizontal thermosyphon (after Lock and Han (989*a*).

region about one tube width from the end[23]. The flow pattern in the end region is thus a complex transition from a vertical wall plume to a vortical pair. This development is shown in Fig. 2.35 for the hot end of a square section tube ($L/D=5$) filled with air. The remnants of the end flow are still evident at $x=X/L=0.143$ when vortices begin to appear in the upper corners. The vortices grow as they progress axially until, midway between the ends, that is $x=0.5$, they fill the entire upper half of the tube, the lower half being occupied by a symmetric pair moving in the opposite direction.

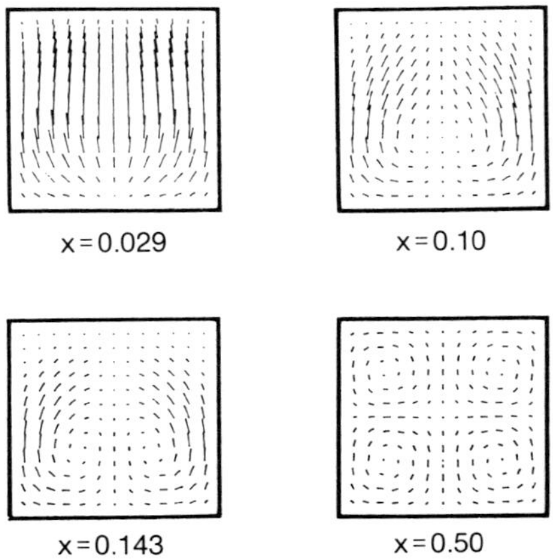

x=X/L

Fig. 2.35. Transverse velocity field in a square, horizontal thermosyphon: $Ra_D=2.5\times10^4$, $Pr=0.71$ (after Lock and Han 1989*a*).

This secondary flow structure exists in any horizontal thermosyphon tube with a cross section of isometric proportions (Lock and Han 1989*a*; Lock and Zhao 1992; Bontoux *et al.* 1986*b*). The vortex pairs simply adjust to the section geometry whether it is symmetric or not. In combination with the primary flow, this secondary motion creates the overall pattern depicted in Fig. 2.36. As indicated, two crescent-shaped vortices mark the division between the end flow and the main flow with its vortex pairs.

It was noted previously that opening one end of a vertical thermosyphon may introduce substantial changes in its behaviour; this is no less true for a horizontal tube. Two-dimensional cavities with one open end have been extensively studied (see, for example, Chan and Tien 1986), but there appear to be few data on the tubular thermosyphon. Figure 2.37 shows the primary flow

[23] In general, the location depends upon the Rayleigh number.

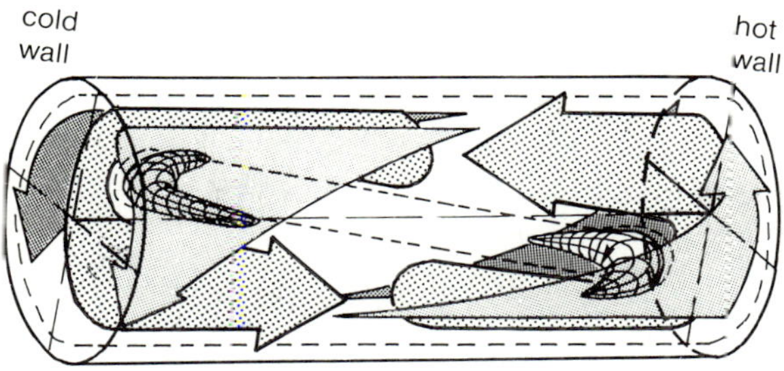

Fig. 2.36. Primary and secondary flow patterns in a circular, horizontal thermosyphon (after Bontoux *et al.* 1986*b*).

pattern for an isothermal, square section tube filled with air, and again with $L/D=5$ (Lock and Zhao 1990*b*). While this clearly illustrates the expected S-shape in the velocity profile, it also reveals a bimodality in the lower (cooler) filament near the open end, for example, $x=0.8$. The transverse velocity field in this region is shown in Fig. 2.38 from which it is evident that the bimodality may be attributed to a pair of secondary vortices. However, in distinction to the closed tube, vortices do not appear elsewhere in the tube. These have a rotation which indicates that they are caused by heating from the side walls. The effect of conduction in the side walls is thus limited to the entry flow; the exit flow is not affected.

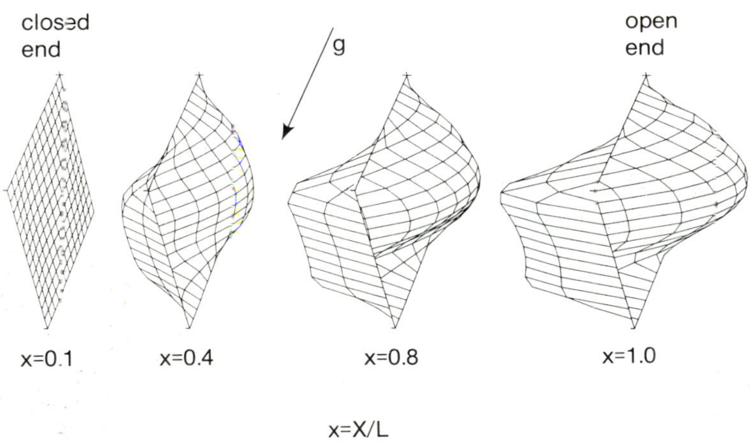

Fig. 2.37. Primary flow, viewed from beneath, in a horizontal, open thermosyphon with a square cross section: $Ra_D=5\times10^4$, $Pr=0.71$, $L/D=5$ (after Lock and Zhao 1990*b*).

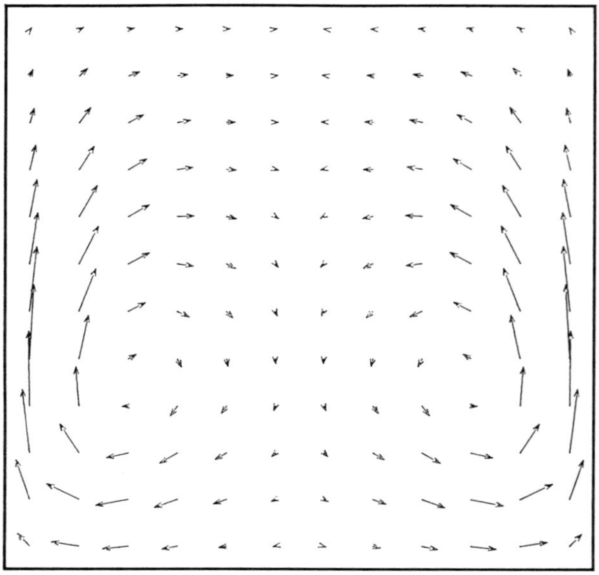

Fig. 2.38. Transverse velocity field in the entry region of a square, open thermosyphon (after Lock and Zhao 1990*b*).

2.4.2 *Characteristics of the tilted thermosyphon*

So far in this section we have dealt with horizontal thermosyphons driven principally by indirect buoyancy forces. Tilting the tube from the horizontal permits direct buoyancy forces to play a greater role which will be reflected in thermal performance. As the tube is tilted from the horizontal, the indirect (longitudinal) effects weaken along with some direct (end) effects while other direct effects strengthen. However, this strengthening occurs over the long walls and therefore reinforces some parts of the primary flow while opposing others. It is thus evident that the progressive development of the primary and secondary flow patterns with tilting will not be simple; at the very least, it must accommodate the gradual transformation from a bifilamental refluent pattern to a core–annulus refluent pattern.

Figure 2.39 illustrates the effect of tilting a square section, closed thermosyphon relative to the horizontal (Lock and Han 1989*a*). For negative tilts, that is, with the hot end higher than the cold end, the flow pattern of the horizontal position persists but with reduced vigour. Under this condition, the

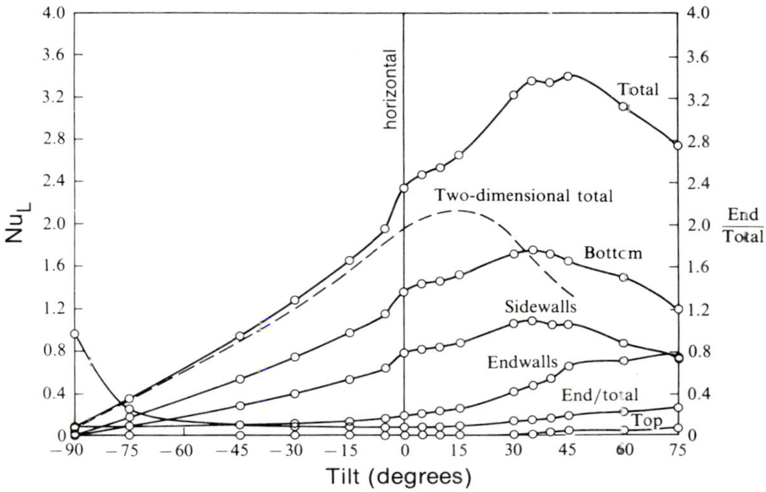

Fig. 2.39. The effect of tilt on heat transfer in a square section, closed thermosyphon: $Ra_D = 2.5 \times 10^4$, $Pr = 0.7$ (after Lock and Han 1989*a*).

device exhibits an anti-gravity capability, albeit a small one, which degenerates to pure conduction in the vertical (inverted) position. The corresponding two-dimensional result, also given in the figure, reveals that secondary motion is not significant until the tube is horizontal. For positive tilts less than 45°, the heat transfer components of the system are not too different from those indicated in Fig. 2.34, but the flow pattern undergoes a substantial, though gradual, change. Around 5°, a stagnation point appears at the bottom corner of the heated end[24]. This is caused by alterations in the direct buoyancy forces operating over the end surface, and over the bottom surface near the end; the former weakens while the latter strengthens, opposing the main filamental flow, which also strengthens. The net result is that a flow reversal eventually occurs at the bottom corner and the filament splits to accommodate it. As the tilt increases, the stagnation point thus created moves towards the axis of the tube, as indicated in Fig. 2.40. When the tube is vertical the stagnation point lies near the axis and the flow exhibits the familiar core–annulus pattern.

A general circulation model for a circular section, closed thermosyphon is shown in Fig. 2.41. This is based on observations, both theoretical and experimental, made on circular and square tubes filled with water or air (Schiroky and Rosenberger 1984; Bontoux *et al.* 1986*a*; Lock and Kirchner 1992; Lock and Han 1989*a*). As indicated, it reconciles the deep core–annulus flow with the mid-length bifilamental flow, the latter dominating the former as

[24] And at the top corner of the cold end.

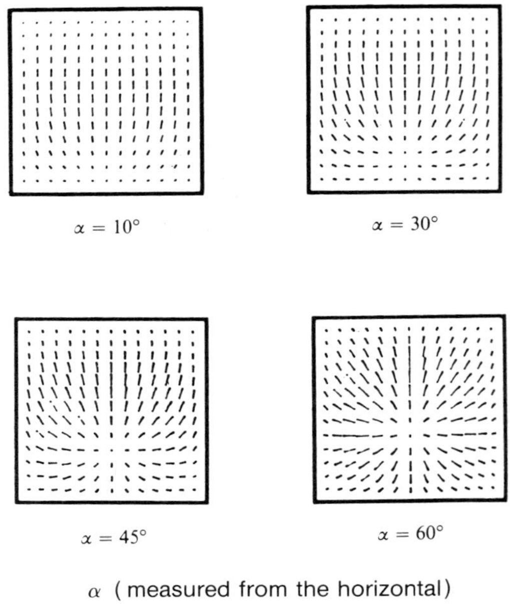

Fig. 2.40. Progressive displacement of the stagnation point at the hot end of a tilted thermosyphon: $Ra_D = 2.5 \times 10^4$, $Pr = 0.71$ (after Lock and Han 1989*a*).

the horizontal position is approached. For increasing tilt angles above the horizontal, the bifilamental flow, which contains the secondary vortices, shrinks in length. Near $\alpha = 45°$, the vortices subside and are virtually absent for angles between this and the vertical. Figure 2.39 indicates that the heat transfer rate then decreases as the vertical position is approached. This type of behaviour is supported by the experimental data of Fig. 2.42 obtained with a water-filled circular tube for which $L_H = L_C$ and $L_H/D = 10$; tilt is here measured from the vertical. In Fig. 2.39, the results refer to the laminar–impeded régime whereas those shown in Fig. 2.42 lie above the impeded régime, probably in the fully mixed turbulent régime.

Theoretical studies of the open thermosyphon (Leslie 1959; Lock and Zhao 1990*b*) suggest that the laminar heat transfer rate increases for small departures from the vertical; this is evidently true in both the impeded and boundary layer régimes. Intermediate tilts are usually characterized by a maximum heat transfer rate. The primary flow field in a square section tube for a tilt (above the horizontal) of 40° is shown in Fig. 2.43. When compared with Fig. 2.37 this reveals the emergence of the core–annulus structure deep in the tube, while the flow nearer the tube mouth remains bifilamental. Also noteworthy at $\alpha = 40°$ is the absence of bimodality near the mouth, and its

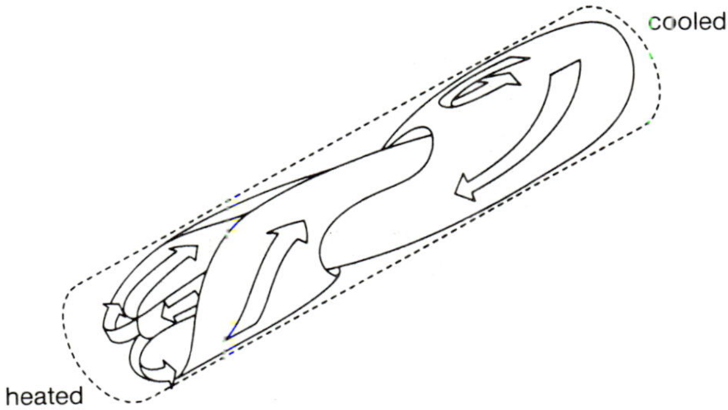

Fig. 2.41. Flow model for the tilted tubular thermosyphon (after Lock and Kirchner 1991).

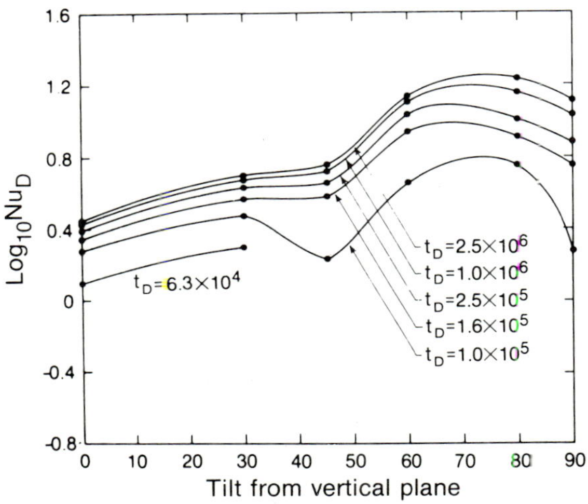

Fig. 2.42. Effect of tilt on the heat transfer rate in a water-filled, closed thermosyphon: $L/D=10$ (after Lock and Kirchner 1991).

apparent re-emergence deeper in the tube, for example, near $x=0.4$. Examination of the secondary flow confirms that a pair of vortices do exist in this region. Their origin, however, no longer lies in conduction from the side walls, but in the nature of the reconciliation between the core–annulus and bifilamental structures suggested in Fig. 2.41.

Many of the flow details mentioned above have been substantiated by visual studies (Martin and Lockwood 1963; Behnia *et al.* 1987). Behnia *et al.* observed the existence of a stagnant length near the closed end of the tube in

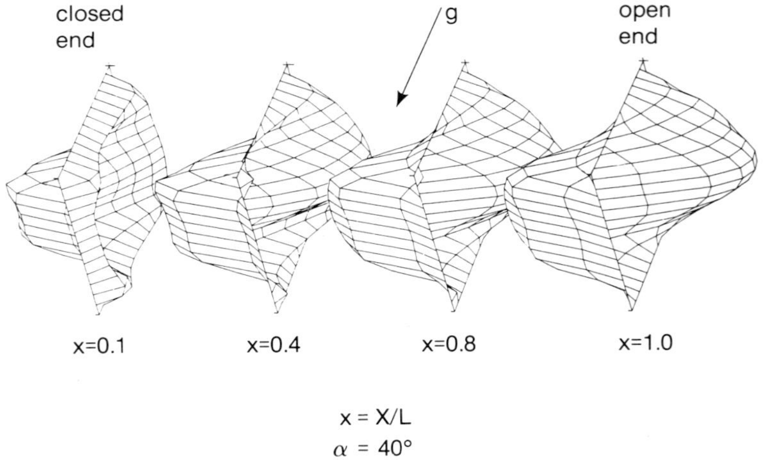

Fig. 2.43. Primary flow, viewed from beneath, in a tilted, open thermosyphon with a square cross section: $Ra_D = 5 \times 10^4$, $Pr = 0.71$, $L/D = 5$ (after Lock and Zhao 1990*b*).

what was presumably the laminar–impeded régime, and noted its decrease with increases in either t_D or tilt relative to the vertical. The paper of Martin and Lockwood contains observations not only on the flow structure but on the effect of mouth conditions. Figure 2.44, for example, shows a dye trace in a water-filled tube under laminar–impeded conditions. Near the entry, the flow is clearly bifilamental while at depth the dye spirals around[25] before heading back up towards the mouth; this pattern is very similar to that suggested in Figs 2.41 and 2.43. Figure 2.45 corresponds to a higher Rayleigh number at which the entering dye trace reveals the presence of a travelling wave instability. The large amplitudes may be the cause of the laminar–turbulent transition in water noted earlier (see Section 2.2).

Nearer the vertical, the influence of tilt is especially difficult to describe. In the closed system, this is partly because the two-filament flow shown in the mid-length region in Fig. 2.41 becomes the coupling region and may contain more filaments of become turbulent. The difficulty is also partly attributable to the ambiguity found in the solutions of the governing equations, and the attendant complexities of the metastable flow patterns and transience (Hart 1985; Lock and Zhao 1990*a*). The first visual observations on near-vertical, closed thermosyphons (Japikse 1969; Japikse *et al.* 1971; Japikse 1973) were made with non-metallic liquids. These showed that the initial effect of tilt depends upon the fluid. For large Prandtl numbers the interdigitations in the coupling region undergo a rearrangement which, in general terms, leads to a

[25] These spirals are not to be confused with the fine heating-wire coils.

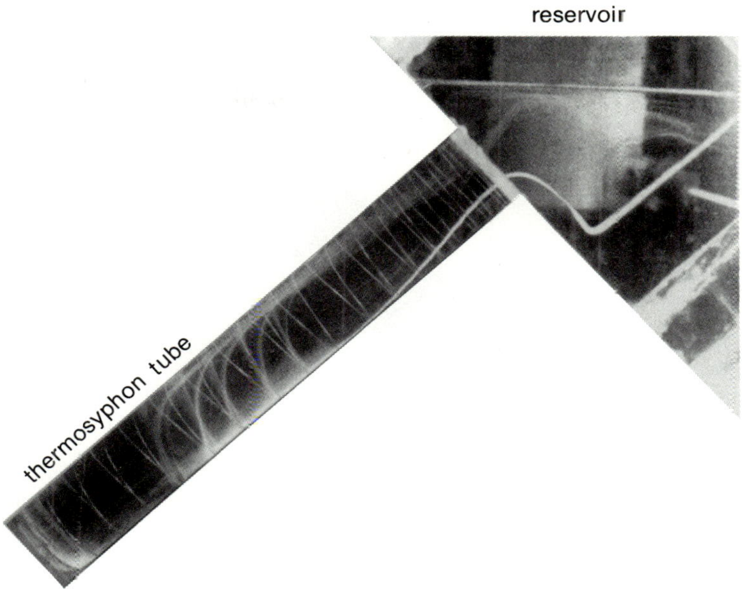

Fig. 2.44. Flow visualization of the laminar–impeded régime in a tilted, water-filled thermosyphon (after Martin and Lockwood 1963).

reduced number of thicker filaments; ultimately, only two filaments are left, one ascending above the other in the tilt plane. For lower Prandtl numbers, the streams are not as well organized initially but do gradually align into two filaments. Regardless of Prandtl number, the two-filament coupling pattern is fully established in non-metallic liquids for tilt angles greater than about 6° from the vertical. The overall circulation pattern then looks like that in Fig. 2.41; further increases in tilt from the vertical simply serve to extend the bifilamental region towards the ends.

The effects of these flow changes are reflected in heat transfer rates. Although no experimental data for closed systems appear to be available with very large Prandtl numbers, it is to be expected that the first 6° of tilt would lead to a decrease in heat transfer rate as the coupling undergoes alterations and the flow path of the general circulation lengthens[26]. After this, the heat transfer rate should increase with tilt, as indicated earlier. For moderate Prandtl numbers, the initial (vertical) coupling mechanism is probably disorganized or bifilamental (Japikse 1973; Crespo *et al.* 1987), in which case the first few degrees of tilt will represent only the beginning of an improved heat transfer trend running over at least 20°.

[26] The maximum length corresponding roughly to a tilt of $\alpha = D/L$.

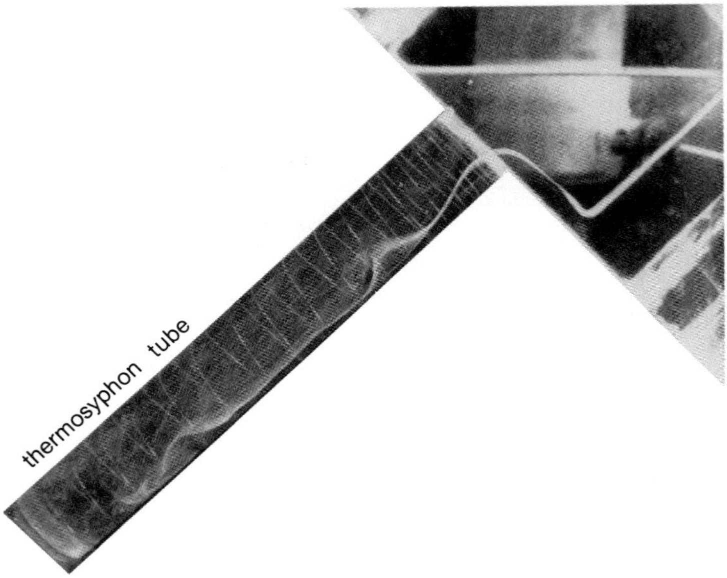

Fig. 2.45. Flow instability in impeded régime of a tilted, water-filled thermosyphon (after Martin and Lockwood 1963).

These comments also apply to open thermosyphons. The effect of Prandtl number on heat transfer rate, for example, is clearly evident in the experimental data shown in Fig. 2.46 (Martin and Cresswell 1957; Martin 1959). Apart from the initial deterioration displayed by glycerine, the heat transfer rate generally increases with tilt from the vertical, at least up to 45°; the effect is greater when measured using the lower surface temperature (solid circles) rather than the upper surface temperature (open circles). The glycerine data are for laminar conditions and thus reflect changes in the entry flow pattern near the vertical: from the complexity of alternations in the ingoing and outgoing streams to the comparative simplicity of a bifilamental flow (Martin and Lockwood 1963). The water data, on the other hand, apparently represent fully mixed turbulence in the presence of which the general flow pattern reflects a shift in balance between core–annulus refluence and bifilamental refluence.

Before considering the behaviour of non-linear thermosyphons, it may be useful to round off the discussion of linear tubes with an examination of the role of cross-sectional geometry, selected examples of which have already been utilized in previous sections. To begin with, it is important to recognize that this geometrical effect has received little attention for either laminar or

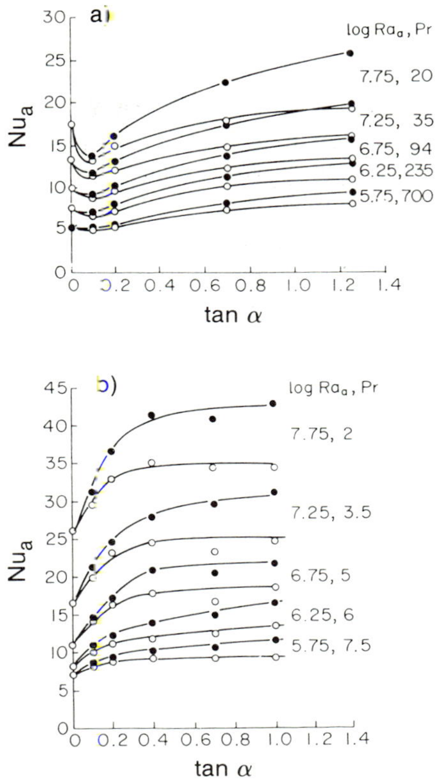

Fig. 2.46. The effect of Prandtl number and tilt on heat transfer in a circular, open thermosyphon: $L/D = 7.5$ (after Martin 1959): (a) glycerine; (b) water.

turbulent conditions, and therefore the following discussion will necessarily be limited. For vertical tubes, two opposing primary effects come into play in the corners of non-circular shapes: a greater tendency for stagnant flow attributable to increased viscous forces and a greater tendency for quicker flow because of heating from two sides. Under laminar conditions, neither of these is overwhelming, at least in the impeded régime (Lockwood and Martin 1964; Lock and Zhao 1990*b*). The hydraulic mean diameter may then offer the best measure of tube width in defining appropriate forms for the Nusselt number and Rayleigh number.

Once the tube is tilted, the flow becomes more complex, and the above tendencies are then accompanied by alterations to the secondary flow attributable to the cross-sectional geometry. Again considering laminar–impeded conditions, numerical studies have been used to examine the flow pattern when a horizontal tube is rotated in intervals about its longitudinal axis (Lock and Han 1989*b*; Lock and Zhao 1992). This work reveals that the basic flow model described earlier is retained, with the details of the secondary flow adapting to the precise shape of the tube. Figure 2.47 illustrates typical flow patterns in the mid-length section. Rotation of a square section tube produces only a slight change in the heat transfer rate but a more substantial change occurs with a (right-angled) triangular section. The performance of the latter is strongly influenced by the hypotenuse, the most vigorous secondary flow occurring with the hypotenuse vertical, as indicated in Fig. 2.47.

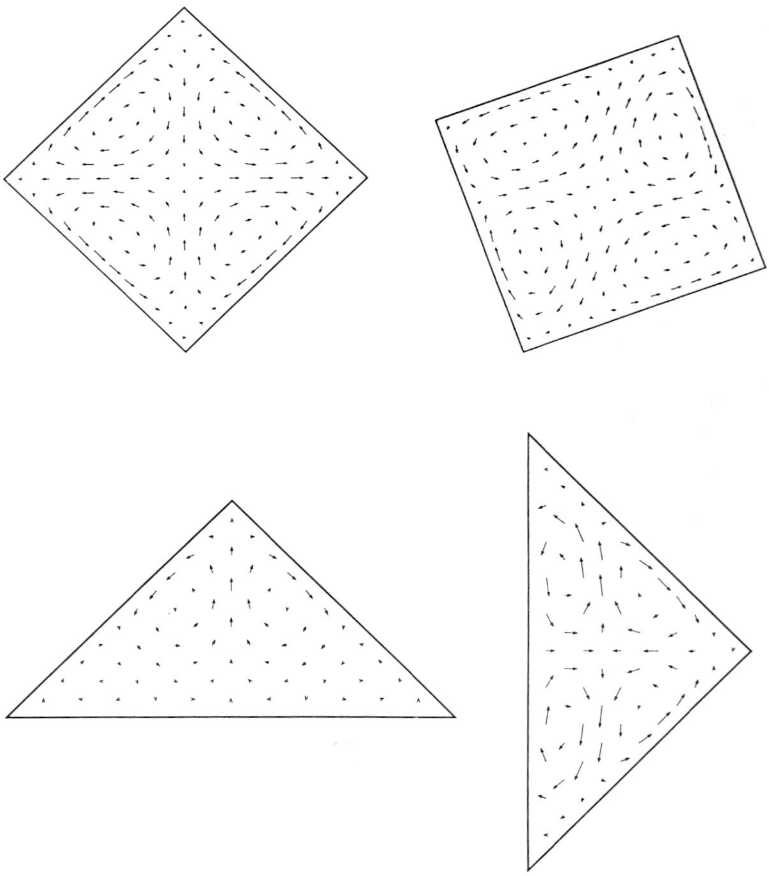

Fig. 2.47. Secondary flow patterns in the mid-length section of square and triangular tubes under laminar–impeded conditions (after Lock and Zhao 1991*a*).

Turbulent flow in non-circular tubes is less easy to discuss, although it is reasonable to expect that secondary circulation would then be reduced in importance. The only published data, for vertical tubes, appear to be those of Lockwood and Martin (1964) who noted the tendency of corners to damp turbulent fluctuations even though their onset occurred earlier. Once turbulence is established, however, its influence is by no means obvious, as the brief discussions in Sections 2.1 and 2.2 would suggest. In a vertical, water-filled tube having a rectangular section of high aspect ratio, for example, the thermal performance is less than or equal to that of the corresponding circular section in which fully mixed turbulence exists. Under these conditions, the turbulent field in the rectangular tube only reaches to the base when the length-based Rayleigh number exceeds 10^{11}; below this value, the lower reaches of the tube are evidently occupied by a *laminar*, impeded flow. With a crescent-shaped cavity, on the other hand, fully mixed turbulence fills the entire tube giving a Nusselt number which exceeds the circular tube value once the length-based Rayleigh number rises above 10^{10}. This bodes well for turbine applications in which the blade is hollow.

2.4.3 *Characteristics of the non-linear thermosyphon*

Turning now to tubes which are not straight, it is immediately obvious that there are too many configurations to consider here in any detail. Figure 2.48 illustrates a number of possibilities, some of which are currently in use[27]. In each of these, the heated section is separated from the cooled section by an adiabatic (unshaded) section which is, in essence, a tube connecting the distributed source of heat to the distributed sink. With the exception of the spiral form, the working sections are all shown as straight vertical or horizontal tubes, but it is clear that particular applications may call for working sections which are curved or inclined.

Until recently, no systematic study of these forms has been attempted and consequently their heat transfer characteristics, together with the internal flow details which give rise to those characteristics, are not widely known. In the limited space available here, coverage will be restricted to only two types, the elbow and cranked thermosyphons, in order to broach a fundamental question common to all; namely, what happens when the flow pattern in one of the working sections is markedly different from that in the other? The insight thus gained will be helpful in assessing the performance of many other non-linear configurations.

In the vertical, cooled section of the elbow thermosyphon shown in Fig. 2.48, it is reasonable to expect the flow to possess a core-annulus structure, while in the horizontal, heated section a bifilamental pattern seems more likely. The form of the reconciliation in the adiabatic bend is suggested

[27] See, for example, Figs 1.11 and 1.12.

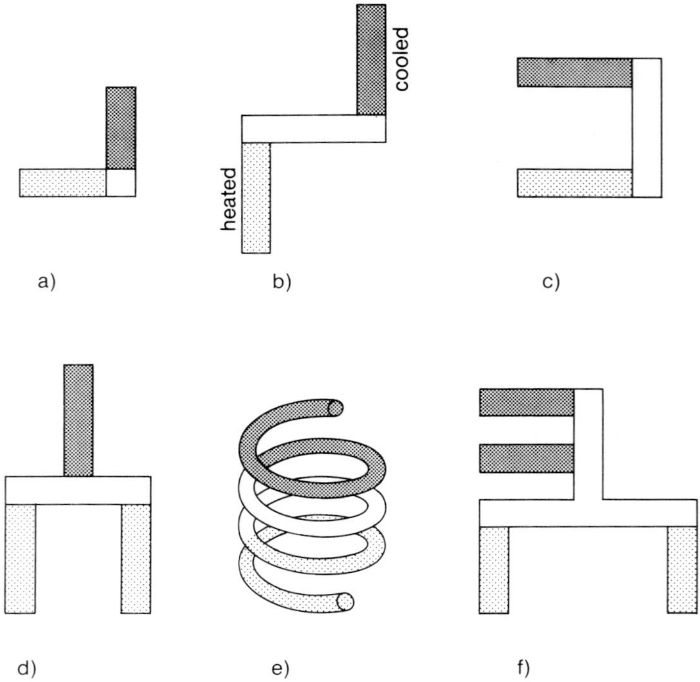

Fig. 2.48. Non-linear thermosyphon configurations: (a) elbow; (b) crank; (c) hairpin; (d) branched; (e) spiral; (f) combination.

by the model developed for inclined tubes, and shown in Fig. 2.41. If the cooled section in that figure were bent upwards into a vertical position, the foreshortened core-annulus structure shown would tend to spread over the entire cooled length. At the same time, if the heated section were raised to the horizontal position, the core-annulus structure there would shrink and eventually disappear. This would leave the upper filament of the heated section feeding the core of the cooled section, forcing the annulus of the cooled section to regroup and feed the lower filament in the heated section.

Figure 2.49 shows the primary flow field, as determined numerically, in a short. water-filled elbow thermosyphon (Lock and Park 1991). These results, for laminar–impeded conditions, clearly show the tongue of hot fluid in the central plane leaving the heated section and entering the core of the cooled section. The side plane data, both taken close to the side walls, illustrate how the upward-flowing core in the cooled section is balanced by a downward-flowing annulus which is gradually transformed into the lower filament in the heated section.

It is found that the elbow thermosyphon, like its linear counterpart, exhibits

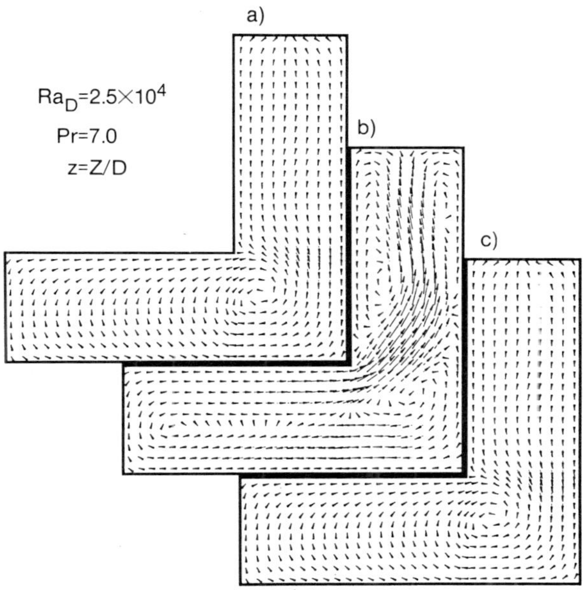

Fig. 2.49. Primary flow field in an elbow thermosyphon with a square cross section $D \times D$: (a) z=0.04; (b) central plane; (c) z=0.96.

conduction, impeded, and boundary layer régimes. The role of geometry, however, is not yet clear; in particular, the appropriate ratio of Rayleigh number and length–diameter ratio under laminar conditions is not obvious, although it presumably lies somewhere between t_D and $t_D(D/L)$, the limiting vertical and horizontal forms, respectively. Some experimental data for water in a small-bore tube (Lock and Ladoon 1991*a*) are shown in Fig. 2.50. For clarity, laminar–impeded data have been omitted from this figure which, at first glance, appears to represent laminar, boundary layer conditions, as suggested by the index 0.22. On closer examination, however, the correlating coordinate is seen to be the same as that displayed in Fig. 2.25 with $L_H = L_C$; namely, $t_D(D/L_H)^3$ which suggests turbulent conditions. Since these data cover the same low Rayleigh number range in which Martin (1955) observed fully mixed flow in a water-filled open thermosyphon, the correlation provided in Fig. 2.50 evidently refers to fully mixed turbulent conditions. By comparison with Fig. 2.25, it is evident that performance was poorer in the small-bore elbow thermosyphon experiments.

In the cranked thermosyphon, as depicted in Fig. 2.48, the reconciliation of the two sectional flow patterns is perhaps easier to predict, at least under laminar–impeded conditions. Given the vertical–horizontal elbow reconciliation revealed above, it is most likely that the heated and cooled sections in the

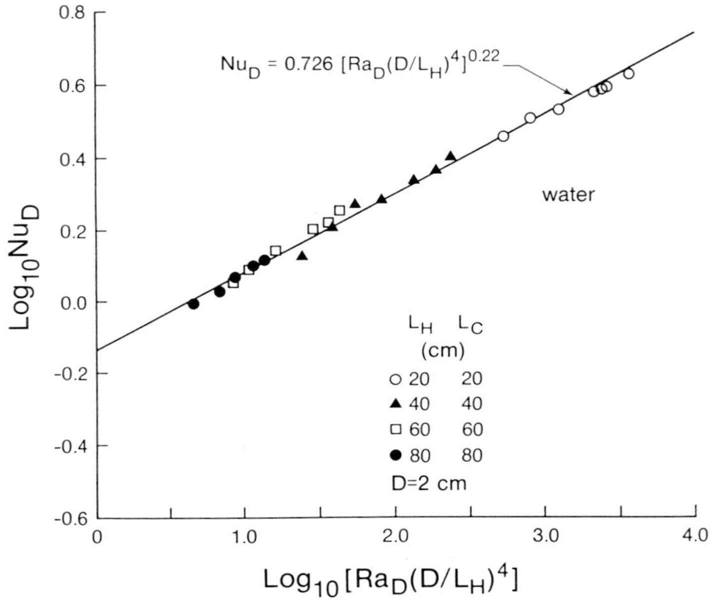

Fig. 2.50. Empirical correlation for an elbow thermosyphon (after Lock and Ladoon 1991*a*).

cranked thermosyphon both have a core-annulus structure, with the two cores being fed from a bifilamental structure in the adiabatic offset length; the rising annulus of the heated section thus transforms into the upper, hotter filament in the offset length and ultimately becomes the core of the cooled section, while the opposite transposition occurs in the lower, cooler filament. Whether or not this circulation pattern is maintained at higher Rayleigh numbers presumably depends upon both the tube geometry and the Prandtl number.

Figure 2.51 shows some water data (Lock and Ladoon 1991*b*) designed to reveal the effect of the offset length L_A. Not unexpectedly, the heat transfer rate decreases as L_A is increased, but not until L_A/D exceeds at least 4 : 1; below this value, the offset length has little effect. Perhaps more importantly, the data show a significant upturn for $Ra_D \gtrsim 10^{7.2}$, indicating that a small offset introduces an improvement over linear performance marked by the dashed line. It is believed that this alteration in behaviour corresponds to the development of a turbulent, boundary layer flow, not only in the offset length but in the vertical sections. Since this occurred in water, for which the conventional wisdom suggets a fully mixed flow, it is the more surprising.

The explanation may perhaps be found in the strong influence exerted by the bifilamental flow in the offset length. This could convert a stable, axisymmetric, annulus–core pattern in each of the vertical sections into an

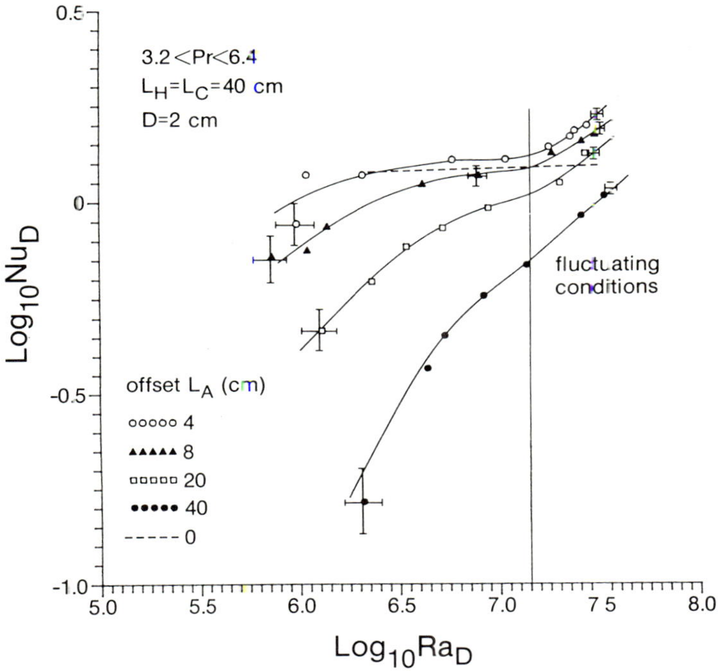

Fig. 2.51. Effect of offset on heat transfer in a cranked thermosyphon (after Lock and Ladoon 1991*b*).

unstable, eccentric, annulus–core pattern in which the tongue of fluid leaving the offset length remains fairly close to the tube wall at the inner corner, thus creating a local recirculation flow of restricted extent; at the same time the flow exiting the vertical section along the outer wall will be strengthened *en route* to the offset length. If the Rayleigh number exceeds a critical value, evidently in the vicinity of $10^{7.2}$, the local recirculating flow near the mouth of the vertical section may collapse, thereby allowing an essentially bifilamental flow in the heated and cooled sections. This speculation is supported by two pieces of evidence: the appearance of wall temperature fluctuations near the heated section mouth for $Ra_D \gtrsim 10^7$, and an exponent of approximately 0.33 in the $Nu_D \sim Ra_D$ relationship.

Further discussion of flow models in non-linear thermosyphons may be found in Lock and Ladoon (1991*c*) and Lock and Fu (1992*a*). Both of these papers deal with the effect of inclination, the former treating the elbow thermosyphon and the latter the cranked thermosyphon. These and other flow models, in which the precise form of the coupling is stipulated, offer two important advantages. Firstly, they indicate if, and when, the performance of a non-linear thermosyphon is likely to exceed that of a linear thermosyphon; and, of course, when it is inferior. Secondly, a prescribed coupling mechanism lends itself to the analysis of complex configurations in which the performance of the system overall may be described in terms of the elements of which it is composed. The treatment outlined in Section 2.3, for example, may thus be extended to a very wide range of tubular thermosyphons.

3
THE EVAPORATIVE, TUBULAR THERMOSYPHON

Above the critical pressure, the tubular thermosyphon behaves as a single-phase system. The same is often true below the critical pressure provided that the fluid specific volume is very much different from the critical value. As Fig. 3.1 indicates, $v \ll v_{cr}$ corresponds to a liquid-filled device while $v \gg v_{cr}$ corresponds to a gas-filled device unless the pressure is very low, that is $P \ll P_{cr}$. Under such single-phase conditions, the descriptions presented in the previous chapter continue to apply. If the critical point is approached in these circumstances, the behaviour of the system changes but not in any fundamental way. The thermal expansion coefficient increases and the circulation becomes more vigorous, but it is still driven by thermal buoyancy forces, directly or indirectly. In the vicinity of the critical point, however, a discontinuity appears in the fluid density if the pressure falls beneath the critical value; a less-ordered vapour phase then coexists with a more-ordered liquid phase at the same pressure and temperature. This molecular-based alteration of fluid properties produces a different type of thermosyphon behaviour, the upper boundary of which is therefore marked by the critical point; the lower boundary is found at the triple point.

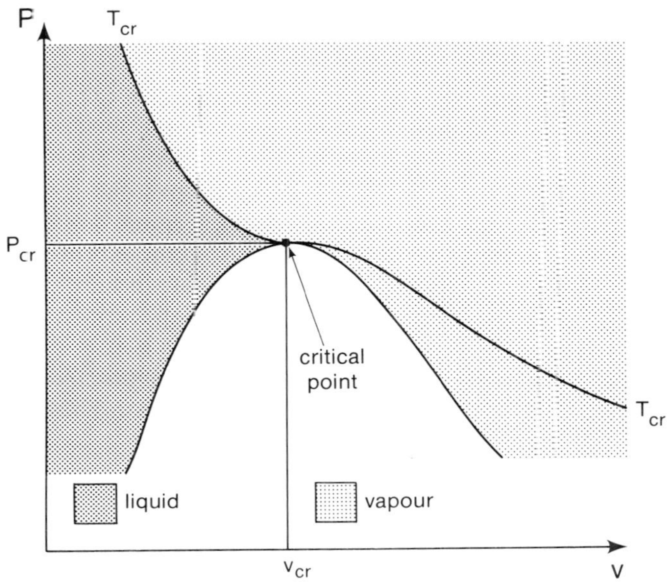

Fig. 3.1. Single-phase and two-phase domains.

Just below the critical pressure, the latent heat λ and the density difference $(\rho_L - \rho_V)$ between coexistent phases are both quite small; thermal buoyancy continues to play an important role, especially in the presence of low viscosities. It is thus to be expected that many of the flow patterns described in Chapter 2 would reappear under what, in all likelihood, would then be turbulent conditions. As the pressure is lowered further, λ and $(\rho_L - \rho_V)$ increase rapidly with the result that Archimedean buoyancy becomes the principal driving force in a circulation which depends increasingly on latent heat transfer. Evaporation now dominates in the heated section while condensation dominates in the cooled section. Temperature gradients continue to generate thermal buoyancy forces but their effect is relegated to a minor role. The flow pattern is no longer that of a homogeneous fluid even though it must continue to follow the dictates of a body force field: less dense fluid tends to rise while more dense fluid tends to fall within the constraints of the tube geometry. The gross features of a buoyancy-driven circulation are thus retained.

Heterogeneity is an intrinsic characteristic of this phase-change heat transfer system. Within either phase, the differential equations governing thermal and hydrodynamic behaviour are unchanged but they are now coupled across the interface separating the phases. The interface thus becomes the distinguishing feature. It is over this interface that crucial heat transfer events occur. Whether at the surface of vapour bubbles, liquid droplets, or over the surface of an extensive liquid film, evaporation and condensation control the rate of heat transfer through the thermosyphon. Latent heat transfer across the liquid–vapour interface thus dictates thermosyphon performance, and the behaviour of the interface itself therefore assumes considerable importance. Broadly speaking, the influence of the interface is felt in two ways: firstly, through its area, relative to the inside surface of the thermosyphon tube; secondly, through its stability, especially when it constitutes the surface of a continuous film flowing over the inside of the tube.

The changing role of the interface separating a rising vapour core from a falling liquid annular film may be anticipated from the expression for overall heat transfer rate which, in its simplest approximate form, is given by

$$\dot{Q} = A_V \rho_V \bar{U}_V \lambda = A_L \rho_L \bar{U}_L \lambda \tag{3.1}$$

where A is the cross-sectional area, ρ the density, $\bar{U}$ the average axial velocity, and λ is the latent heat of evaporation. The average axial heat flux density, written in terms of the vapour, is therefore given by

$$\bar{\dot{q}} = \rho_V \bar{U}_V \lambda \tag{3.2}$$

from which it follows that the greater the vapour density, latent heat, and vapour velocity, the greater will be the heat transmitted. Precisely at the critical point this expression is invalid because it ignores sensible heat transfer, but below the critical point it provides very valuable insight. Just beneath the critical pressure, ρ_V is not too different from ρ_L and hence $\bar{U}_V \simeq \bar{U}_L A_L/A_V$ which suggest that vapour velocities are not likely to exceed liquid velocities by much, if at all. Under such conditions, the liquid–vapour interface may not be subjected to large shear stresses but may still become unstable because the surface tension is very small in the vicinity of the critical point. On the other hand, if the pressure is lowered substantially beneath the critical value the surface tension and liquid viscosity both increase significantly and thus tend to promote a stable liquid film. By then, however, $\rho_V \ll \rho_L$, and the heat transfer rate would decrease rapidly with falling pressure according to eqn (3.2) were it not for two countervailing tendencies: an increase in λ and an increase in $\bar{U}$. Well below the critical pressure, changes in λ become quite small, and it is therefore the vapour velocity which must compensate for the vapour density if the same heat flux is to be maintained.

Now the Clausius–Clapeyron equation

$$\frac{dP}{dT} = \frac{\Delta s}{\Delta v}$$

applied to a liquid and its vapour in equilibrium well below the critical pressure has the approximate solution

$$P \simeq P^{\circ} \exp \frac{\lambda(T - T^{\circ})}{RTT^{\circ}}$$

where T°, P° are reference saturation conditions. Since this assumes the vapour to be an ideal gas, it implies that

$$\rho_V = \frac{P}{RT} \simeq \frac{P^{\circ}}{RT} \exp \frac{\lambda(T - T^{\circ})}{RTT^{\circ}}$$

which is a steep function of vapour temperature T. It therefore follow from eqn (3.2) that reductions in the saturation temperature within a thermosyphon carrying a given heat load will create large increases in the vapour velocity and thus produce large interfacial shear stresses. Typically, low pressures and temperatures lead to interfacial alterations which range from surface waves to complete flow reversal. The changing condition of the interface will be called upon repeatedly to explain thermosyphon behaviour in this chapter. But first it is necessary to understand the fundamentals of boiling and condensation in tubes.

3.1 Condensation in vertical tubes

Cluster theory is perhaps the easiest way to describe the condensation process in both microscopic and macroscopic terms. The growth of liquid droplets from a homogeneous vapour phase, for example, may be viewed as the consequence of a critical balance being reached by individual molecular clusters. Initiated through chance encounters, the clusters may grow through the retention of subsequently encountered molecules but may also decay through fragmentation and escape mechanisms. At any given time, clusters are distributed over a wide range of sizes, with the most probable size depending upon the vapour temperature and pressure.

When the pressure is above, or the temperature below, the saturation value, the vapour is said to be supersaturated or supercooled, respectively. In such a metastable state, the vapour chemical potential lies above the corresponding value for the liquid, and this introduces a maximum in the cluster Gibbs functions when plotted against cluster size. The cluster is thus unstable in the vicinity of this optimal or critical size which, when it is small enough to approach the most probable cluster size, precipitates spontaneous and irreversible growth with the addition of only one more molecule to the cluster. That is, with a sufficient degree of supercooling or supersaturation the critical clusters suddenly grow into small droplets.

The presence of foreign molecules, particles, and surfaces may greatly influence the condensation process. This is felt in two ways: promoting or impeding the formation of critical clusters; and altering the condensation rate at the liquid–vapour interface. More will be said about these influences later. At this point it is sufficient to note that the tube wall, as a foreign material, may be chosen so that it is either wetted or not wetted by the condensate. Without wetting, the condensate beads on the surface in what is described as *drop* condensation; surface tension is then important. Wetting, on the other hand, induces the condensate to cover the surface in the process known as *film* condensation. The latter is by far the most common and may also be influenced by surface tension.

3.1.1 *Unconfined condensation*

Before making a detailed examination of condensation inside a tube, let us first consider the simpler situation when the vapour is unconfined and quiescent. This is depicted in Fig. 3.2 for a vertical flat plate completely wetted by the condensate. The plate surface temperature T_C is taken to be less than the saturated vapour temperature T_{sat}. Under laminar conditions, the two-dimensional problem thus defined may be phrased in terms of the normalized variables

$$x = \frac{X}{X^c}, \quad y = \frac{Y}{Y^c}, \quad u = \frac{U}{U^c}, \quad v = \frac{V}{V^c}, \quad \text{and} \quad \phi = \frac{\theta}{\theta^c}$$

which incorporate the natural scales designated by the superscript c. Following the technique used in the previous chapter, these may be substituted into the primitive forms of the governing equations to uncover the

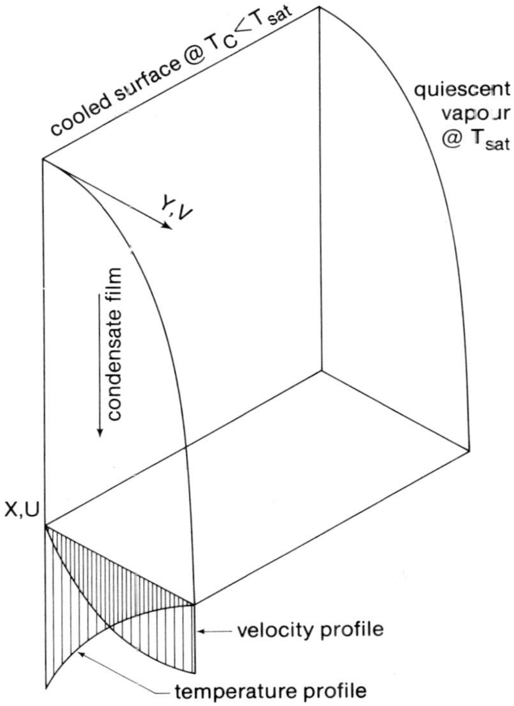

Fig. 3.2. Film condensation on a vertical plate.

basic physics applying to the *film*. Thus, the longitudinal equation of motion for the film becomes[1]

$$\left(\frac{\rho(U^c)^2}{X^c g\Delta\rho}\right)\left(u\,\frac{\partial u}{\partial x}+v\,\frac{\partial u}{\partial y}\right)=1+\left(\frac{\rho v U^c}{(Y^c)^2 g\Delta\rho}\right)\frac{\partial^2 u}{\partial y^2} \tag{3.3}$$

where $\Delta\rho=\rho_L-\rho_V$, while the energy equation becomes

$$\left(\frac{U^c(Y^c)^2}{\kappa X^c}\right)\left(u\,\frac{\partial\phi}{\partial x}+v\,\frac{\partial\phi}{\partial y}\right)=\frac{\partial^2\phi}{\partial y^2}. \tag{3.4}$$

These must be solved subject to the following hydrodynamic and thermal boundary conditions:

$$u(0,\,y)=\phi(0,y)=0$$

$$u(x,\,0)=\phi(x,\,0)=0$$

$$\frac{\partial u}{\partial y}\,(x,\,\delta)=\phi(x,\,\delta)-1=0 \tag{3.5}$$

[1] Ignoring the pressure departure and longitudinal diffusion.

in which $\phi=(T-T_C)/(T_I-T_C)$, where T_I is the interface temperature, and $\delta=Y^c$ is the film thickness scale. Also at the interface,

$$\lambda\dot{m}=k\left(\frac{\partial T}{\partial Y}\right)_\delta-k_V\left(\frac{\partial T_V}{\partial Y}\right)_\delta \tag{3.6}$$

where $\dot{m}$ is the condensation rate per unit interfacial area and the subscript V represents the vapour. Hence

$$\left(\frac{\lambda\dot{m}Y^c}{k\theta^c}\right)=\left(\frac{\partial\phi}{\partial y}\right)_\delta-\left(\frac{k_V\theta_V^c Y^c}{k\theta^c Y_V^c}\right)\left(\frac{\partial\phi_V}{\partial y}\right)_\delta$$

in which the conductivity ratio k_V/k is typically small. Here it is assumed that the vapour is saturated and hence the superheat ratio $\theta_V^c/\theta^c=0$, so that

$$\left(\frac{\lambda\dot{m}Y^c}{k\theta^c}\right)=\left(\frac{\partial\phi}{\partial y}\right)_\delta. \tag{3.7}$$

Since $T_I=T_{sat}$ and $\dot{m}X^c\simeq\rho U^c\delta$, where $X^c=L$ is the surface height, this balance between the latent heat transfer and the lateral heat conduction in the film implies that

$$\frac{\rho\lambda U^c\delta^2}{k(T_{sat}-T_C)L}=O(1). \tag{3.8}$$

Similarly, the balance of the viscous and Archimedean buoyancy forces in the equation of motion requires that

$$\frac{\rho\nu U^c}{\delta^2 g\Delta\rho}=O(1). \tag{3.9}$$

which, together with eqn (3.8), yields[2]

$$\delta=L\left(\frac{Ja}{Ar_L Pr}\right)^{1/4}$$

$$U^c=\frac{\kappa}{L}Ja\left(\frac{Ar_L Pr}{Ja}\right)^{1/2}. \tag{3.10}$$

Comparing these with the corresponding scales for natural convection on a vertical plate, that is eqns (2.7) and (2.8), it is evident that the non-dimensional group

[2] Keeping in mind that these coefficients may be set arbitrarily at unity, as noted in Chapter 2.

$$\frac{Ar_L Pr}{Ja} = \frac{\lambda g L^3 \Delta\rho}{\nu k(T_{sat} - T_C)}$$

plays a role similar to the Rayleigh number. In this group lie the Prandtl number $Pr = \nu/\kappa$, the Archimedes number $Ar_L = gL^3\Delta\rho/\rho\nu^2$, and the Jakob number $Ja = c_p(T_{sat} - T_C)/\lambda$. Given the structure of the Archimedes number, which is not unlike the Grashof number, it is perhaps surprising to find that $Ar_L Pr$ does not completely specify the vigour of the flow, as $Gr_L Pr$ would in natural convection. The additional feature is provided by the Jakob number, which is a measure of the sensible heat of the film in relation to the latent heat. This ratio is important in the description of heat transfer across any interface where condensation or evaporation are occurring.

Using these natural scales, eqns (3.3) and (3.4) may be rewritten

$$\frac{Ja}{Pr}\left(u\frac{\partial u}{\partial x} + v\frac{\partial u}{dy}\right) = 1 + \frac{\partial^2 u}{\partial y^2}$$

$$Ja\left(u\frac{\partial \phi}{dx} + v\frac{\partial \phi}{\partial y}\right) = \frac{\partial^2 \phi}{\partial y^2} \qquad (3.11)$$

which immediately reveal that in the common circumstances where $Ja \ll 1$, the temperature profile will be almost linear, as originally assumed by Nusselt (1916). It is also evident that the velocity profile will be closely approximated by a parabola in viscous liquids, that is when $Pr \gg 1$. The heat transfer rate under these conditions may be estimated from

$$\dot{Q} \simeq \lambda\rho U^c \delta$$

or

$$\dot{Q} \simeq \frac{k\theta^c L}{\delta},$$

both of which reveal that

$$Nu_L = O\left(\frac{Pr\,Ar_L}{Ja}\right)^{1/4}. \qquad (3.12)$$

It is interesting to compare this with eqn (2.12) which applies to a laminar, boundary layer flow driven by thermal buoyancy. The analogy is obvious and suggests that comparisons with natural convection behaviour may prove useful under the more complex conditions found in a thermosyphon tube.

Velocity and temperature profiles in a condensate film have been obtained numerically by Sparrow and Gregg (1959) using a similarity analysis. They demonstrated that

$$Nu_L = C\left(\frac{PrAr_L}{4Ja}\right)^{1/4}, \tag{3.13}$$

which is a more precise form of eqn (3.12). The function

$$Nu_L\left(\frac{4Ja}{PrAr_L}\right)^{1/4} = C(Ja, Pr)$$

is shown in Fig. 3.3 from which it is evident that the simple Nusselt prediction ($C=1$) is retained for non-metallic liquids providing $Ja \leq 0.1$, but is inaccurate for liquid metals; this was heralded by eqns (3.11). However, both of these predictions ignore the effect of surface ripples which cause the slightly higher heat transfer rates observed experimentally.

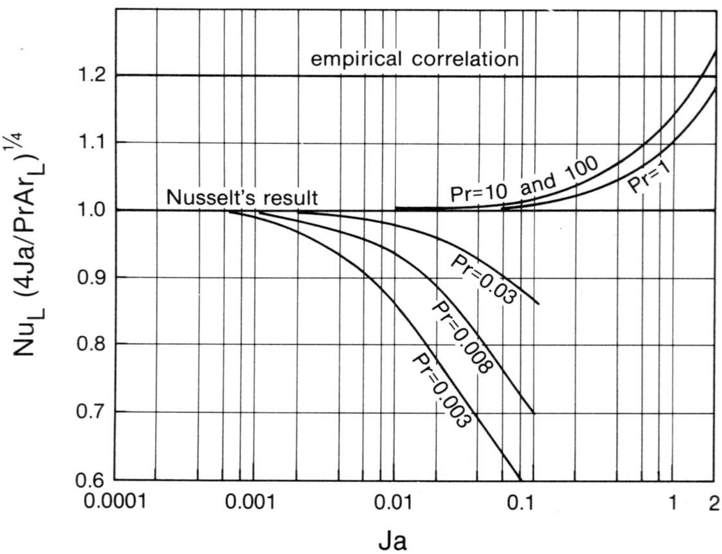

Fig. 3.3. Prediction of laminar film condensation heat transfer rate (after Sparrow and Gregg 1959).

Equation (3.13) may be written in a number of ways, depending upon its particular use. For example, it may be transformed into a heat transfer coefficient and averaged over the height L, in which the case result is

$$\bar{h} = 0.943\left(\frac{g\lambda k^3 \Delta\rho}{\nu L(T_{sat} - T_C)}\right)^{1/4}, \tag{3.14}$$

an equation attributable to Nusselt[3]. Alternatively it may be written in terms of the film Reynolds number defined by

$$Re_{\mathrm{f}} = \frac{\delta \bar{U}}{\nu} = \frac{\Gamma}{\mu} \tag{3.15}$$

where Γ is the local mass flow rate per unit width of wall. For a height L, the overall energy balance requires that

$$\bar{h}_{\mathrm{L}}(T_{\mathrm{sat}} - T_{\mathrm{C}}) = \frac{\lambda \Gamma}{L} \tag{3.16}$$

and hence eqn (3.14), averaged over this height, may be rearranged using eqn (3.16) to give

$$\bar{h}_{\mathrm{L}} = C_1 \left(\frac{\rho g k^3 \Delta\rho}{\mu^2}\right)^{1/3} \left(\frac{\mu}{\Gamma}\right)^{1/3} \tag{3.17}$$

where C_1 is an empirical coefficient. Thus, if the average film Nusselt number is redefined by

$$\bar{Nu}_{\mathrm{f}} = \frac{\bar{h}_{\mathrm{L}}}{k}\left(\frac{\mu^2}{\rho g \Delta\rho}\right)^{1/3},$$

in which $(\mu^2/\rho g \Delta\rho)^{1/3}$ is a length corresponding to $Ar = 1.0$, it follows from eqn (3.17) that

$$\bar{Nu}_{\mathrm{f}} = C_1 Re_{\mathrm{f}}^{-1/3} \tag{3.18}$$

for a laminar condensate whereas for a turbulent condensate film it is found that

$$\bar{Nu}_{\mathrm{f}} = C_2 Re_{\mathrm{f}}^{1/3}. \tag{3.19}$$

These relations are shown in Fig. 3.4. Using Nusselt's laminar theory, $C_1 = 0.926$ whereas the empirical value is $C_1 = 1.11$; the empirical turbulent result with $C_2 = 0.021$ is attributable to Grigull (1952).

A film Reynolds number of around 380 thus serves to distinguish between laminar and turbulent conditions. However, this inversion of behaviour does not occur simply and abruptly at a single transitional Reynolds number; the onset of turbulence has been noted around $Re_{\mathrm{f}} = 300$ (Kutateladze and Styrikovich 1960; Fulford 1964). For much lower values, the laminar film develops surface waves, and it is these which eventually produce an increase in laminar heat transfer rate by varying the local film thickness; their net effect is to increase the average temperature gradient (Sofrata 1980; Uehara *et al.*

[3] To match the 20% increase in Nusselt number found empirically, and suggested in Fig. 3.3, the coefficient in eqn (3.14) must be replaced by 1.13 (McAdams 1954).

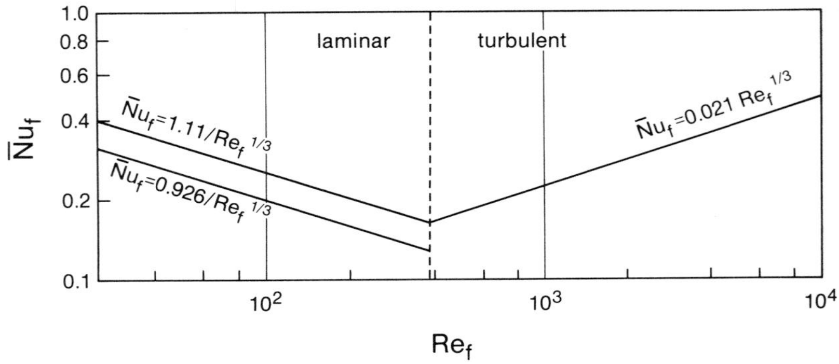

Fig. 3.4. Effect of Reynolds number on film condensation heat transfer.

1982; Suzuki *et al.* 1983). This wave behaviour may be attributed firstly to gravity and secondly to capillarity, and can only be fully understood in terms of the Froude number Fr and Weber number We. In falling water films, for example, gravity waves appear when $Fr = O(1)$ while the onset of capillary waves occurs when $We = O(1)$ (Fulford 1964). In general, therefore, film condensation on a vertical surface requires a description

$$\bar{N}u_f = f(Re_f, Pr, Fr, We) \tag{3.20}$$

if it is to be capable of covering a wide range of conditions including, for example, the formation of rivulets.

The description must also be capable of accommodating the surfaces effects of microgeometry and composition. Before a film can exist, any vapour approaching the surface must first nucleate. If the surface is clean and microrough, the vapour will tend to nucleate rapidly at many sites, thereby facilitating wetting of the surface; a macroscopic film quickly forms and grows. On the other hand, if the surface is microsmooth and is exposed to a non-wetting agent, either as a coating or in the vapour itself, the condensation process is radically altered by the growth of drops (Umur and Griffith 1964; LeFevre and Rose 1966; Graham and Griffith 1973; Brodowicz and Czaplicki 1982). Surface tension then assumes a primary importance.

The precise details of drop condensation are not yet fully understood but it appears that either or both of two distinct mechanisms are in operation. One of these postulates the prior existence of a microscopic condensate film which becomes unstable as it thickens and thus breaks up into droplets (Ruckenstein and Metiu 1965; Sugawara and Katsuka 1966). The other explanation simply assumes that droplets grow from nuclei formed at specific surface sites (McCormick and Westwater 1965). In either event, the microdroplets first formed are observed to grow at the expense of the condensate around them;

this is attributable to surface tension effects. Initially, the accretion may be caused by mobility in the thin film but eventually it is the result of coalescence which creates a smaller number of comparatively large drops. On a vertical surface, these drops become increasingly pendulous until their weight can no longer be held by surface tension forces. As they begin to run down, they accrete the drops ahead of them and thus accelerate, leaving behind a clean surface on which the entire process begins again. Film condensation typically creates heat transfer coefficient of the order of 10^4 W/m^2 K. Drop condensation heat transfer coefficients may be an order of magnitude higher than this, but non-wetting conditions are difficult to maintain in practice.

One important limitation of eqn (3.20) is to be found when the condensing vapour is intimately mixed with other vapours, for example air, which condense at much lower temperatures. In this situation the other gases dilute the condensing vapour and limit the rate of condensation by introducing an additional resistance; in approaching the interface, the vapour must diffuse through the gas, and upon reaching the interface the vapour molecules will be at a lower pressure and temperature than in the vapour bulk[4]. The diffusive resistance may be considerable (Sparrow and Eckert 1961; Sparrow and Lin 1964; Mori and Hijikata 1973), and its determination may be complicated by the fact that the heat and mass fluxes are generally coupled and must therefore be treated in accordance with the dictates of irreversible thermodynamics (De Groot and Mazur 1962). However, as noted in Chapter 1, the presence of incondensable gases may be turned to advantage in the isothermal operation of tubular thermosyphons.

3.1.2 *Refluent condensation*

To a first approximation, condensation inside a vertical tube is similar to that on a plane vertical surface wrapped into the form of a short, and usually circular, cylinder. Providing the length–diameter ratio is not much greater than unity, all of the descriptions and results discussed above continue to apply. This behaviour therefore provides a useful point of departure in exploring the various flow régimes which occur under the refluent conditions introduced by closing the top of the tube. For any given length–diameter ratio, a wetted tube surface ensures a falling film, the thickness of which increases with distance from the closed upper end. At the same time, a replenishing core of vapour moves upwards through a 'channel' of increasing cross-sectional area. In this situation the vapour and liquid flow may each be either laminar or turbulent, thus giving rise to four possible régimes in which the liquid–vapour interface may often be regarded as a passive surface on which condensation takes place.

[4] For the same reason, the gas partial pressure is higher at the interface; a continuing build-up of gas is thus prevented.

Under laminar conditions at least, a comparison with single-phase tubular behaviour provides some insights. The Nusselt result represented by eqn (3.12), for example, may only be applied when the film is thin in relation to the tube diameter. According to eqn (3.10) this condition would limit the result to

$$\frac{2\delta}{D} = \frac{2L}{D}\left(\frac{Ja}{Ar_{\mathrm{L}}Pr}\right)^{1/4} \ll 1$$

or

$$\frac{Ar_D PrD}{JaL} \gtrsim O(10^5),$$

analogous to the t_D limit discussed in Chapter 2. Below this limit the liquid film may become thick enough to experience a substantial drag caused by vapour moving in the opposite direction; the lower the magnitude of the group $Ar_D Pr/Ja$, the greater the film thickness and the greater the impeding effect of the vapour. In contrast to the natural convection system, however, a decrease in this group may be brought about by an *increase* in the driving temperature difference; the opposite is true in natural convection where the Rayleigh number is the controlling group. This again illustrates the significance of the interface, as reflected in the Jakob number. Continued decreases in $Ar_D Pr/Ja$ eventually lead to dramatic alterations at the interface, as will be seen later.

When the liquid and vapour flow are both laminar, the system is amenable to numerical analysis (Lee and Mital 1971; Spendel 1982, 1984) from which a number of observations may be made. Perhaps the most obvious of these is the alteration of conditions at the liquid–vapour interface. In particular, the absence of shear stress reflected in the (unconfined) boundary condition $\partial u/\partial y = 0$ is no longer justified. This apparently minor change destroys the similarity boundary layer forms used in obtaining Fig. 3.3 and requires solutions to the partial differential equations in their original elliptic form or, at the very least, in the parabolic forms given by eqns (3.3) and (3.4). Since the longitudinal mass flow of both vapour and liquid decreases with height it is to be expected that the interfacial shear stress would also decrease with height; in fact, it is found that the stress not only decreases but changes sign thereby signalling a flow reversal in the upper end of the tube. Corresponding to this reversal is a change in sign of the longitudinal pressure gradient: that is, the pressure increases with height near the tube mouth but decreases near the closed end. These effects introduce only small changes from the overall heat transfer rate calculated using Nusselt's laminar theory. The numerical data of Spendel (1984), for example, indicate that refluence decreases the local heat transfer rate near the mouth but introduces a compensating increase near the closed end. Figure 3.5 displays typical results, with Nu_N designating the Nusselt prediction. From Fig. 3.5(a), it is evident that performance generally improves with increases in the Jakob number, at least in this range. Figure 3.5(b), on the other hand, indicates that the Froude number $Fr = U_0^2/gD$, where D is the tube diameter and U_0 is the mean vapour entry velocity, produces little overall effect despite significant changes near the tube ends. For a given entry Reynolds number, this finding may be interpreted in terms of the tube diameter; increasing Fr then implies a decrease in tube diameter thereby producing a greater departure from the local Nusselt prediction.

For laminar flow throughout, the interfacial shear stress is caused mainly by momentum transfer through condensation and less by viscous effects (Chen *et al.* 1984). Once the *vapour* becomes turbulent, however, it appears that the reverse may be true, thus signalling a change in behaviour which may begin while the liquid film is still laminar ($Re_f < 300$). Such conditions were studied by Ho and Tien (1982) with the vapour Reynolds numbers[5] in the range 3247 to 6345. They found that turbulent vapour may increase or decrease the heat transfer rate relative to the classical laminar value but the changes were not substantial (<16%). However, when the condensate *film* becomes turbulent, the laminar form represented by eqn (3.18) no longer applies and must be replaced by a turbulent relation such as eqn (3.19). Laminar, transitional, and turbulent data have been reported by Gross and Hahne (1986) who clearly demonstrate the progressive effect of film Reynolds number between two asymptotic forms which, although developed for a vertical cylinder, are similar to eqns (3.18) and (3.19) as displayed in Fig. 3.4. This effect will be discussed further in Section 3.4.

The onset and development of turbulence are both associated with increased condensation rates, that is internal circulation rates, which, if they continue to rise, eventually lead to fundamental changes in interfacial conditions. As noted earlier, these changes may be attributed to a thickening film and the rising vapour velocity which accompanies it when the Jakob number is progressively increased. In a vertical tube, for example, interfacial waves begin to grow until they represent a physical impedance to the vapour flow which, since it is returning under the influence of a pressure gradient, then begins to exert a major influence on the buoyancy-driven film. By analogy with the phenomenon of separation, the pressure gradient in the vapour tends to reduce the downward film velocity near the wall, while the surface waves are subjected to the buffeting effect of the constricted vapour flow. At some point, the Archimedean buoyancy will be balanced by the pressure gradient, the film will be thick enough to create a physical blockage, and the surface waves will disintegrate. This point is usually known as the *flooding* point. Under these conditions, the situation is not unlike that found in cocurrent, annular,

[5] The traditional value of 2500 at transition is presumably modified at a free liquid surface.

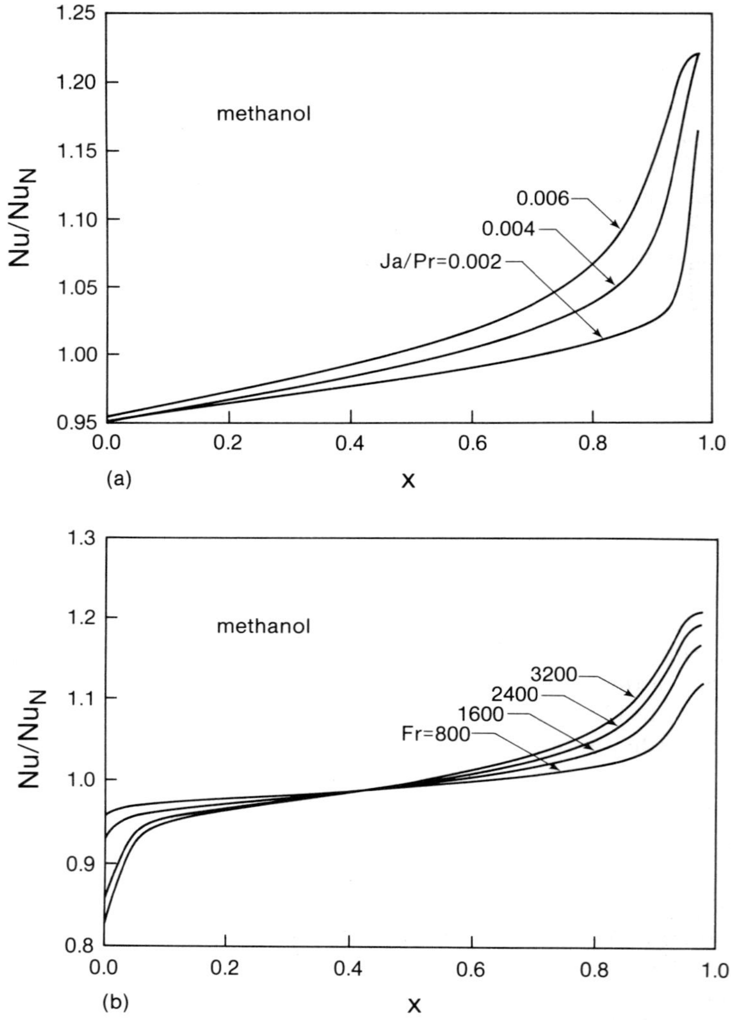

Fig. 3.5. Comparative heat transfer rate versus distance from the tube mouth (after Spendel 1984): (a) effect of Jakob number; (b) effect of vapour Froude number.

gas–liquid systems with the gas flowing upwards. These have been reviewed by Whalley (1987) who provides a simple analysis based on the insight that the Archimedean buoyancy force acting downwards on a large wave at the flooding point is essentially balanced by the alteration in the vapour pressure distribution created by the wave's own volume. This assumption leads to the result

$$Fr_D = \frac{\rho_V U_V^2}{gD\Delta\rho} = \frac{\pi}{4} \tag{3.21}$$

where D is the tube diameter. Viscous effects are ignored, thus implying that the inertial forces which give rise to the local pressure alterations are of the same order as the buoyancy forces. Now the magnitude of the inertial forces in relation to the surface tension forces is given by the Weber number, while the corresponding ratio of the buoyancy forces to the surface tension forces is given by the Bond number; that is, $Fr_D = O(We_D/Bo_D)$, and hence eqn (3.21) suggests that

$$We_D = O(Bo_D) \tag{3.22}$$

may provide a useful criterion for flooding. This suggestion will be taken up again later.

Figure 3.6 provides a one-dimensional representation of events in the condensate film as the vapour velocity is increased to the point of flooding and beyond; the interface is shown straight but would, in general, by wavy. Interfacial shear and the pressure distribution over shallow waves act increasingly to retard the liquid flow near the interface until it becomes stationary and flooding occurs; by that time the growth of interfacial waves has become substantial. Further increases in the vapour velocity eventually reduce the net flow in the film to zero, a condition described as *holdup*. The study of Bezrodnyi and Volkov (1982*a*) offers support to this explanation of flooding and holdup in a vertical condenser tube. Although executed in an adiabatic section, their experiments suggest flow régimes similar to those observed in diabatic, cocurrent systems. Initially, the friction factor of the vapour decreases with Reynolds number in a conventional manner. This is the régime where interaction beween the vapour and liquid streams, manifest through interfacial shear stress, is minimal. It persists until waves and spray characterize interfacial conditions. Beyond this, Bezrodnyi and Volkov report a rapid rise in friction factor leading first to flooding and then holdup, particularly in small-bore tubes.

Flooding behaviour is associated with wave behaviour and therefore has its origin in the stability of the liquid–vapour interface. At first thought, the situation brings to mind the classical Kelvin–Helmholtz stability problem, but there are important distinctions as well as similarities. The classical problem deals with incipient 'gravity' waves on a plane, *horizontal* interface separating two phases having a velocity differential ΔU. A balance between the inertial and Archimedean buoyancy forces leads to the stability criterion (Lamb 1932)

$$Fr_\Lambda = \frac{2\pi\rho(\Delta U)^2}{\Lambda g\Delta\rho} = O(1) \tag{3.23}$$

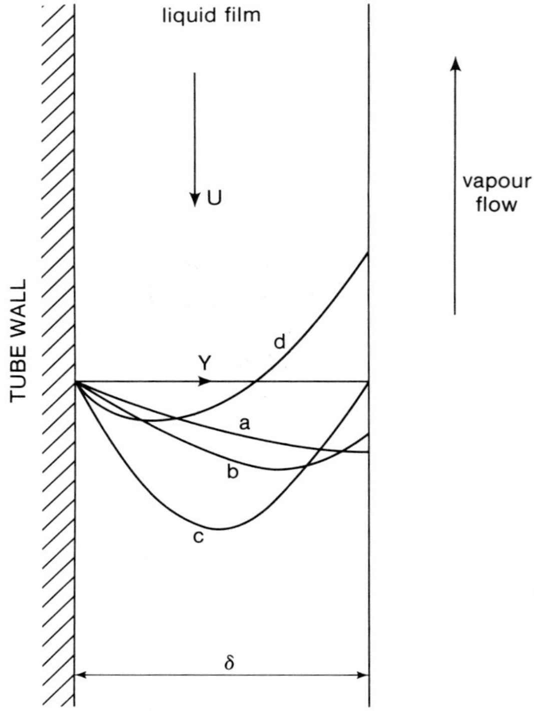

Fig. 3.6. Effect of vapour velocity on liquid film velocity profile: (a) $U>0$, $(\partial U/\partial Y)_\delta = 0$; (b) $U>0$, $(\partial U/\partial Y)_\delta \neq 0$; (c) $U_\delta \simeq 0$, flooding; (d) $\bar{U}=0$, hold up.

where Λ is the wavelength. This result is independent of the surface tension between the phases, and implies that

$$We_\Lambda = \frac{\rho^2(\Delta U)^4}{g\sigma\Delta\rho} = Ku^4$$

where Ku is the Kutateladze number; it contains no characteristic length. The growth of such waves thus occurs while

$$Ku = O(We_\Lambda^{1/4}) = O(Bo_\Lambda^{1/4})$$

since $Bo = We$ when $Fr = 1$. Should the waves grow until the surface tension forces balance the Archimedean buoyancy forces,

$$Bo_\Lambda = \frac{g\Lambda^2\Delta\rho}{4\pi^2\sigma} = O(1), \tag{3.24}$$

and it then follows that conditions on the nominally horizontal interface may be described by

$$Ku = We_\Lambda = Bo_\Lambda = Fr_\Lambda = O(1). \tag{3.25}$$

On the other hand, the growth of capillary waves on the surface of the annular condensate film flowing down inside a *vertical* tube is signalled by

$$We_\Lambda = \frac{\Lambda \rho_V U_V^2}{2\pi\sigma} = O(1) \tag{3.26}$$

where U_V plays the role of ΔU above. This result is independent of buoyancy and, by reasoning similar to the above, implies that growth takes place while

$$Ku = O(Fr_\Lambda^{1/4}) = O(Bo_\Lambda^{-1/4}). \tag{3.27}$$

Should this growth continue until Archimedean buoyancy forces acting downwards are roughly balanced by inertial (pressure) forces acting upwards, then $Fr_\Lambda = O(1)$ and eqn (3.25) is again produced.

This result may be generalized by recognizing that the wavelength Λ may not always be the appropriate length scale. In 'large-bore' tubes, that is those in which the liquid film is never thick enough to allow wave amplitudes comparable with the tube radius, it is to be expected that the tube diameter D would exert little influence on flooding. This suggests that Λ is then indeed the appropriate length scale, and may be determined from the condition $Bo_\Lambda = O(1)$, for example. It need only be added that the wavelengths observed in these circumstances are typically much less than the tube diameter and therefore the diameter-based, dimensionless quantities are not of order 1; for example, $Bo_D \gg 1$. On the other hand, D (or $D/2$) appears to be a more appropriate length scale when wave amplitudes are comparable with the tube radius. For such 'small-bore' tubes,

$$Ku = We_D = Bo_D = Fr_D = O(1). \tag{3.28}$$

This extends the particular result given by eqn (3.22), suggesting that it is a necessary, though not sufficient, condition for flooding in 'small-bore' tubes.

It thus appears that the large waves and spray which characterize flooding occur when the surface tension forces are of the same order as the Archimedean buoyancy forces (acting on the liquid) and the inertial forces (acting on the vapour). The latter are also balanced by the vapour pressure distribution, thus implying that viscous forces are not important. This indicates that even though interfacial shear plays a role at low condensation (circulation) rates, and may influence the onset of surface waves, if has little effect on flooding and holdup, during which inertial forces in the vapour shape the behaviour of the system.

By dramatically reducing circulation, flooding represents an upper limit for the heat transfer rate which may be estimated from

$$\dot{Q} \simeq \rho_V A U_V \lambda$$

where A is the cross-sectional area of the vapour, that is, essentially that of the tube. Prenger (1984) has suggested that this may be expressed in non-dimensional terms by taking

$$Q_E = \frac{\dot{Q}}{\rho_V A \lambda^{3/2}} = \frac{U_V}{\lambda^{1/2}}$$

or, using eqn (3.26),

$$Q_E = (2\pi E_t)^{1/2} \tag{3.29}$$

where

$$E_t = \frac{\sigma}{\rho_V \lambda \Lambda}.$$

Prenger followed Kutateladze (1969) and took the wavelength as

$$\Lambda = \left(\frac{\sigma}{g\Delta\rho}\right)^{1/2} \tag{3.30}$$

which is equivalent to fixing the Bond number at $1/4\pi^2$. Data for various fluids studied experimentally under flooding conditions are shown in Fig. 3.7 from which the correlation was found to be

$$Q_E = 2.99 E_t^{0.545},$$

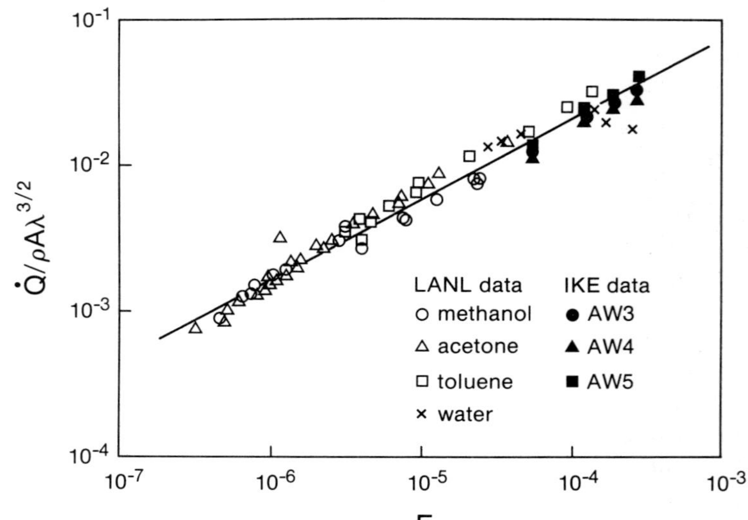

Fig. 3.7. Maximum heat transfer rate during flooding for various fluids in a vertical tube (after Prenger 1984).

thus substantiating the form of eqn (3.29) and lending support to the physical arguments leading to eqns (3.25) and (3.28). Since Q_E is typically small, it is clear that

$$U_V^2 \ll \lambda$$

thus revealing that while kinetic energy in the vapour controls the flooding limit it contributes little to the energy flux.

Much less is known about drop condensation in a vertical tube, although Xin and Xia (1986) have confirmed that heat transfer coefficients are much greater than those for film condensation. Their results, which evidently lend support to the theory of LeFevre and Rose (1966), suggest that refluent conditions lead to a reduction in heat transfer coefficient attributable to the vapour shear which impedes the descent of drops.

Finally, it is worth incorporating all of the above information into a general flow model which applies, in particular, to the more common mode of film condensation. This is shown schematically in Fig.3.8 which provides an

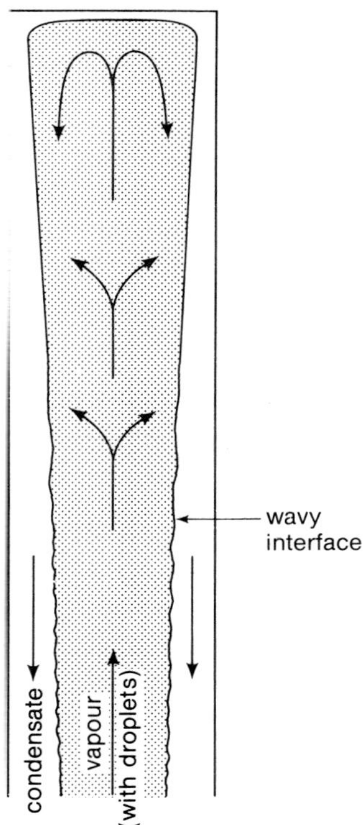

Fig. 3.8. Model of condensation in a vertical tube closed at the top.

overall description of refluent conditions in a vertical tube closed at the top. It will be recalled that for length–diameter ratios not too much greater than unity, interaction between the liquid film and the vapour core is unimportant and behaviour closely follows classical film theory whether the film is laminar or turbulent; in practice, the film may break into rivulets. As the heat flux and length–diameter ratio increase, interfacial conditions become increasingly important: firstly, because of the growth of capillary waves and the generation of spray which enters the vapour core and impedes its upward movement; secondly, because developed interfacial waves near the mouth eventually create flooding and, ultimately, holdup.

3.2 Evaporation in vertical tubes

Evaporation is not simply the reverse of condensation, notwithstanding the dictates of equilibrium thermodynamics; nor is the initiation of a vapour bubble a nucleation phenomenon, strictly speaking, because the parent liquid molecules move apart rather than come together to form a new, and coherent, structure, None the less, the basic tenets and consequences of cluster theory continue to apply. A critical cluster size is necessary for the spontaneous appearance and growth of bubbles in a superheated liquid; and foreign materials, the tube surface in particular, are capable of exerting a considerable influence on the nucleation process. At the liquid–vapour interface, condensation and evaporation are essentially reversible, and this is true, in particular, when the interface forms the free surface of a falling film; the Nusselt film analysis applies exactly during such evaporation. However, when the vapour phase is dispersed throughout the liquid phase the heat transfer process is more complex.

3.2.1 *Unconfined boiling*

Even though boiling within a tube differs from that on the outside, the geometry of the surface does not alter the general sequence of boiling events illustrated schematically in Fig. 3.9. This curve—commonly called the boiling curve—reveals the effect of increasing the temperature difference between the heated surface and the liquid which is to be boiled. It is evident that the heat flux is a strong function of the driving temperature difference.

Assuming the bulk of the liquid to be initially below its saturation temperature T_{sat}, the immediate effect of raising the surface temperature T_H is to establish the natural convection currents discussed extensively in the previous chapter. During this process, evaporation will likely occur at the liquid free surface where the role of the convection currents is simply to provide a continual source of warmer liquid. However, evaporation rates in this natural convection régime (a–b) are much smaller than those typical of boiling, as the figure indicates.

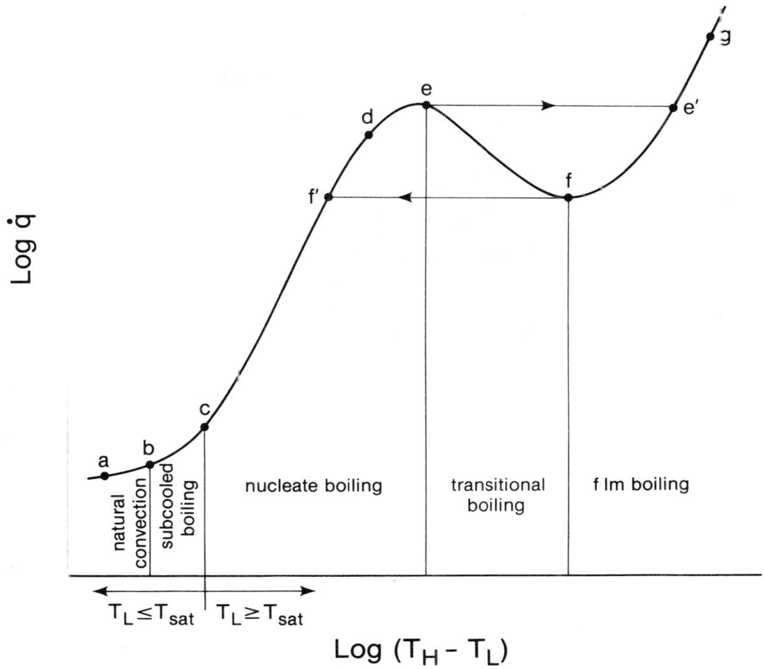

Fig. 3.9. The boiling curve (modified from data of Farber and Scorah 1948).

If the heated surface temperature continues to rise it will eventually exceed the saturation temperature (at that pressure) thereby introducing a small degree of superheat in the liquid, the precise amount depending upon any resistance to the formation of vapour bubbles at the surface. Figure 3.10 uses a *T–S* diagram to illustrate the corresponding states. When $T_H \leq T_{sat}$, the bulk liquid temperature T_L will lie on the subcooled section of the isobar $P_L = P_{sat}$: that is, $T_L < T_H \leq T_{sat}$. Vapour bubbles will not form until $T_H \geq T_V$, where

$$T_V \simeq T_{sat} + \frac{2\sigma R T_{sat}^2}{\lambda r_V P_V} \tag{3.31}$$

in which R is the gas constant of the vapour, σ is the interfacial (surface) tension, λ is the latent heat of vaporization, and P_V is the pressure inside a vapour bubble of radius r_V.

Bubbles[6] form first at those sites which are most favourable, as determined by the microgeometry of the surface and the chemical composition of the

[6] These are not to be confused with bubbles of dissolved gas which are also released by heating; however, nucleation sites sometimes contain trapped pockets of undissolved gas.

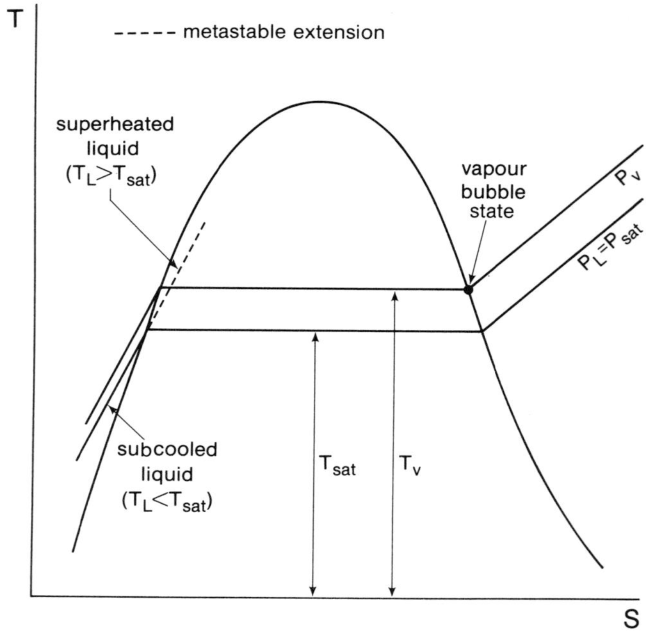

Fig. 3.10. Liquid and vapour states.

boiling fluid and the surface itself. As a general rule, bubbles begin in microscopic cavities of radius slightly larger than a critical value r_{cr} derived from eqn (3.31). That is

$$r_{cr} = \frac{2\sigma R T_{sat}^2}{\lambda P_V (T_L - T_{sat})}. \tag{3.32}$$

In any such cavities, where the surface temperature exceeds T_L, bubbles will form and grow until they more than fill the cavity. The subsequent spill of vapour beyond the microcavity is constrained by surface tension acting at the three-phase intersection between the vapour, its parent liquid, and the heated surface. Eventually, the attached vapour bubbles balloon to the point where the buoyancy force produced by evaporation overcomes the attachment force. The bubbles then detach themselves in a manner which depends upon the orientation of the surface and the vigour of the neighbouring circulation.

Once bubbles begin to form and detach, the heat transfer rate increases substantially. As Fig. 3.9 shows, the boiling curve begins to steepen (b–c) as bubbles form in numbers which gradually increase with the surface temperature, that is as more (smaller cavity) sites are activated. This is the subcooled boiling régime. Initially, these bubbles are small, almost spherical, and do not grow following detachment. If the bulk of the liquid is below the saturation temperature, $T_L < T_{sat} < T_V$, the bubbles will collapse as they move away from the surface, giving up latent heat to warm the bulk further. The bulk temperature T_L may thus be raised to, and may slightly exceed, the saturation temperature at which point (c) the boiling curve steepens yet again in what is often referred to as the nucleate boiling régime (c–d–e).

The growth of a detached bubble in the nucleated régime may be represented in the radially symmetric form shown in Fig. 3.11. The temperature and pressure are essentially uniform throughout the bubble and are related to each other through an equation of state; both are above liquid saturation values. A heat balance at the vapour–liquid interface yields

$$\rho_V \lambda \frac{dr_V}{dt} = h(T_L - T_V) \qquad (3.33)$$

where h is the bubble heat transfer coefficient and ρ_v is the vapour density.

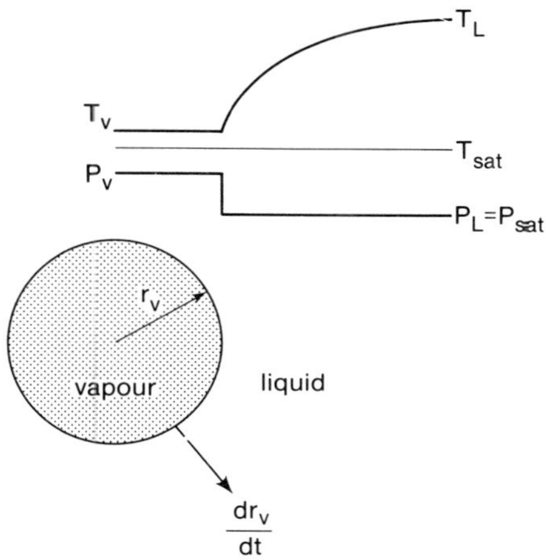

Fig. 3.11. Evaporative growth of a bubble.

Nucleate boiling is characterized by very large heat transfer coefficients at the heated surface, for example 10^3–10^4 W/m^{2}K. These are attributable partly to the necessary supply of latent heat but are largely the result of vigorous convection near the surface: rapid displacement of liquid by the growing bubble and rapid replacement of the detached bubble by liquid; both of these events induce macroscopic mixing processes equivalent to an extremely high level of turbulence. As Fig. 3.9 reveals, the vigour of the macroscopic mixing process increases as the surface temperature is raised above the saturation temperature. Bubbles increase in number through site activation attributable to a decreasing value of r_{cr}, given by eqn (3.32), and through increasing frequency at any given site. The bubbles also increase in size by growing faster, according to eqn (3.33), and by subsequent coalescence.

When increases in surface temperature are particluarly great (d–e) the bubbles become very large and very numerous; vapour columns appear and lead to a very unstable transitional régime (e–f) in which the heat flux decays as the columns disappear with the eventual formation of a continuous blanket of vapour over the heated surface. Beyond f, heat is transferred through a thin vapour film, principally by conduction and radiation; agitation of the liquid bulk is then much reduced. Should the transition régime be traversed with the heat flux fixed at the value first reached at e, the route followed will be from e to e′ which is an irreversible and rapidly traversed path. Upon reducing the heat flux, the backward route e′–f–f′ is also an irreversible path. Thus, boiling in and around the transitional régime is hysteretic if control is exerted through the heat supply rate. On the other hand, if the surface temperature is used to exercise control, the path f′–e–f–e′ is reversible. The condition e is often referred to as *burnout*, the term originating in the discovery of the path e–e′ when the surface temperature exhibits a large upward excursion. While this is no longer an accurate description under reversible, temperature-controlled conditions, it will be retained throughout this book to signify the local maximum. It is important to note that the attainment of this limiting or critical heat flux may or may not produce unstable hysteretic behaviour.

The above description refers to *pool* boiling, in which the bulk motion of the unconfined liquid is unimportant in comparison with the vigorous internal mixing created by bubble behaviour. Under such conditions, it is to be expected that the *bubble* heat transfer rate would depend upon the bulk liquid properties, the driving temperature difference, and on the stirring action created by the bubbles: that is,

$$Nu_b = f(Pr, Ja, Re_b) \tag{3.34}$$

where Nu_b, Re_b refers to the bubbles. The Jakob number and Reynolds number may be defined by

$$Ja = c_{pL}\frac{(T_H - T_{sat})}{\lambda} \quad \text{and} \quad Re_b = \frac{U_b r_V}{\nu_L}$$

in which U_b is the mean (superficial) vapour velocity estimated from a balance between the surface heat supply rate and the convective (latent heat) flux of the vapour; that is,

$$\dot{q} = \rho_V U_b \lambda \tag{3.35}$$

links heater performance to bubble dynamics. The bubble radius may be estimated from a static force balance; thus, we define

$$r_V = \left(\frac{\sigma}{g\Delta\rho}\right)^{1/2} \tag{3.36}$$

as the effective radius[7] following detachment. Using eqns (3.35) and (3.36), it follows that

$$Nu_b = \frac{\dot{q}}{k_L(T_H - T_{sat})}\left(\frac{\sigma}{g\Delta\rho}\right)^{1/2},$$

$$Re_b = \frac{\dot{q}}{\nu_L \lambda \rho_V}\left(\frac{\sigma}{g\Delta\rho}\right)^{1/2},$$

and hence

$$Nu_b = \frac{Pe_b}{Ja}\frac{\rho_V}{\rho_L}, \tag{3.37}$$

in which $Pe_b = Re_b Pr_L$ is the bubble Peclet number.

The relationship between the Jakob number and the bubble Reynolds number in pool boiling was first revealed by Rohsenow[8] (1952) in the form

$$Ja \simeq C_{sf}\left(Re_b \frac{\rho_V}{\rho_L}\right)^{1/3} Pr_L^{5/3} \tag{3.38}$$

thus permitting eqn (3.37) to be written in the form

$$Nu_b \simeq \frac{1}{C_{sf}}\left(\frac{\rho_V}{\rho_L}\right)^{2/3}\left(\frac{Re_b}{Pr_L}\right)^{2/3} \tag{3.39}$$

and hence

$$\dot{q} \simeq \frac{\mu_L c_{pL}^3}{\lambda^2 C_{sf}^3 Pr_L^5}\left(\frac{g\Delta\rho}{\sigma}\right)^{1/2}(T_H - T_{sat})^3. \tag{3.40}$$

Equation (3.39) embodies the fundamental role of bubble dynamics in the boiling process but has little practical value unless the velocities and diameters of the bubbles are known, at least on average. On the other hand, eqn (3.40),

[7] This expression for radius has the same structure as the critical wavelength associated with instability of the liquid–vapour interface in the transitional regime (Bellman and Pennington 1954). It also appeared as eqn (3.30) for the wavelength associated with flooding, and is rooted in the Bond number, the ratio of buoyancy force to surface tension force. Thus when $Bo_r = 1$, the radius is given by eqn (3.36).

[8] The exponent of the Prandtl number in eqn (3.38) is approximately 5/3 for most liquids; for water, however, it is 1.0, altering eqns (3.39) and (3.40) accordingly.

which reflects the steepness of the boiling curve in the nucleate régime, may be applied directly to pool boiling provided that C_{sf} is known. This coefficient is a function of the chemical properties of both the heated surface and the boiling liquid but also depends upon the microgeometry of the surface. Table 3.1 lists representative values indicating that in general $2 \times 10^{-3} < C_{sf} < 3 \times 10^{-2}$.

In concluding this brief introduction to pool boiling it is important to point out that many heat transfer correlations other than eqn (3.40) have been developed, most being limited to particular conditions. The correlation of Cooper (1984), based on the reduced pressure, appears to be an exception which avoids the need to specify the liquid–surface characteristics, at least within experimental uncertainty. For design purposes, this has the considerable merit of simplicity, but it provides little physical insight. It is also worth noting that various attempts (Webb 1981) have been made to enhance the boiling process by attaching promoters to the bare heated surface. Tong *et al.* (1983), for example, have studied the effect of a neighbouring perforated plate and found a significant performance improvement in the low-temperature difference range, while Yabe *et al.* (1984) report a similar benefit using a fibrous polypropylene membrane.

3.2.2 *Refluent evaporation*

The above discussion applies in general terms to pool boiling on any surface which is isometric, or nearly so. It is therefore to be expected that boiling in a vertical tube would exhibit the precise behaviour described above only if there were no superimposed forced flow and the length–diameter ratio was not too much greater than 1.0. In short tubes, the boiling curve closely follows the classical shape with an equation of the type (3.40) applying in the nucleate régime. The upper limit of thermal performance is then dictated by the burnout phenomenon. For tubes with larger length–diameter ratios the situation changes substantially.

It will be recalled from previous discussion that the result of increasing the length–diameter ratio during condensation or natural convection in a vertical tube closed at one end is to introduce a choking or impeding effect; wall effects then propagate across the entire cross section of the tube thus creating a return core flow of significant strength. A similar tendency may be found during boiling. Savchenkov *et al.* (1982), for example, investigated the flow structure of simulated boiling in a vertical tube, closed at the bottom, and noted well-defined circulation cells which they attributed to impeding conditions. However, a direct comparison with thermally buoyant flows is difficult to establish because bubbling may not conform to the same overall flow pattern. In particular, the bubbles do not always rise uniformly near the wall to produce a downward replenishing core of liquid flowing about the tube axis. With a ring bubbler, for example, it has been noted (Lock and Kirchner 1989*a*) that bubbles have a tendency to migrate radially inwards thereby creating a

TABLE 3.1. *Representative values of the surface–liquid coefficient* C_{sf}

Liquid–surface combination	C_{sf}
Water–copper	0.013
Water–platinum	0.013
Water–brass	0.0060
Water–emery-polished cooper	0.0128
Water–ground and polished stainless steel	0.0080
Water–chemically etched stainless steel	0.0133
Water–mechanically polished stainless steel	0.0132
Water–emery-polished and paraffin-treated copper	0.0147
Water–scored copper	0.0068
Water–Teflon pitted stainless steel	0.0058
Benzene–chromium	0.010
n-Butyl alcohol–copper	0.003 05
Ethyl alcohol–chromium	0.027
Isopropyl alcohol-copper	0.002 25
n-Pentane–emery-polished copper	0.0154
n-Pentane–emery-polished nickel	0.0127
n-Pentane–lapped copper	0.0049
n-Pentane–emery-rubbed copper	0.0074

cell near the closed, lower end of the tube; the extent of this cell varies with the vapour fraction. Above the cell, the circulation is in the opposite direction; that is, a rising core of higher vapour fraction is then balanced by a descending annulus of lower vapour fraction. This behaviour is discussed in more detail in Section 5.3.

The effect of length–diameter ratio is also manifest through substantial alterations to the bubbles themselves. With the tube initially filled with liquid, an increasing heat flux generates increasingly higher vapour fractions which, in the confined space of the tube, lead to bubble interactions. A number of investigations (Larkin 1971; Bezrodnyi 1978; Savchenkov *et al.* 1982) have shown that the flow may change from bubbly to frothy to cellular-frothy to plug flow, and on up to a pulsing churn flow. In general terms, these régimes are similar to those established for one-dimensional, two-phase flow in the absence of phase change (Baker 1954; Wallis 1969; Hewitt and Hall-Taylor 1970), but the details reflect important differences: the continual formation and detachment of bubbles at the tube wall, the effect of returning condensate, and the inherent refluent pattern. If the heat flux is too high, burnout will occur. The cause is not obvious but the mechanisms which have been suggested for unidirectional flow conditions (Hewitt 1978) are likely to be operative. These include: film boiling as the result of bubble proliferation; nucleation site overheating, which prevents rewetting, especially at very high heat fluxes; and vapour pocketing accompanied by a very thin liquid film which evaporates rapidly to prevent rewetting, particularly at low heat fluxes.

When the tube is not completely filled with liquid initially the situation changes yet again. A reduced height in the liquid pool standing above the closed base may be a design choice or the result of liquid having been transferred to the condensing section, or perhaps both. Within the reduced pool, the liquid boils in the manner described above, and it is only necessary to bear in mind that as the liquid level falls the effect of tube geometry within the pool decreases and behaviour approaches that of classical pool boiling. Above the pool, evaporation takes place both at the pool surface and over the surface of the returning film as it runs down over the heated wall; a vapour core flows in the opposite direction. This situation thus combines the complexities of falling film hydrodynamics with the thermodynamics of phase change.

The hydrodynamic behaviour of the film is largely determined by conditions in the condensing section upstream. This was discussed in detail in the previous section from which it will be recalled that the film may be laminar or turbulent, smooth or wavy, and may descend in annular or rivulet form. The rate of heat transfer through the film may thus vary significantly over various flow régimes, even when considering simple evaporation from the free surface. Any subcooling of the condensate may carry over[9] into the evaporating section, potentially raising the local critical heat flux (Ivey and Morris 1962). Subcooling also has the effect of delaying the onset of nucleation until

$$T_{\mathrm{H}}-T_{\mathrm{sat}} \simeq \left(8\sigma T_{\mathrm{sat}}\frac{\dot{q}}{\lambda\rho_{\mathrm{V}}k_{\mathrm{L}}}\right)^{1/2}, \tag{3.41}$$

assuming that the temperature profile in the condensate film is linear (Davis and Anderson 1966). For higher heat transfer rates, boiling may occur in the film itself thereby causing disruptions; in particular, vapour bubbles burst and spray water droplets into the upward-moving vapour. This behaviour is not unlike that observed in the annular flow régime in some non-evaporative two-phase systems (Wallis 1969) except that in the vertical thermosyphon tube the liquid annulus and vapour core do not move in the same direction. Annular flow under cocurrent conditions occurs when the vapour fraction is higher than that in the churn flow régime; in a boiling context, this leads to higher heat transfer coefficients (Chen 1966). However, evaporation then accelerates the flow, the liquid annulus in particular, whereas under annular refluent conditions the new vapour tends to decelerate the liquid annulus and may

[9] Depending upon conditions in the intermediate adiabatic section, if any.

even lead to a flow reversal. This behaviour was discussed earlier and is an intrinsic feature of the thermosyphon.

It is known that nucleate boiling in the returning condensate film contributes to choking phenomena which limit the permissible heat transfer rate. A vigorously boiling film sprays droplets into the vapour core, the resulting entrainment adding to the momentum exchange which ultimately leads to a flooding condition. Below this flooding limit, the fall of liquid pool levels imposes yet another performance constraint. Should the level drop too far, and especially if the pool vanishes entirely, the lower extremity of the liquid film may evaporate completely and thus expose a patch of bare tube wall. This limit is appropriately called *dryout*. In its simplest form, incipient dryout reflects the fact that the liquid phase is distributed as a film over the entire inside surface of the tube; the pool is absent. In general, however, it must be borne in mind that in the vicinity of the flooding limit the impeding effect of the vapour return may enable a weaker film to evaporate completely before reaching the top of the liquid pool; that is, dryout may occur in the presence of a pool. Dryout is closely associated with the phenomena of quenching and rewetting in which the heated surface temperature ahead of a liquid tongue increases sharply to create localized boiling. Such behaviour gives rise to a sputtering effect in which small droplets are ejected ahead of the tongue to produce precursory cooling (Sun *et al.* 1975). Similar behaviour accompanies dryout.

For most boiling applications, the limits of performance are associated with the maximum permissible heat transfer rate but noteworthy behaviour may also be found at low heat transfer rates, especially those which are encountered during startup in long tubes. Under these conditions, the boiling process may be pulsatile, resulting in the *geyser* effect (Larkin 1971; Bezrodnyi and Alekseyenko 1977*a*,*b*; Casarosa *et al.* 1983). As its name suggests, this cyclic effect is divided into two parts: a period of eruptive discharge of vapour and liquid followed by a quiet period during which boiling is absent. In the latter, heat is transferred to the liquid by conduction from the wall so that convection is limited to a weak natural circulation during which nucleation is delayed. This superheating process occurs mainly beneath the subcooled upper layers of the pool and may continue only until the Jakob number is great enough to allow a convulsion in which bubble growth stimulates convection which promotes more rapid bubble growth, each reinforcing the other in the presence of superheat which, at the onset of eruption, is abnormally high. The result is a very rapid boiling process during which the absorption of latent heat tends to cool the liquid being displaced by large bubbles. The vigour of the action increases the local heat transfer coefficient with the wall and thereby accelerates the reduction of local tube wall temperature by the withdrawal of sensible heat. With temperatures falling towards the saturation temperature, the boiling process eventually subsides while the expelled mixture, externally condensed and cooled, prepares to flow back down into the pool and thus return the system to its earlier quiet condition.

The frequency of this process extends over a significant range in which three separate, but overlapping, flow régimes may be identified (Liu and Ang 1988). For very low frequencies, circulation in the liquid is especially weak and the high superheat characteristically creates a single vapour bubble; high frequencies, on the other hand, are associated with a frothy or churned mixture, the superheat being very small. Between these extremes is a range of intermediate frequencies (and superheats) in which intermediate behaviour is often described as pulse boiling.

The features outlined in this section, taken together, lend themselves to the development of a general model of evaporation in a tube closed at the bottom and open at the top to a stream of condensate running down the tube wall. This is illustrated schematically in Fig. 3.12. When the tube is completely filled with a pool of liquid, behaviour follows the dictates of the classical boiling curve modified by two major effects which are more pronounced as the length of the tube is increased in relation to the diameter. The first effect consists of cellular circulation patterns driven by the combination of bubble behaviour and condensate return. The second effect, which is rooted in bubble growth and interaction, results from a progressive change in flow régime as bubbles grow and gradually coalesce into larger bubbles which may occupy the entire cross section; this effect influences the limiting performance for either high or low heat fluxes.

When the liquid pool does not extend over the entire heated section, a replenishing film is observed above the pool. The behaviour of this film is governed partly by its thickness and velocity (i.e. the circulation rate) and partly by surface tension forces acting at its free surface. It is also strongly influenced by internal boiling. Should this boiling become too vigorous, either or both of two limiting events may occur: the film may completely evaporate before it reaches the pool, or the rate of liquid droplet ejection into the upward-moving vapour core may seriously impede the circulation. The full understanding of these events, and indeed of refluent evaporation in general, requires a wedding with refluent condensation in the confines of the same tube. The next two sections provide an analysis of complete thermosyphon tubes: their thermal behaviour, performance characteristics, and limits.

3.3 The vertical, evaporative thermosyphon

It will be recalled from the Introduction that one of the earliest systematic studies of phase change in a closed, vertical thermosyphon tube was carried out by Bayley (1955*a*) who attempted to simulate the conditions which might exist in a hollow gas turbine blade (Cohen and Bayley 1955). Since that time, interest in the evaporative device has grown steadily as new applications have

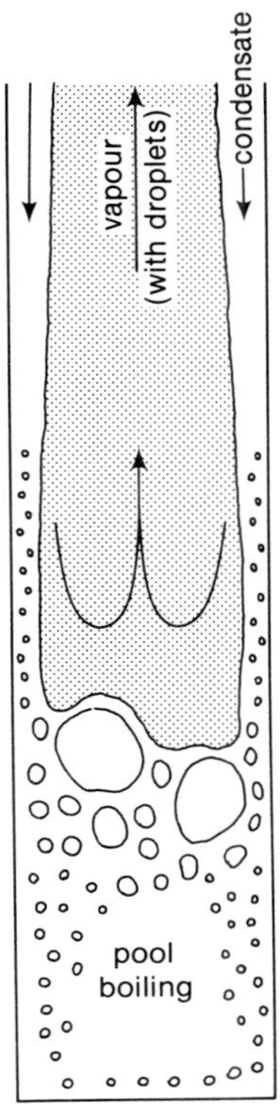

Fig. 3.12. Model of evaporation in a vertical tube closed at the bottom.

arisen in connection with nuclear reactor cooling, permafrost aggradation, waste heat exchangers, etc. Consequently, a great number of independent investigations has since been undertaken, each contributing to our understanding of evaporative behaviour. As with single-phase behaviour, the characteristic effectiveness of the device is attributable to a large convective flux created by thermally induced buoyancy forces, but the evaporative mode clearly differs in three important respects: the buoyancy, being Archimedean, is greater by a factor $\Delta\rho/\rho\beta\Delta T$ which may be as high as 10^3; the circulation pattern within the tube is essentially the same in the heating and cooling sections, thereby removing the problem of flow reconciliation at their junction; and the existence of a liquid–vapour interface, continuous or dispersed, greatly influences the limits of performance. These differences tend to restrict comparisons with the single-phase system to the overall features of refluent behaviour which are, however, central to our understanding.

While it is possible to construct both open and closed versions of the evaporative thermosyphon, most interest lies in the closed form, which offers greater flexibility in design; most notably in the permissible heat flux and temperature ranges. Consequently, this section focuses on closed tube behaviour. Using the discussion of evaporation and condensation presented in the previous two sections, it is possible to construct a composite model of the closed device. This is shown in Fig. 3.13 which is essentially the superposition of Figs 3.8 and 3.12, with the tacit assumption that the joining of the two open ends, with or without an intermediate adiabatic length, does not materially alter the behaviour in either section; that is, the evaporative and condensing systems are coupled only through the requirements of continuity. As a first

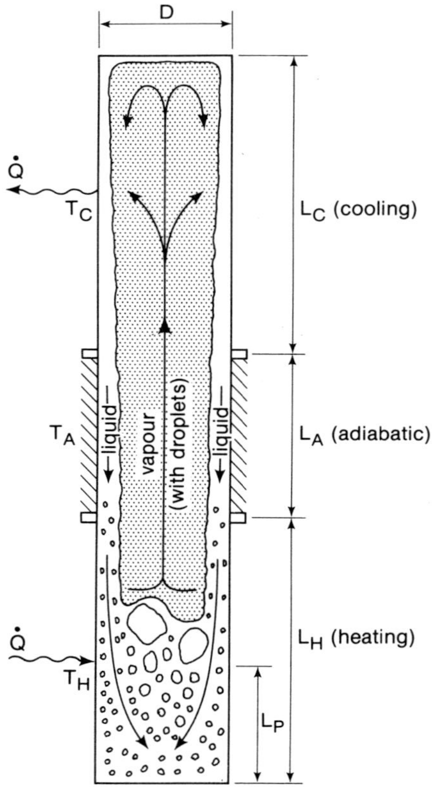

Fig. 3.13. Model of evaporative tubular thermosyphon standing vertical.

approximation, this assumption provides a satisfactory starting point, but it is now evident from earlier discussion that under certain circumstances the processes and events taking place in the evaporator may exert a considerable influence on the condenser, and vice versa. It is worthwhile keeping this physical model in mind during the ensuring discussion of thermal characteristics and their dependence on the filling charge, tube geometry, and system pressure. It is also important to recognize at the outset that the rate at which heat is transmitted through the system is a function of many variables. The previous sections, for example, suggest a complex relation described by

$$Nu = f\left(Ar,\ Ja,\ Fr,\ We,\ Pr,\ \frac{P}{P_{cr}},\ \frac{L_H}{D},\ \frac{L_C}{D},\ \frac{L_A}{D},\ \frac{L_P}{D}\right) \tag{3.42}$$

where L_P is the height of the liquid filling in the absence of evaporation. Not surprisingly, this relation has not yet been fully mapped, and hence the following treatment is limited to an exploration of the effects of the principal variables.

3.3.1 *Thermal model for a vertical tube*

Shiraishi *et al.* (1982) suggested that the physical model of Fig. 3.13 may be converted into a heat transfer model represented by the overall relation

$$\dot{Q} = \frac{T_H - T_C}{R} \tag{3.43}$$

where R is the overall thermal resistance constructed from component resistances in the heating and cooling sections. In the heating section, for example, heat transfer occurs in both the pool and film such that

$$\dot{Q} = \dot{Q}_P + \dot{Q}_F$$

where

$$\dot{Q}_P = \bar{h}_P A_P \Delta T_P$$

and

$$\dot{Q}_F = \bar{h}_F A_F \Delta T_F$$

are the heat transfer rates to the pool and the evaporating film, respectively; $\bar{h}$ is the average heat transfer coefficient and the heat transmitting areas are given by

$$A_F = \pi D(L_H - L_P)$$

and

$$A_P = \pi D L_P,$$

ignoring the tube end. In general the radial temperature difference between the heated wall and the interior fluid is a function of height, but may be approximated by a constant value for each of the lengths over which heat transfer occurs: in the pool, uniform heating and hydrostatic head together suggest a linear decrease in wall and liquid temperature with height, but the temperature difference remains constant; across the liquid film, the wall temperature and (saturated) vapour temperature remain fixed in both the condenser and evaporator.

Using Nusselt's laminar analysis of film behaviour above the pool and an empirical correlation based on nucleate boiling in the pool itself, Shiraishi *et al.* were able to test this model. A representative set of their data for water at various pressures is shown in Fig. 3.14 for $L_H/D=7.57$, $L_A/D=13.51$, and $L_C/D=12.16$. The agreement is seen to be good when the liquid volume fills the evaporator, but less so when the filling is halved, perhaps because the lower reaches of the condensate film were not laminar.

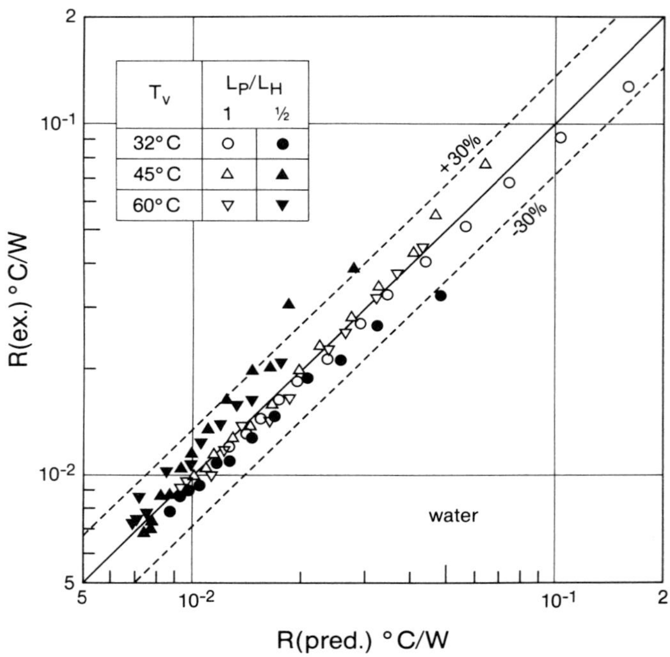

Fig. 3.14. Comparison of the predicted overall thermal resistance with experimental results for water (after Shiraishi *et al.* 1982).

A comparison of the temperature drops occurring in a similar system is shown in Fig. 3.15 using the data of Hahne and Gross (1982) obtained with $L_H/D=12.5$, $L_A/D=25$, $L_C/D=12.1$, and a filling of refrigerant R115 occupying 45% of the overall tube length. The data indicate that the condenser contributed most to the overall thermal resistance but this is not an invariable result, as will be seen later; as expected, the adiabatic length offers the least resistance. Figure 3.16 reveals that the condenser heat transfer coefficient, along with the effective thermal conductivity of the entire system, increases steadily with heat flux, but only beyond the point where the condensate film undergoes a transition from from laminar to turbulent flow, that is near $\dot{Q} = 1$ kW.

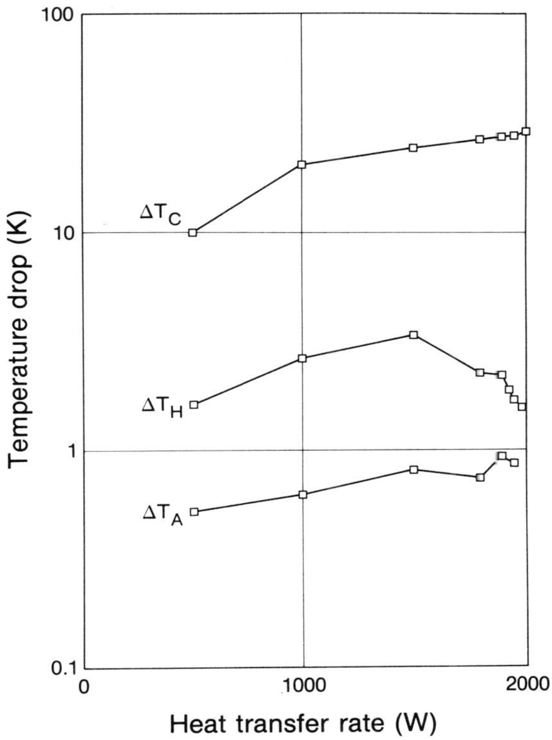

Fig. 3.15. Effect of heat flux on the temperature drop in the heating, adiabatic, and cooling sections (after Hahne and Gross 1982).

These observations suggest that the physical model of Fig. 3.13 will be amenable to theoretical analysis if it is cast in the idealized form shown in Fig. 3.17. As indicated, the cooling, adiabatic, and heating sections may be incorporated as separate entities. Above the pool, three vertically averaged film thicknesses are chosen to represent the three sections of the film. Under steady, axisymmetric conditions this idealization lends itself to one-

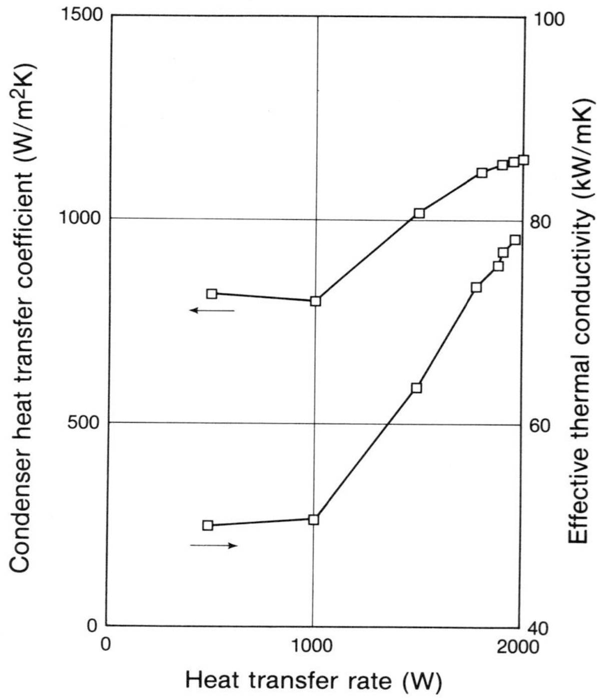

Fig. 3.16. Effect of heat flux on effective thermal conductivity and condenser heat transfer coefficient (after Hahne and Gross 1982).

dimensional analysis based on the conservation laws applied to the liquid film and the vapour core (Dobran 1984, 1985; Reed and Tien 1987).

Assuming that $\delta \ll D/2$ in Fig. 3.18, the change in Γ, the liquid mass flux per unit circumference, is

$$\Gamma_{X+\Delta} - \Gamma_X = \dot{m}\left(1 - \frac{2\delta}{D}\right)\Delta$$

over the length X to $X+\Delta$: $\dot{m}$ is the condensation rate per unit surface area of the film. Thus for the condenser,

$$\Gamma_{LC} = \dot{m}_C\left(1 - \frac{2\delta_C}{D}\right)L_C. \tag{3.44}$$

Secondly, if $\rho_V \ll \rho_L$, a force balance on the length Δ is given by

$$\pi D[(\Gamma\bar{U}_L)_{X+\Delta} - (\Gamma\bar{U}_L)_X] - \dot{m}U_I\pi(D-2\delta)\Delta$$
$$= \pi(D-\delta)\delta[\rho_L g\Delta - (P_{X+\Delta} - P_X)] - \pi D\tau_W\Delta - \pi(D-2\delta)\tau_I\Delta$$

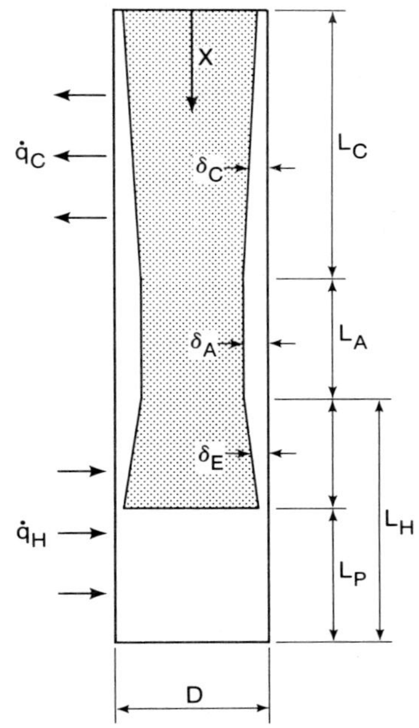

Fig. 3.17. Idealized model of vertical tubular thermosyphon.

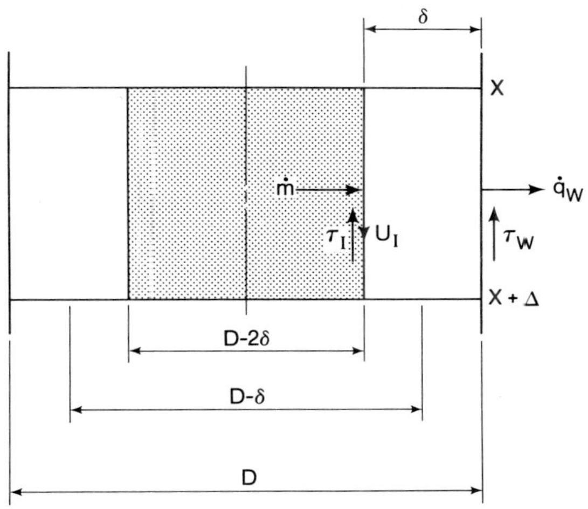

Fig. 3.18. Fluxes and forces on an elementary condensate ring of thickness δ and length Δ.

where $\bar{U}_L$ is the average liquid velocity, U_I the interfacial value, and τ_W, τ_I are the flanking shear stresses at the wall and the interface, respectively. Applied to the condenser, this equation reduces to

$$g\rho_L\delta_C\left(1-\frac{\delta_C}{D}\right)=\tau_W+\tau_I\left(1-\frac{2\delta_C}{D}\right) \tag{3.45}$$

if the Froude number $Fr=\bar{U}_{LC}^2/gL_C \ll 1$, implying that inertial forces may be neglected. Thirdly, a film energy balance may be written

$$\dot{m}(\lambda+c_p\theta^c)\pi(D-2\delta)\Delta-\pi D\dot{q}_W\Delta=\pi D[(c_p\bar{\theta}\Gamma)_{X+\Delta}-(c_p\bar{\theta}\Gamma)_X]$$

in which $\bar{\theta}=\bar{T}-T_W$ is the average liquid temperature difference and $\theta^c=T_V-T_W$. Defining $\phi=\bar{\theta}/\theta^c$, this equation applied to the condenser yields

$$(1+Ja)\left(1-\frac{2\delta_C}{D}\right)-\frac{\dot{q}_W}{\dot{m}_C\lambda}=\frac{Ja(\phi\Gamma)_{LC}}{\dot{m}_C L_C} \tag{3.46}$$

in which $Ja=c_p\theta^c/\lambda$ is the Jakob number. From eqn (3.44) it is clear that $\Gamma_{LC}=O(\dot{m}_C L_C)$, and since $\phi=O(1)$ it follows that eqn (3.46) may be reduced to the simple balance $\dot{q}_W=m_C\lambda(1-2\delta_C/D)$ if $Ja\ll 1$, which is typical.

Equations similar to (3.44), (3.45), and (3.46) may be derived for the liquid films in the adiabatic and evaporative sections. Likewise, conservation equations may be formulated for the vapour core, but these add little information except that the force balance reveals magnitudes which are smaller than the gravitational and shear terms in the film equations. This particular feature implies that the interfacial shear stress must be introduced through an empirical friction factor for the tubular 'channel' through which the vapour flows. Similarly, the wall shear stress may be represented by an expression obtained from a Nusselt analysis applied to the liquid film. The description of the idealized model is completed with an energy balance on the pool

$$(1+Ja)=\frac{\dot{q}_P L_P}{\lambda\Gamma_P}+Ja\phi_P \tag{3.47}$$

and the specification of the mass of the filling charge M by

$$M=\rho_L\pi D\left[\frac{DL_P}{4}+\left(1-\frac{\delta_E}{D}\right)(L_E-L_P)+\left(1-\frac{\delta_A}{D}\right)L_A+\left(1-\frac{\delta_C}{D}\right)L_C\right]. \tag{3.48}$$

The model has the advantage of describing the thermosyphon in algebraic terms and thus offers a simple interpretation of low-pressure characteristics when the interface is cylindrical and clearly defined. For example, it may be used to predict the heat transfer rate, the film Reynolds number and film thickness, the liquid pool depth and vapour temperature, assuming saturated conditions. Beyond these, it has the ability to indicate the dryout and flooding points. In accord with earlier discussion, the former is specified either by $L_P=0$ or $\Gamma_E=0$, while the latter, reflecting incipient choking, may be defined by $d\delta/d\dot{Q}\rightarrow\infty$.

Figure 3.19 illustrates a prediction for the laminar film thickness in the adiabatic section compared with experimental data, both obtained by Reed and Tien (1987). As indicated, the agreement is generally good and reveals two important facts: firstly, the Nusselt theory, which neglects interfacial shear stress, becomes increasingly inaccurate as the film Reynolds number (or heat transfer rate) increases; secondly, the flooding Reynolds number is slightly overestimated, because of uncertainty in fluid properties and the neglect of interfacial alterations, the existence of entrainment in particular. Using the same thermosyphon, the flooding Reynolds number is shown plotted against vapour temperature in Fig. 3.20. The one-dimensional model evidently provides a satisfactory indication of how the thermal performance increases with saturation temperature and pressure in this range of conditions. It has been extended to include entrainment (Casarosa and Dobran 198) and transience (Reed and Tien 1987).

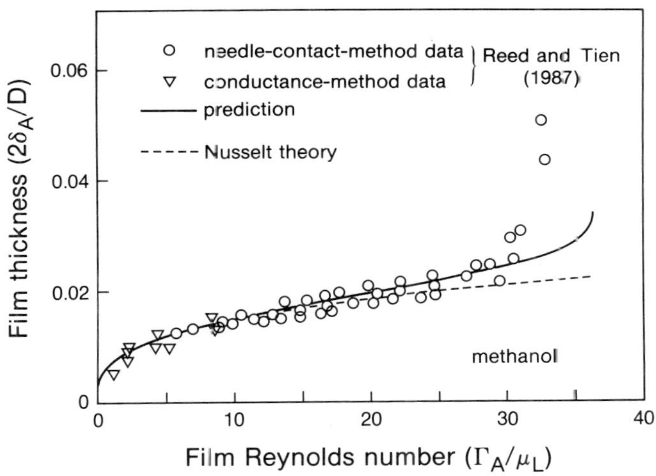

Fig. 3.19. Laminar film thickness in a vertical thermosyphon as a function of Reynolds number: $D=1.42$ cm, $L_C=1.08$ m, $L_A=0.305$ m, $L_H=1.03$ m, $T_C=36$ °C (after Reed and Tien 1987).

3.3.2 *The influence of pressure*

At very low pressures, when the vapour density is much reduced and vapour velocities are correspondingly higher, the emphasis in the idealized model shifts from the condensate film to the vapour core. This introduces new limits

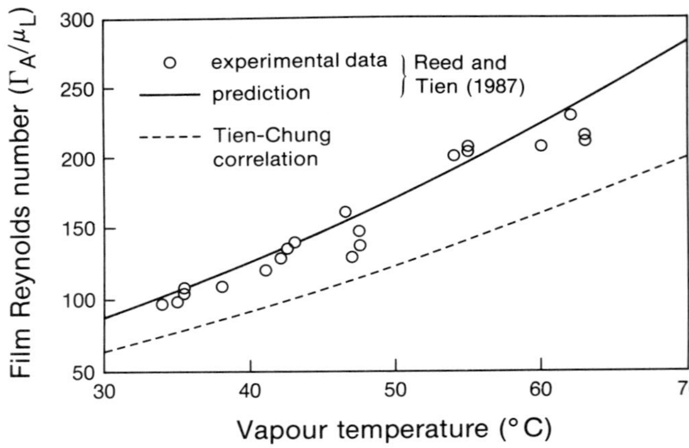

Fig. 3.20. Flooding Reynolds number as a function of vapour temperature in a vertical thermosyphon (same device used in Fig. 3.19).

imposed through the attainment of the sonic velocity or a zero pressure (Busse 1973). The sonic limit occurs in situations where the inertia of the vapour stream is so great that it dominates the viscous effects in the longitudinal equation of motion; compressibility is important. Referring to Fig. 3.21, which ignores both the pool and the film described in Fig. 3.17, this equation of motion may be represented by

$$\bar{P}_X - \bar{P}_{X+\Delta} = (\overline{\rho U^2})_{X+\Delta} - (\overline{\rho U^2})_X$$

in which $\bar{P}_X$ is the average vapour pressure at X, now measured upward from the tube base, and ρ, U are the density and axial velocity, respectively. Consistent with the neglect of viscosity, U may be interchanged with $\bar{U}$, the average value across the core, while $\bar{\rho}$ may be interchanged with ρ if two further conditions apply: firstly, that the equation of state may be expressed by $P/\rho = RT = \text{constant}$[10]; secondly, that lateral pressure gradients may be neglected. The above equation may then be rewritten

$$P_o - P(X) = \rho(X) U^2(X) \tag{3.49}$$

where the subscript o refers to the closed end of the evaporator. Now as noted previously, the axial heat flux $\dot{Q}_a$ is essentially the latent heat flux and hence

$$\dot{Q}_a = \frac{\pi}{4} D^2 \bar{\dot{q}}_a = \frac{\pi}{4} D^2 \lambda \overline{\rho U}. \tag{3.50}$$

Using eqn (3.49),

[10] Justified *a posteriori* rather than *a priori* (Busse 1973).

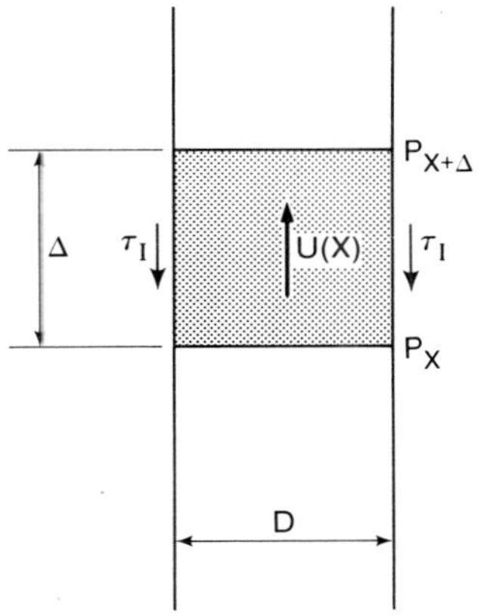

Fig. 3.21. Forces acting on a cylindrical vapour core.

$$\bar{\dot{q}}_a(X) = \lambda(\rho_o P_o)^{1/2}\left[\frac{P}{P_o}\left(1-\frac{P}{P_o}\right)\right]^{1/2}$$

which exhibits a maximum value[11] $\dot{q}_s$ given by

$$\dot{q}_s = 0.5\lambda(\rho_o P_o)^{1/2} \tag{3.51}$$

corresponding to a maximum vapour velocity

$$U_s = (RT_o)^{1/2}, \tag{3.52}$$

again using eqn (3.49). Under the assumed isothermal conditions this is the sonic velocity[12], which is independent of the tube geometry. Acceleration of the vapour caused by evaporation from the surrounding film, which otherwise plays a passive role, is therefore restricted to subsonic conditions; that is, the axial heat flux is limited by choking at the evaporator outlet, thereby adding another variable, the Mach number, to eqn (3.42).

At the other, very low-pressure extreme, the limit is imposed by the inability of the pressure gradient to overcome the dominant viscous forces, in which case the longitudinal equation of motion, again derived from Fig. 3.21, is given by

$$\frac{dP}{dX} = -\frac{4\tau_I}{D} \tag{3.53}$$

[11] More precisely, lateral variations in U may be accommodated by introducing the ratio $\overline{(U^2)}/(\bar{U})^2$; however, this is typically close to 1.0.

[12] Under the more usual isentropic conditions, $U_s = (\partial P/\partial \rho)_{isen} = (\gamma R T_o)^{1/2}$, but the result is essentially the same.

where τ_I is the interfacial shear stress. Compressibility is now unimportant. Under laminar conditions,

$$\tau_I = -\mu\left(\frac{\partial U}{\partial R}\right)_{D/2} = \frac{8\mu\bar{U}}{D}$$

and hence

$$\frac{dP}{dX} = -\frac{32\mu\bar{U}}{D^2}$$

or, using the axial heat flux eqn (3.50) and the isothermal equation of state,

$$P\frac{dP}{dx} = -\frac{32\mu RT_o}{\lambda D^2}\bar{\dot{q}}_a$$

which integrates to give

$$\bar{\dot{q}}_m(X) = \frac{\lambda D^2 \rho_o P_o}{64\mu L_{eff}}\left[1 - \left(\frac{P}{P_o}\right)^2\right] \tag{3.54}$$

where

$$\bar{\dot{q}}_m L_{eff} = \int_0^X \bar{\dot{q}}_a \, dX.$$

In this equation, L_{eff} is the effective length[13] corresponding to $\bar{\dot{q}}_a$ reaching its highest value $\dot{q}_m$ at X. From eqn (3.54) it is evident that this value increases until $P = 0$ at the evaporator outlet when

$$\dot{q}_v = \frac{\lambda D^2 \rho_o P_o}{64\mu L_{eff}} \tag{3.55}$$

defines the viscous limit of heat transfer. Unlike the sonic limit, this varies inversely with the heated length–diameter ratio of the tube.

In developing a more detailed model of the sonic and viscous limits, Busse and Prenger (1984) have proposed a more succinct description using the radial Reynolds number defined by

$$Re_r = \frac{\bar{V}_I D}{2\nu}$$

where $\bar{V}_I$ is the mean radial velocity at the interface: thus, $\bar{V}_I = \dot{m}_I/\rho_I$, where $\dot{m}_I$ is the evaporation (condensation) rate. This enables the average axial heat flux density at the evaporator outlet to be rewritten in the form

[13] For example, uniform evaporation over an evaporator length L_H requires that $\bar{\dot{q}}_a$ increase linearly from zero up to $\bar{\dot{q}}_m$, in which case $L_{eff} = L_H/2$.

$$\bar{\dot{q}}_a = \frac{16\mu L_{eff} Re_r}{D^2}$$

and therefore the sonic limit may be simply restated as

$$\frac{\dot{q}_s}{\lambda(\rho_o P_o)^{1/2}} = 0.482,$$

a more accurate asymptote than the value of 0.5 given in eqn (3.51), whereas the viscous limit reduces to

$$\frac{\dot{q}_v}{\lambda(\rho_o P_o)^{1/2}} = \frac{1}{2} Re_r^{1/2}.$$

These asymptotes are shown in Fig. 3.22 along with a numerical solution for the intermediate range which is seen to extend over $0.1 \lesssim Re_r \lesssim 10$.

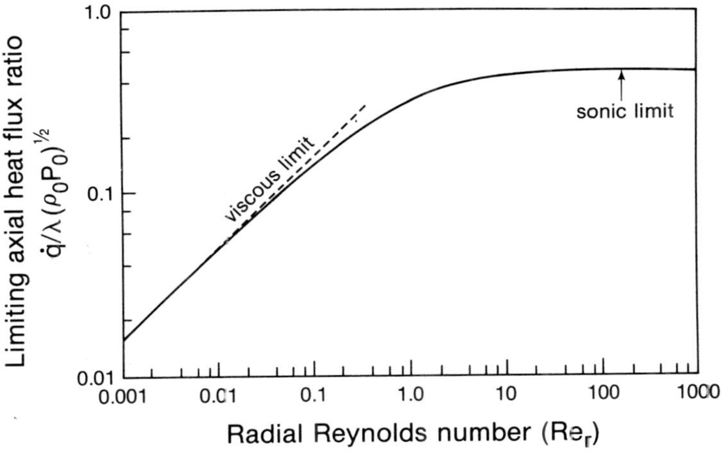

Fig. 3.22. Limiting axial heat flux ratio in a very low-pressure thermosyphon (after Busse and Prenger 1984).

At very high pressures, the effect of the vapour is felt at the interfaces it creates with the liquid. It is the behaviour of these interfaces, dispersed in the evaporator pool and continuous over the condenser film, which then controls the characteristics and limits of the thermosyphon. As a general rule, the heat transfer rate increases monotonically with saturation pressure, implying that the thermal resistance must exhibit a corresponding decrease, but this tendency is altered substantially near the critical pressure P_{cr}. When $P \to P_{cr}$, the dramatic reduction in surface tension promotes the formation of smaller bubbles while bubble frequency and the number of nucleation sites both increase; these trends follow from eqn (3.32). Even so, simple nucleate boiling in which $\dot{q} \sim (\Delta T)^3$ or $h \sim (\dot{q})^{2/3}$ is not observed in the pool. Gross and Hahne (1985) found that instead, $h \sim \dot{q}$, which suggests that $\dot{q}$ and ΔT are independent of each other; increases in heat transfer rate are evidently attributable to increases in nucleation rate. Decreases in buoyancy felt as $\rho_V \rightarrow \rho_L$ are offset by the increased vigour of small bubble growth and detachment. The net increase in boiling heat transfer coefficient with pressure is illustrated in Fig. 3.23 which reveals that $h \sim \exp(P/P_{cr})$, except very near the critical point. The sudden reduction shown only on the lowest power curve at $P/P_{cr} \simeq 0.96$ was apparently caused by the film boiling transition; that is, the burnout limit had been reached. This limit was observed to shift to lower pressures when the heat flux was increased, and hence offers support to the above suggestion that $\dot{q}$ controls the vapour nucleation rate.

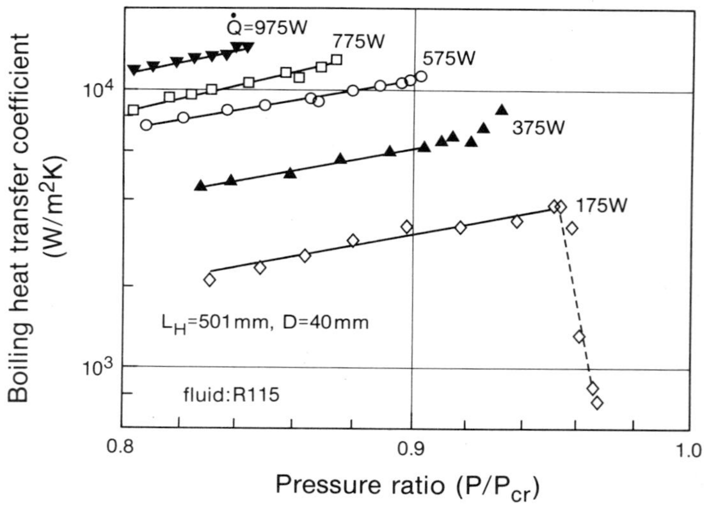

Fig. 3.23. Boiling heat transfer coefficients in a vertical thermosyphon near the critical point (after Gross and Hahne 1985).

Very different behaviour occurs in the condensing section. This is illustrated in Fig. 3.24 which reveals that condensation heat transfer coefficients are comparable to those found at lower pressure and are again less than the boiling heat transfer coefficients. The effect of heat transfer rate (or film Reynolds number) on h is to reproduce the laminar–turbulent transitional minimum discussed in Section 3.1; this is particularly noticeable for $P/P_{cr} \lesssim 0.9$. However, the most striking feature of the curves, their marked rise, stands in sharp contrast to the expectations of the Nusselt's laminar film theory which suggests a gradual decrease in heat transfer coefficient attributable to decreases in λ and $\Delta\rho$ as $P \to P_{cr}$. The explanation appears to lie in interfacial stability, as reflected in eqn (3.29) which indicates a limiting heat flux proportional to $\lambda(\Delta\rho)^{1/4}$ corresponding to the development of interfacial waves. This suggests a maximum (flooding) heat transfer rate which decreases as P/P_{cr} increases; that is, the flooding point will occur at lower pressures if the heat supply rate (circulation rate) is raised. The figure appears to confirm this. The precise effect of the waves is unknown but, at the very least, they can be expected to induce or augment turbulence in the presence of a small viscosity.

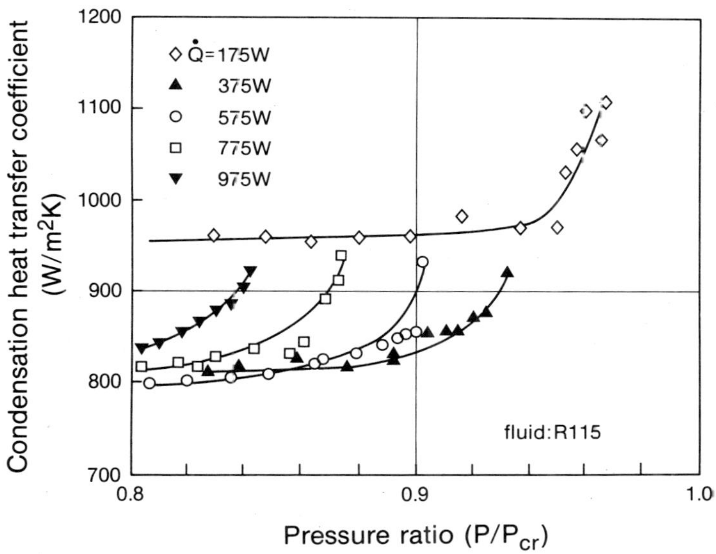

Fig. 3.24. Condensation heat transfer coefficients in a vertical thermosyphon near the critical point (after Gross and Hahne 1985).

3.3.3 *The influence of filling*

It has been noted that the behaviour of the tubular thermosyphon is often determined by various interfacial phenomena which are associated with flooding, dryout, and burnout. Other than pressure, the principal design variables which influence these phenomena are the tube geometry and the filling. Perhaps the most crucial of these is the filling, which exerts an influence partly through fluid properties but mainly through its magnitude, the charge, the importance of which was heralded earlier. The effect of this quantity has been studied by many research workers using experimental data, visualization techniques, and analysis. The pioneer analysis of Dobran (1984), for example, uses the equation of motion of a thin liquid film, constrained by a fixed charge, to explore the relation between the heat transfer rate and the average condensate[14] film thickness $\bar{\delta}$. The result is shown in Fig. 3.25 for a representative fluid and geometry; the Bond number $Bo = gD^2\rho_L/\sigma$. At the

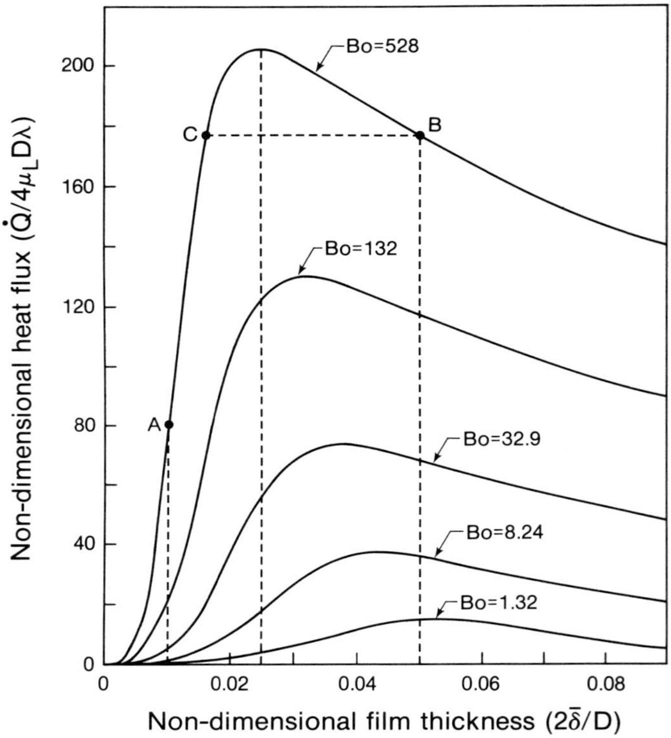

Fig. 3.25. Heat flux–film thickness relation in a vertical thermosyphon (after Dobran 1984).

onset of pool dryout, continuity demands that the non-dimensional film thickness $2\bar{\delta}/D$ be equal to half the filling ratio (here defined as the liquid volume divided by the total thermosyphon volume), and hence the behaviour of a given system may be followed by tracing the appropriate curve in the figure up to this limiting thickness. When this limit lies to the left of the maximum on the curve, the heat transfer rate increases monotonically with film thickness up to the onset of dryout (e.g. at A), beyond which point the system performance would deteriorate rapidly in practice. When the limit lies to the right of the maximum, the heat transfer rate is no longer a single-valued

[14] Without distinguishing the cooling, heating, and adiabatic sections.

function of $\bar{\delta}$, as suggested when dryout begins at B. This ambiguity in the solution of the equation of motion may signal the oscillating force balance sometimes observed near the maximum heat transfer rate. According to Dobran, at one extreme (C) a thinner film above a liquid pool is subject to a relatively low interfacial shear stress; at the other (B) a higher interfacial shear stress heralds incipient dryout, or flooding[15]. Figure 3.25 thus suggests why oscillatory behaviour near the limiting heat transfer rate does not occur for filling ratios which are less than the optimal value. Such values evidently lead to a pool dryout limit. On the other hand, higher values are associated with the oscillations observed during flooding and thermal blocking[16].

Experimental and visual studies generally confirm the above suggestions (Imura *et al.* 1983; Shiraishi *et al.* 1984; Fukano *et al.* 1986). It has been found that very small charges lead to the pool dryout limit, but conditions in the evaporating film vary considerably, from surface evaporation to nucleate boiling with the bursting of vapour bubbles accompanied by the production of liquid droplets. Under these variable conditions, the maximum heat flux increases with the charge, as suggested by Fig. 3.25. However, beyond a certain value of charge, the void ratio of the pool increases substantially, with the resulting two-phase mixture climbing up the thermosyphon as the heat transfer rate increases. This type of behaviour frequently leads to the flooding limit and, inevitably, to film dryout. The maximum heat transfer rate is largely unaffected by any further increases in charge, although it might be mentioned that dryout is then limited to the upper reaches of the evaporator in the presence of a vigorously boiling pool beneath; that is, holdup disconnects the returning condensate from the pool without eliminating the latter. This leads to the general observation that holdup becomes a cause of threshold behaviour while dryout is the limiting effect. Figure 3.26 illustrates the two filling regions: for small charges, the heat transfer limit increases as some power of the filling ratio; for large charges, the heat transfer limit remains constant.

Given the effect of charge, it is clear that its 'optimal' value should be large enough to avoid pool dryout. This corresponds to a filling ratio $v_R = V_{liq}/V_{evap}$ which is just large enough to create a flooding limit with the evaporator just filled with a churning two-phase mixture. Using the limiting (flooded) heat flux density

$$\dot{q}_C = 0.16 \frac{\lambda D}{L_H} [\sigma g \rho_V^2 (\rho_L - \rho_V)]^{0.25} \left(\frac{\rho_L}{\rho_V}\right)^{0.13} \quad (3.56)$$

[15] It may be recalled from Section 3.1 that flooding is not attributable to interfacial shear directly, but interfacial waves are not incorporated in this model.

[16] A state beyond the flooding limit where the system pressure evidently fluctuates with time but the heat transfer rate does not (Casarosa and Dobran 1988).

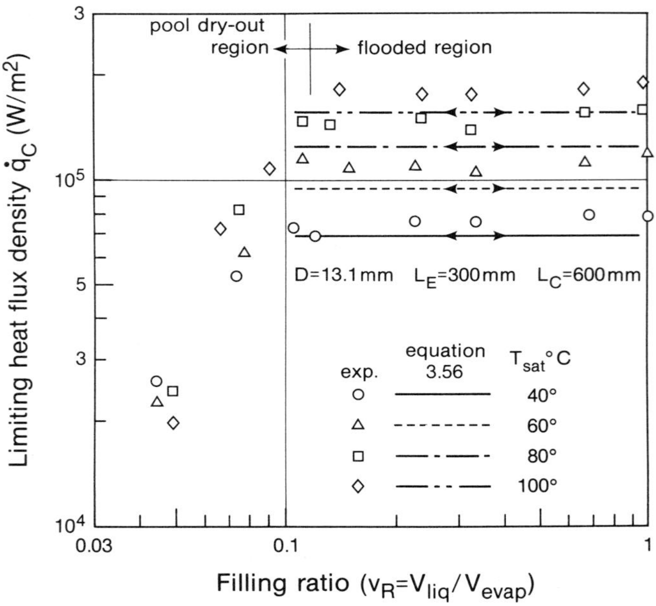

Fig. 3.26. Effect of filling ratio on limiting heat flux in a water-filled thermosyphon (after Imura *et al.* 1983).

shown in Fig. 3.27, Imura *et al.* (1983) have proposed that the optimum filling ratio be given by

$$v_{\mathrm{R}} = v_{\mathrm{o}} + f + \frac{\rho_{\mathrm{V}}}{\rho_{\mathrm{L}}}\left(f + \frac{L_{\mathrm{C}} + L_{\mathrm{A}}}{L_{\mathrm{C}}}\right) \tag{3.57}$$

where

$$f = \frac{4}{D}\left(\frac{0.8L_{\mathrm{C}} + L_{\mathrm{A}}}{L_{\mathrm{C}}}\right)\left(\frac{3\mu_{\mathrm{L}} L_{\mathrm{H}} \dot{q}_{\mathrm{C}}}{\lambda g \rho_{\mathrm{L}}^2}\right)^{1/3}.$$

The first term on the right-hand side of eqn (3.57) represents the volume fraction of *liquid* which occupies the evaporator when it is filled with a churning two-phase mixture. This has been estimated at $v_0 = 0.2$–0.33 (Larkin 1971; Harada *et al.* 1980; Imura 1980) but is not a constant; it may be expected to vary with tube geometry, fluid properties, and pressure. The second and third terms represent, respectively, the liquid and vapour contained in the length of tube above the evaporator; these terms are especially important if the condenser and adiabatic sections are much longer that the evaporator. Equation (3.57) thus defines the boundary between the two regions identified

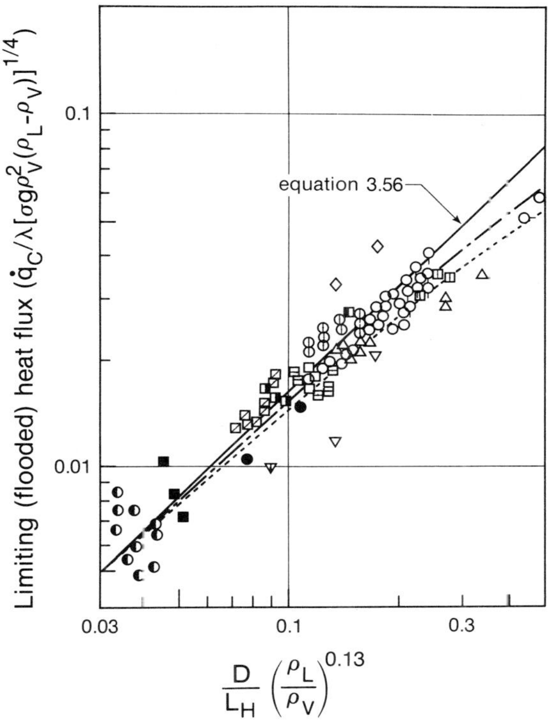

Fig. 3.27. Limiting (flooded) heat flux correlation (after Imura *et al.* 1983)

in Fig. 3.26. The calculated boundary indicated by the arrows is in good agreement with the data. It is worth noting the inverse effect of the length–diameter ratio in eqn (3.56); the effect carries over into eqn (3.57).

The behaviour of the condensate film while the heat transfer rate is gradually increased up the flooding limit has been studied visually by Weber-Carstanjen and Busse (1984) who observed a complex sequence of events. Although undertaken with grooved evaporator and condenser tubes separated by a wick-lined adiabatic section, their study provides observations which are useful in understanding vertical thermosyphon behaviour. They found, for example, that dry patches may exist between rivulets on the evaporator wall even when the mass circulation rate is great enough to ensure a free flowing film on the condenser wall. The interaction between the ascending vapour and the falling film begins in the lower reaches of the condenser and spreads upwards with increases in the heat transfer rate. To begin with, surface ripples appear, but soon the film converts into the rivulets reported by many workers. As the heat transfer rate increases, the rivulets exhibit complex behaviour: at first they acquire their own capillary waves; then they thicken appreciably; and finally, the largest rivulets grow by tributarial accretion. For sufficiently high heat transfer rates, the large rivulets in the condenser oscillate vertically in sympathy with the development of the oscillatory film dryout in the lower reaches of the evaporator. As the maximum heat transfer rate is approached, the further development and break up of surface waves signals entrainment followed by flooding during which the evaporator is invariably dried out. Beyond this limit, where flow oscillations become pronounced, the heat transfer rate falls sharply.

In an attempt to explain the oscillatory flow behaviour, Weber-Carstanjen and Busse (1984) proposed a process in which the force balance between the descending film and the ascending vapour core is cyclic, much as suggested earlier in Fig. 3.25 between points B and C. Starting from a moment when interruption in the returning condensate film marks the simultaneous existence of holdup and film dryout, the process begins with the consequence of a reduced evaporation rate. Vapour velocity falls and the condensate is then able to redistribute itself as a thin annular film of descending liquid which immediately begins to thicken as the evaporation rate is restored. Growth of the film at its thickest point, the mouth of the condenser, continues in the presence of a rising vapour velocity. The subsequent appearance and growth of large waves leads to the creation of a 'hanging lip' which protrudes further and further into the vapour stream causing an additional increase in the flow resistance. Eventually, the protruding 'lip' is blown back into the condenser, its momentum being transferred to the film; this is sufficient to cause an interruption in the film, thereby recreating a partial dryout in the evaporator. Although idealized, this cyclic hypothesis helps explain why flow oscillations are not often observed with liquid charges which are very small; the film may never be thick enough to produce a 'lip'. The hypothesis is also consistent with observations on cocurrent, gas–liquid systems which also exhibit 'lip' formation and oscillatory behaviour near the flooding limit (Whalley 1987).

Under transient conditions, the evaporative thermosyphon exhibits both similarities and differences with respect to the steady state discussed extensively above. The imposition of (transient) surface temperatures, for example, tends to bypass the well-known hysteretic behaviour associated with imposed surface heat fluxes; and it is not unreasonable to expect the recurrence of the various boiling régimes, from bubbly to plug to annular flow. The work of Kaminaga *et al.* (1990) generally confirms these similarities, but also draws attention to the distinctive aspects of transience.

The character of transience is best revealed through the system time constant or, more precisely, through the sequence of events leading to boiling and its consequences. By and large, once heat is supplied the evaporator wall temperature rises until boiling begins; it then falls with the sudden increase in heat transfer coefficient, but the details are strongly dependent upon the overall temperature difference (or applied heat load) and the filling charge. A large temperature difference, for example, tends to promote stable nucleate boiling and an orderly, monotonic approach to steady conditions, whereas a lower temperature difference may lead to pulse boiling, as described in Section 3.2; the magnitude of the heat load thus influences behaviour. Similarly, if the charge is small enough to permit a large superheat, this may also promote pulse boiling with accompanying slugs of liquid being delivered to the condenser. A small charge and a low heat load thus tend to create pulse boiling which may be pronounced enough to produce dryout and prevent a normal steady state from ever being reached. Kaminaga *et al.* noted that to prevent such a failure, the charge must be roughly double the value that produces optimal performance under steady conditions.

Finally, brief mention must be made of the effect of changing the composition of the filling; specifically, through the addition of incondensable gases. In practice, such additions may be part of a deliberate attempt to have the thermal conductance of the thermosyphon adjust sympathetically to changes in heat load; or they may simply represent an undesirable leakage. In either event, the incondensable introduces two new features: condensing vapour must now diffuse through the incondensable, which thereby adds a resistance to condensation; and the total accumulation of the incondensable, the gas load, creates an inactive region at the closed end of the condenser. Hijikata *et al.* (1984) analysed this problem by treating the incondensable gas, under steady conditions, as a stagnant plug whose extent and shape were dictated by two-dimensional diffusion in the presence of condensation on the neighbouring annular film. Their results demonstrated that the fraction of the condenser shut down by the incondensable plug increased with the gas load, as would be expected; they also showed that the lower the condensation rate, the less significant was the impeding effect of the incondensable, diffusion then being mainly in the axial direction (i.e. one-dimensional).

Earlier treatments of laminar diffusion commonly assumed the interface between the vapour and the incondensable gas to be planar (Marcus 1972; Edwards and Marcus 1972). This simplified model has also been used by Takuma *et al.* (1984) to demonstrate that turbulence in the vapour core thickens the diffusion front. Their experimental data support their theoretical predictions, some of which are shown in Fig. 3.28 in the form of condenser wall temperature profiles. These typify the effect of the incondensable in shutting down the upper reaches (closed end) of the condenser, where the wall temperature falls to the level of the external coolant temperature T_C, here equal to 20 °C. The three different working fluids have different saturation vapour pressures and temperatures, and therefore they produce different shut-off lengths; their vapour Reynolds numbers were also different, being 19 985, 2507, and 962 for R113, ethanol, and water, respectively. The diffusion front

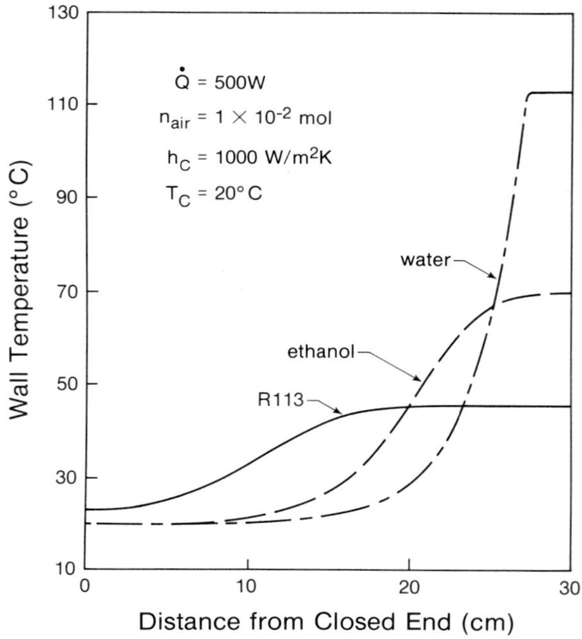

Fig. 3.28. Condenser wall temperature for three fillings in the presence of air (after Takuma *et al.* 1984).

for the laminar water vapour flow is noticeably narrower than those of the turbulent vapour flows, particularly on the vapour side.

3.4 Tilted and modified evaporative thermosyphons

The above discussion has been limited to vertical tubes whose performance may be described by eqn (3.42). Despite the wide range of this equation, it omits important effects which occur when the body force field behind the buoyancy forces is not uniformly parallel to the tube axis. As we shall see in the next chapter, this situation may arise through Coriolis accelerations when the tube rotates about axes which are parallel or normal to its own. More commonly, however, it results from tilting of the tube in the earth's gravitational field. This section is devoted principally to tilted and bent tubes operating under conditions similar to those discussed above; typically, when the length–diameter ratio is large and the filling is non-metallic. The discussion is thus concerned with the influence of a lateral component of gravity on thermal behaviour. As in the previous section it will focus on the *modus operandi*, performance characteristics, and limiting behaviour of the device.

3.4.1 *Thermal model for a tilted tube*

It is apparent from the behaviour of the single-phase thermosyphons considered in Section 2.4 that the effect of the lateral component of gravity is generally complex, and is perhaps best understood in terms of its role in the heated and cooled sections taken separately. Its effect on film condensation is suggested in Fig. 3.29 when the tube is tilted at an angle α from the vertical; $g \sin \alpha$ thus acts normal to the tube axis. The secondary circulation indicated is not unlike that occurring in the single-phase system except for the existence of the liquid–vapour interface; thermal buoyancy augments Archimedean buoyancy but is typically small by comparison. It is evident that the secondary flow results in a film thickness which varies azimuthally, being thinnest along the upper surface of the tube and thickest along the bottom. This feature implies a radial heat flux which decreases from the upper to the lower region of the film.

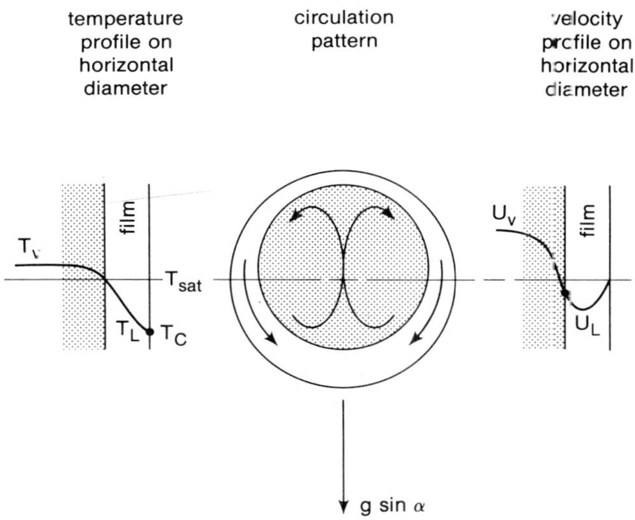

Fig. 3.29. Secondary circulation in film condensation in a tilted tube.

In the evaporator section, the discussion must be divided into two parts: the pool and the film, if any. Figure 3.30 suggests schematically that nucleate boiling in the pool induces a vapour bubble flow which roughly follows the wall in moving up towards the upper surface of the tube where the number of bubbles is greatest. While this secondary circulation may not produce much change in the high heat transfer rates associated with bubble growth and detachment, the azimuthal variation in void fraction may be expected to play a significant role because it can create a departure from simple nucleate boiling conditions along the upper surface of the tube. Further away from the closed end of the tube, evaporation takes place in the returning condensate film. This

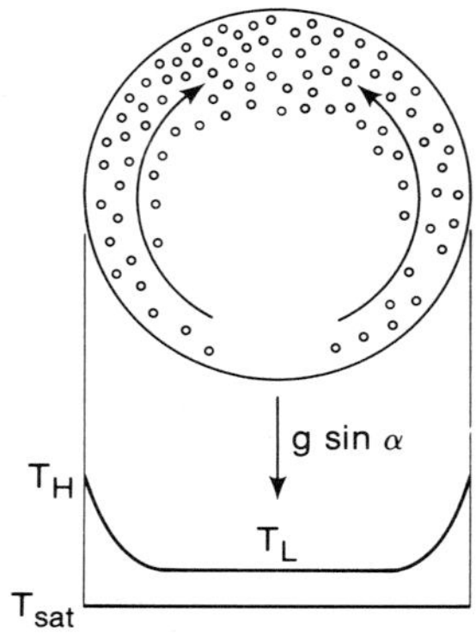

Fig. 3.30. Secondary circulation in pool boiling in a tilted tube.

is depicted in Fig. 3.31, the left-hand side of which suggests the development of the temperature profile in the eccentric film as it moves axially from the condenser exit (where it may be subcooled, especially on the lower surface) to the top of the boiling pool (where it may be superheated, especially on the upper surface). The figure also suggests the corresponding variation in void fraction caused by nucleate boiling which tends to reverse the liquid secondary flow pattern evident in Fig. 3.29. In general, the void fraction will be higher on the upper surface where the film is often thin enough to permit numerous small bubbles to burst, thus spraying liquid droplets into the vapour core. Indeed, the entire circumference of the tube may be covered by such a thin film or a series of rivulets.

Combining all of the above secondary effects caused by the lateral acceleration $g \sin \alpha$, it is possible to construct a model of the primary flow in a tilted thermosyphon. Figure 3.32 presents this model and attempts to illustrate both the similarities with the vertical tube, as represented by Fig. 3.13, and the significant differences. Principal among the latter are the eccentricity of the condensate film and the alteration of the boiling régime. Close to the vertical, eccentricity is small but it grows with tilt until, near the horizontal, most of the condensate return is contained in a thick, slowly moving rivulet which essentially converts the annular refluent pattern to a bifilamental form in which the direct effect of Archimedean buoyancy is

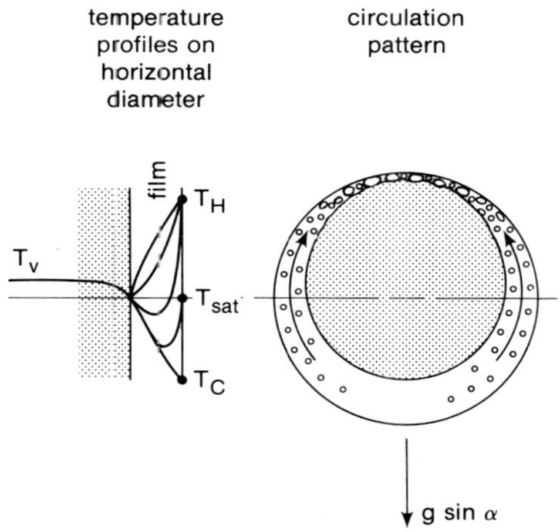

Fig. 3.31. Secondary circulation in a boiling film in a tilted tube.

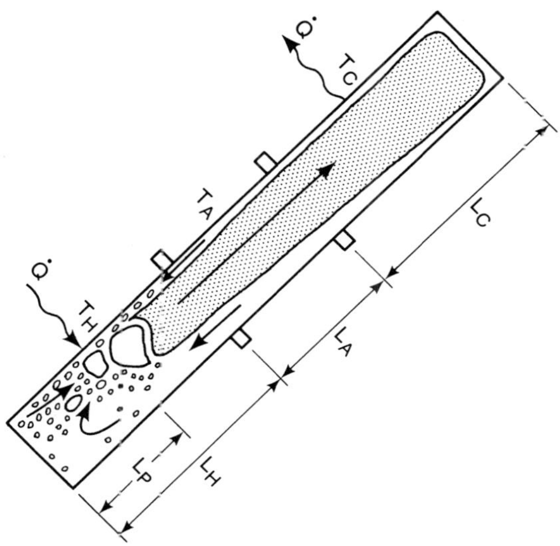

Fig. 3.32. Model of the tilted, evaporative tubular thermosyphon.

replaced by an indirect effect. Likewise, the nucleate boiling régime changes little near the vertical but an increasing tilt creates a shorter spiralling path for vapour bubbles until, nearer the horizontal, the high void fraction along the upper surface of the evaporator produces large vapour pockets and a modified two-phase flow régime in the presence of a very thin condensate film. It is thus to be expected that flow behaviour in the vicinity of the upper surface of the evaporator would vary rapidly as the horizontal position is approached; from slug flow to churning flow to stratified flow. This general model will now be considered in greater detail.

3.4.2 *The influence of tilt, charge, and pressure*

When the charge is a sufficiently large fraction of the evaporator volume (e.g. $v_R > 35\%$) evaporation in the lower reaches of the condensate film is less important, and thermal behaviour in the vertical position is largely governed by the condenser, where the heat transfer coefficient is often lower. Tilting the tube from the vertical improves performance initially because it introduces the secondary circulation incorporating a thinner average film with a progressively shorter path length; this creates a lower (laminar) Reynolds number. The condenser heat transfer coefficient thus rises while the evaporator heat transfer coefficient changes little, depending upon the relative importance of vapour pocketing and the development of a bifilamental flow structure. Large tilts, however, substantially reduce the axial component of gravity and therefore reduce the Archimedean buoyancy driving the primary flow. An optimum tilt angle may thus be created, as illustrated in Fig. 3.33. In general terms, this behaviour is reminiscent of single-phase behaviour: strengthening secondary effects initially improve performance but weakening primary effects eventually exert control. Should the condensate film be turbulent, however, the reduction in film Reynolds number which accompanies tilting will reduce the condenser heat transfer coefficient; under these conditions, the effect of tilt may be to decrease performance progressively. Near the flooding limit, a deteriorating performance with tilt suggests that film flow reversal occurs earlier as the tilt increases. This is illustrated by the data of Pioro (1986) who observed that

$$\dot{q}_{\max} = C\left(\frac{D}{L_H}\right)^{0.9} (\sin\beta)^{0.3} \tag{3.58}$$

where $\dot{q}_{\max}$ is the maximum heat flux density based on cross-sectional area, $\beta = 90° - \alpha$, and C is an empirical coefficient depending on the fluid (e.g. $C = 3.17 \times 10^6$ W/m^2 for water). Apart from revealing the inverse effect of L_H/D, this expression also illustrates the rapid deterioration in performance as the horizontal position is approached; this deterioration is clearly evident in Fig. 3.33. It thus appears that the maximum heat flux density which might be

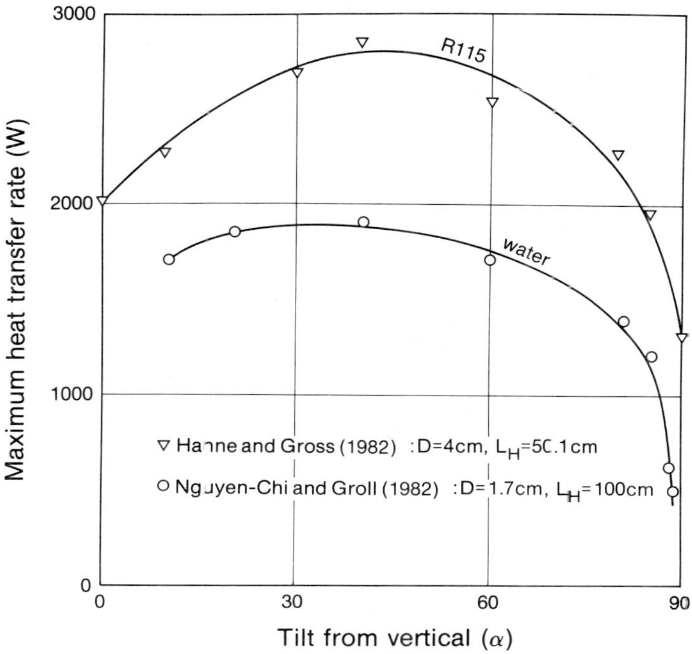

Fig. 3.33. Effect of tilt on maximum heat flux in two thermosyphons.

expected for non-metallic liquids in evaporative thermosyphons is in the range 10^4–10^5 W/m^2, based on the heated surface area. The use of the cross-sectional area provides a very different perspective, as may be seen from the data in Fig. 3.33.

Results such as these typify large charges (i.e. $v_R > 35\%$) with their characteristic flooding limit. When the charge is reduced, the system behaviour changes substantially, as demonstrated by the work of Negishi and Sawada (1983) using water. With $v_R = 25\%$, for example, evaporation and condensation take the forms shown in Fig. 3.34. In the vertical position, nucleate pool boiling occurs near the base but substantial bubble growth and interaction higher up tend to create a two-phase plug flow which maintains a high heat transfer rate across the thin liquid film thereby spread over the evaporator wall. In the upper reaches of the evaporator, this film ends in the bursting of bubbles generating a fine spray. At the same time, condensation above (film or drop) produces shifting rivulets which supply the pool below.

The initial effect of tilt from the vertical is to reduce the axial component of gravity and thereby enable vapour plugs to expel the liquid ahead of them further and further into the condenser as the tilt increases. At the low vapour

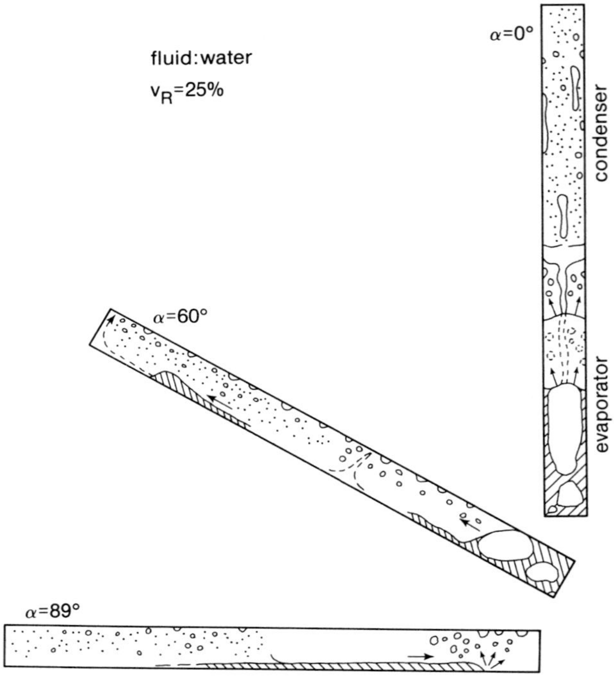

Fig. 3.34. Boiling and condensation patterns in a tilted thermosyphon with a small charge (after Negishi and Sawada 1983).

pressures typical of water vapour, the liquid tongue blown up the tube and the returning rivulet of condensate both oscillate along the lower surface of the condenser. This 'forced convection' augmentation of heat transfer in the condenser balances the general effect of a reduced primary circulation until the tube is within a few degrees of the horizontal. The situation then changes dramatically with the disappearance of any pool and the attendant absence of large vapour bubbles. A condensate rivulet formed principally from accumulated drops flows almost unimpeded into the evaporator where it ends in a disruption of bursting bubbles beyond which is a dryout region; the latter extends increasingly further from the base as the horizontal position is approached.

The overall effect of tilt on heat transfer is represented by the data shown in Fig. 3.35. As with the vertical tube, the effect of charge on the tilted thermosyphon is very small for $v_R > 35\%$ but becomes increasingly significant when v_R is lowered beneath 25%. Figure 3.36 provides an explanation for the marked decrease in heat transfer experienced at $v_R = 5\%$, when no pool exists at the base of the evaporator. In the vertical position, a combination of drop and film condensation again generates rivulets which trickle down the evaporator wall where they evaporate firstly at their free surface and finally through nucleate boiling. This second mechanism is clearly the more effective. In each rivulet, vapour bubbles continually grow and burst, spraying the opposite wall with droplets which evaporate in a process akin to drop condensation in reverse. Should the rivulets become too thin, they decompose into beads which boil and spray in much the same manner. These evaporative processes evidently create a cooling effect which varies little with tilt until the horizontal position is approached. The behaviour of a single rivulet running along the lower surface of the evaporator is then limited by the sluggish and erratic advance of the toe.

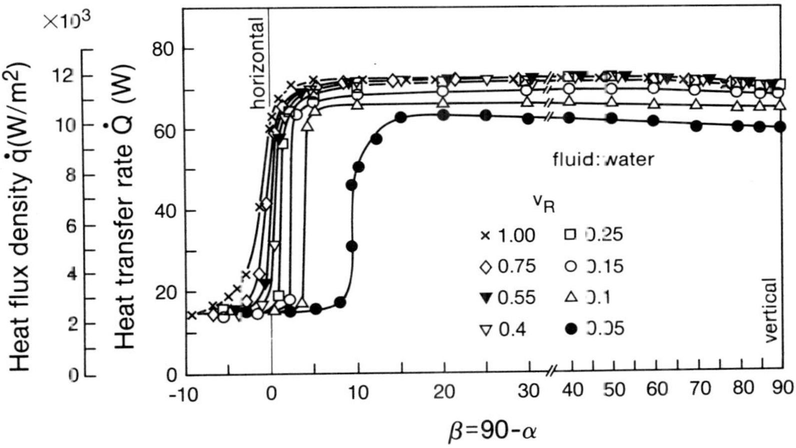

Fig. 3.35. The effect of tilt on heat transfer in a thermosyphon filled with water (after Negishi and Sawada 1983).

The formation of the droplets is reminiscent of the spray formation mechanism on water waves (Blanchard and Woodcock 1957), while their flight paths through the vapour bring to mind the saltation process (Owen 1964; Francis 1973). Air bubbles found in the foam of breaking waves, for example, break at their thinnest (uppermost) points, allowing the escaping air to rupture the bubble film further and thereby produce and disperse a cloud of very fine droplets. Immediately after, however, surrounding water rushes in to fill the bubble crater left behind, and it is this process which produces an unstable jet that breaks into one or more larger droplets having a diameter of about one-tenth that of the original bubble (Mason 1971). Under these circumstances, the droplets are ejected to heights of the order of 20 cm, and it is therefore reasonable to expect that a similar vigour in the bursting of vapour

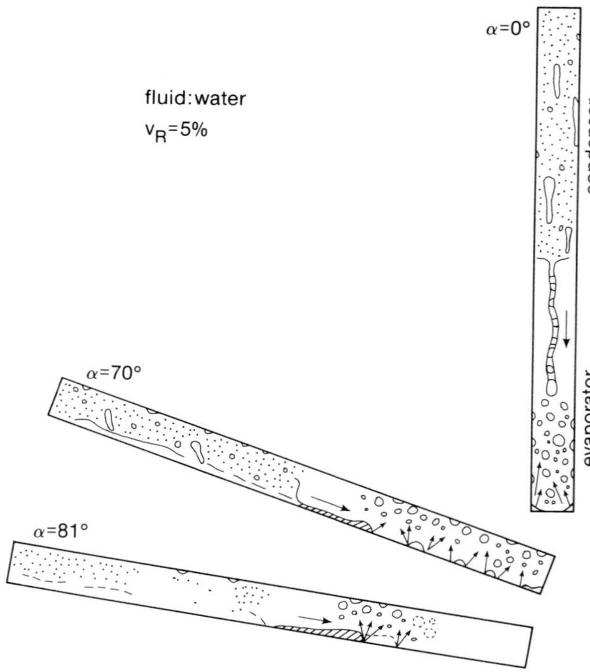

Fig. 3.36. Boiling and condensation patterns in a tilted thermosyphon with a very small charge (after Negishi and Sawada 1983).

bubbles in a thermsyphon would create a form of spray cooling on areas of the evaporator which would otherwise dry out. More generally, it is very likely that bubble bursting will play a role in the entrainment process which precedes flooding.

Figures 3.34 and 3.36 also help illustrate why thermal performance decays so rapidly as the horizontal position is aproached. For all but very large tilt angles, the direct Archimedean buoyancy $g \cos \alpha \Delta\rho/\rho$ drives the flow axially. As $\alpha \to \pi/2$, however, two important changes occur: the axial driving force is controlled principally by variations in condensate film thickness; and pooling in the bottom of the evaporator may leave its upper surface unwetted. The first of these feature is analogous to indirect thermal buoyancy discussed in the previous chapter, and is readily described in terms of the velocity and thickness scales of the film. For a two-dimensional liquid film moving with velocity U in the X direction, for example, the longitudinal (horizontal) equation of motion reduces to

$$0 = -\frac{\partial P}{\partial X} + \mu \frac{\partial^2 U}{\partial Y^2} \tag{3.59}$$

if inertia is neglected ($Pr \gg 1$) along with longitudinal viscous terms: Y is measured vertically upwards. Under the same conditions, the lateral (vertical) equation of motion reduces to

$$0 = -\frac{\partial P}{\partial Y} - g\rho.$$

which integrates to give

$$P = -\rho g Y + \rho g \delta + P_{\mathrm{I}}(X, Y)$$

where δ is the film thickness and P_{I} (X, Y) is the vapour pressure at the interface. Hence

$$\frac{\partial P}{\partial X} = g\Delta\rho \frac{\partial \delta}{\partial X}.$$

Introducing the normalized variables

$$x = \frac{X}{X^{\mathrm{c}}}, \quad y = \frac{Y}{Y^{\mathrm{c}}}, \quad u = \frac{U}{U^{\mathrm{c}}}, \quad \text{and} \quad p = \frac{P}{P^{\mathrm{c}}}$$

it is evident that

$$0 = -\left(\frac{P^{\mathrm{c}}(Y^{\mathrm{c}})^2}{\mu X^{\mathrm{c}} U^{\mathrm{c}}}\right) \frac{\partial p}{\partial x} + \frac{\partial^2 u}{\partial y^2}$$

$$\frac{\partial p}{\partial x} = \left(\frac{g Y^{\mathrm{c}} \Delta\rho}{P^{\mathrm{c}}}\right) \frac{\partial \gamma}{\partial x} \tag{3.60}$$

where $\gamma = \delta / Y^{\mathrm{c}}$. It thus follows that

$$\frac{g Y^{\mathrm{c}} \Delta\rho}{P^{\mathrm{c}}} = O(1) \tag{3.61}$$

and

$$\frac{P^{\mathrm{c}}(Y^{\mathrm{c}})^2}{\mu X^{\mathrm{c}} U^{\mathrm{c}}} = O(1). \tag{3.62}$$

To these must be added the consequence of balancing lateral heat conduction with the latent heat transfer, as given in eqn (3.8), for example. Thus

$$\frac{\rho \lambda U^{\mathrm{c}}(Y^{\mathrm{c}})^2}{k X^{\mathrm{c}} \theta^{\mathrm{c}}} = O(1).$$

Hence the thickness and velocity scales for a film of length L are given by

$$Y^c = \delta = L\left(\frac{Ja}{Pr\,Ar_L}\right)^{1/5}$$

$$U^c = \frac{\kappa Ja}{L}\left(\frac{Pr\,Ar_L}{Ja}\right)^{2/5} \tag{3.63}$$

respectively. The heat transfer rate may then be expressed by

$$Nu_L = C\left(\frac{Pr\,Ar_L}{Ja}\right)^{1/5}. \tag{3.64}$$

Equations (3.63) and (3.64) are worth comparing with those obtained earlier in this chapter, for a vertical surface, and with those corresponding to thermal buoyancy acting indirectly over a horizontal surface. An analogy is clearly evident. For laminar condensation (or evaporation) from a vertical film, the key exponent of the central group $Pr\,Ar_L/Ja$ is 0.25, the same as the Rayleigh number exponent for laminar, boundary layer natural convection over a vertical surface. Likewise, the exponent for a horizontal surface is 0.20 for both types of buoyancy. An important distinction, however, is found in the effect of temperature difference on the film thickness. Under phase-change conditions, the thickness increases as the temperature difference increases, regardless of the film inclination. It is therefore to be expected that increasing the heat transfer rate in the tilted thermosyphon will increase the flow impedance and may eventually lead to a flow reversal, as suggested earlier.

It is also evident that film evaporation in a horizontal tube is weaker than in a vertical tube. The relatively sluggish movement of the film is accompanied by its tendency to run along the bottom surface of the tube. In the evaporator, this tendency, reinforced by the pocketing of vapour bubbles along the upper surface, inevitably leads to a stratified flow in which the upper (vapour) layer transfers relatively little heat from the tube wall. Under these circumstances, the wall may cease to be isothermal, at least in the azimuthal direction, unless the boiling is vigorous enough to quench or, in the presence of bursting bubbles, spray the unwetted portion. Even so, the heat transfer rate in a horizontal evaporator may be reduced substantially, as illustrated by Fig. 3.35; this is especially true for small charges.

For large charges (e.g. $v_R > 75\%$), the effect evident in Fig. 3.35 is lessened but it increases again when the heat flux approaches the maximum value. This is clear from Fig. 3.37, based on the work of Terdtoon *et al.* (1990) who split the near-optimal régime shown into two parts. Here it is simply defined as the region in which improvement (or deterioration) in condenser performance is roughly balanced by deterioration (or improvement) in evaporator performance. In the near-horizontal régime, which extends over rougthly 30°, the thermosyphon is limited principally by the evaporator while it is the condenser which generally limits performance in the near-vertical régime. It must be emphasized, however, that the extent of these régimes is not unique. As Fig. 3.37 indicates, they are influenced by fluid properties, but we have seen that they also depend upon the charge, the geometry, and on the proximity of the heat flux to its maximum value. Perhaps the only generalization which may be permitted is that the optimal value will commonly fall on a shallow rise of the heat transfer curve in the range $20° < \alpha < 60°$; that is, $30° < \beta < 70°$. Even this may not be absolutely true according to the data of Pioro (1986).

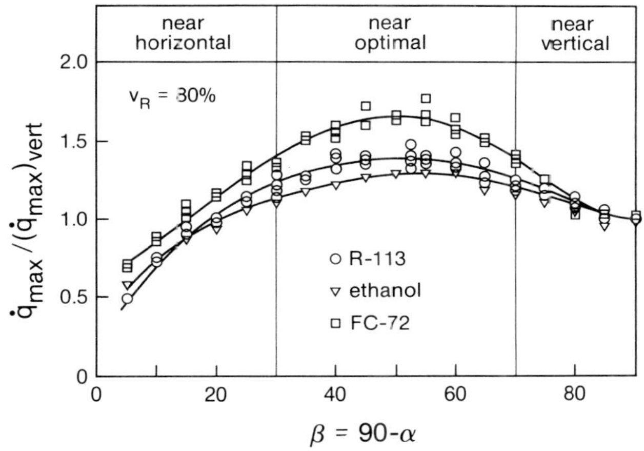

Fig. 3.37. Tilt régimes of evaporative thermosyphon under limiting conditions (after Terdtoon *et al.* 1990).

The effect of charge on maximum heat flux at various tilt angles is perhaps best interpreted in terms of the particular mechanism which leads to the limit of performance. With a substantial charge, this limit is typically set by flooding, at least in low-pressure thermosyphons. However, the situation changes when the charge is lowered too far (e.g. $v_R < 25\%$) and the pool is either much reduced in size or absent. Negishi and Sawada (1983) observed that a small pool can create conditions in which the expansion of vapour pockets may keep the heated surface covered in a thin liquid film. Even when this film is incapable of covering the entire evaporator surface, its breakup through a quenching process may still produce effective cooling. If this ceases, the tube reaches the dryout point. Thus, the initial effect of tilt from the vertical when the charge is small is to enhance the impulsive action of vapour pockets and thereby increase the heat transfer limit. Should the tilt angle become too great, however, film dryout may occur either in the association with hold up or as the sole result of the returning condensate tongue evaporating completely before it reaches the base of the evaporator. As observed earlier, film dryout is characteristic of small charges, regardless of tilt, but it is now evident that it will limit the effective operation of the thermosyphon to a tilt range which shrinks as the charge diminishes.

Turning now to the effect of pressure, it is important to recall that large vapour bubbles are characteristic of low-pressure thermosyphons. At higher pressures (and temperatures), a reduced bubble size is generally associated with higher heat transfer rates, as noted in the previous section. For a tilted thermosyphon, this is well illustrated in the work of the Gross and Hahne (1985, 1986) using a large charge and the same value of $L_H/D = 12.5$ employed in their earlier work (Hahne and Gross 1982). In the evaporator, the heat transfer coefficient varies with pressure in much the same manner as shown previously in Fig. 3.23 for a vertical tube. The effect of tilt is illustrated in Fig. 3.38 which reveals a modest decrease in heat transfer coefficient as α increases from 0° to 60°; it is likely that this tendency will steepen as $\alpha \rightarrow 90°$.

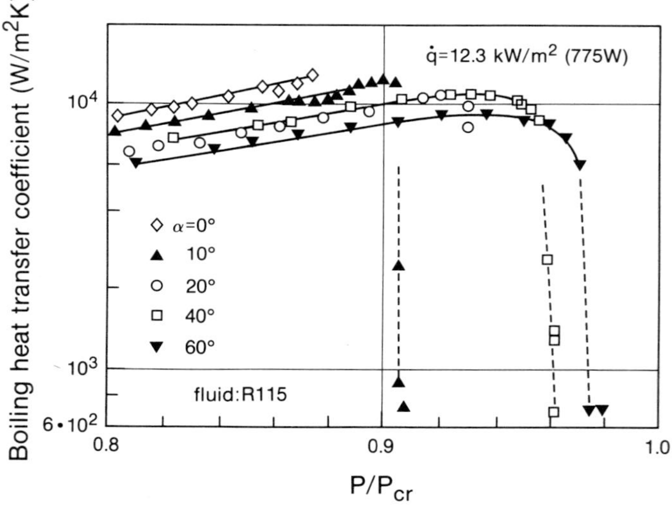

Fig. 3.38. Effect of tilt on the boiling heat transfer coefficient at high pressures (after Gross and Hahne 1985).

In the condenser, the existence of a laminar film is again capable of producing an optimum tilt at which increases in the secondary flow effects are balanced by decreases in primary flow effects; this is particularly noticeable at low heat fluxes. Condenser data for the same heat flux density used in Fig. 3.38 are presented in Fig. 3.39 which reveals that tilt has a substantial effect on the heat transfer coefficient. When the pressure is not too high (e.g. $P/P_{cr} < 0.9$), a laminar condensate film creates an increase in h with increasing tilt up to $\alpha \simeq 50°$. For higher pressures (e.g. $P/P_{cr} > 0.9$), a turbulent condensate film evidently produces a decrease in h with increasing tilt. Once the transition from laminar to turbulent flow occurs it may bring about a substantial improvement in performance. In a detailed study of this transitional region, Gross and Hahne (1986) have produced the correlation shown in Fig. 3.40 and given by

$$Nu = [(fNu_l)^2 + (Nu_t)^2]^{1/2} \tag{3.65}$$

where Nu_l, Nu_t are the asymptotic forms indicated for the laminar and turbulent films, respectively. Re_α is the corresponding film Reynolds number for a vertical surface multiplied by a function of tilt $F(\alpha)$ which is equal to 1.0 when the tube is vertical and given by

$$F(\alpha) = \frac{2.87D}{L_C \sin \alpha} \tag{3.66}$$

for $\alpha \geq 10°$. Note again the inverse effect of length–diameter ratio. The function f, chosen to fit the data, is given by

$$\frac{1}{f} = 1 - 0.63\left(\frac{P}{P_{cr}}\right)^{3.3}. \tag{3.67}$$

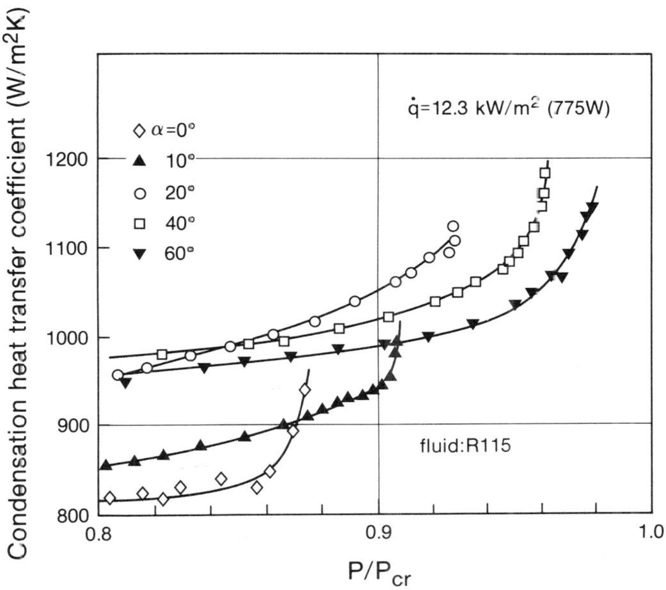

Fig. 3.39. Effect of tilt on the condensation heat transfer coefficient at high pressures (after Gross and Hahne 1985).

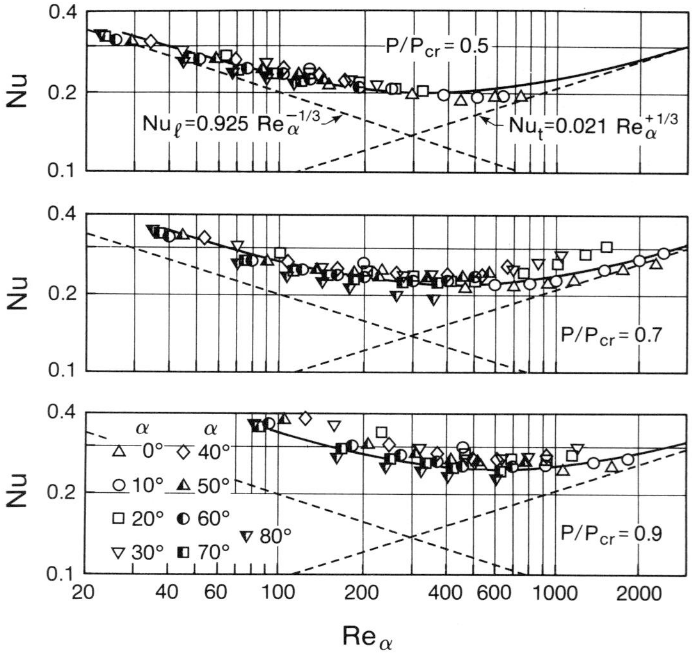

Fig. 3.40. Effect of pressure and Reynolds number on heat transfer in a tilted thermosyphon condenser (after Gross and Hahne 1986).

The transitional region is seen to extend over an order of magnitude in Reynolds number with the intersection of the asymptotes occurring near $Re_\alpha = 290$, consistent with the vertical plate estimate discussed in Section 3.1. As the pressure is increased, the experimental data lie increasingly above the laminar asymptote, indicating that interfacial instability and consequent wave phenomena become increasingly important in laminar behaviour as $P \to P_{cr}$, presumably because of a reduction in surface tension.

The effect of high pressure on the *maximum* heat transfer rate in the tilted thermosyphon is best interpreted in terms of the burnout typical of limiting behaviour at very high pressures (Gross and Hahne 1985). In the vertical position, this limit evidently occurs when $P/P_{cr} \lesssim 0.9$ with the primary circulation in the evaporator being poorly defined. Tilting the tube tends to create a bifilamental primary flow with fresh condensate flowing down along the lower surface of the tube and vapour bubbles streaming upwards along the upper surface. Moreover, flow velocities increase with decreasing λ as $P \to P_{cr}$ if the heat load is fixed. These effects apparently lead to burnout being delayed to higher pressures. Figures 3.38 and 3.39 demonstrate that this delaying tendency continues with increasing tilt, at least up to $\alpha = 60°$. On the other hand, at very low pressures, when the viscous and sonic limits are experienced, the effect of tilt may be ignored because the appropriate force balances on the vapour do not contain a gravitational term of any significance. Strictly speaking, the axisymmetric model no longer applies in the tilted position but the slight change in vapour core geometry and the effect of an azimuthal variation in interfacial shear stress do not alter the essential description of limiting conditions at very low pressures.

It has been noted that flooding is characteristic of limiting performance in low-pressure thermosyphons, but the precise details depend upon several factors, not least among which are the pressure and tilt angle. Lowering the pressure in long, fully charged evaporators creates large vapour pockets in the upper reaches where they eventually lead to condenser holdup accompanied by film dryout. The principal effect of tilt is to modify the primary flow pattern on which the flooding behaviour is superimposed (Nguyen-Chi and Groll 1982). As the tube is tilted from the vertical, a deeper but narrower rivulet gradually forms along the lower surface of the condenser. Since it is on the reduced interfacial area of this rivulet that instability is introduced, it is reasonable to expect that the initial effect of tilting would be to increase the limiting heat transfer rate associated with flooding. As the tilt is increased further, however, the direct Archimedean buoyancy force driving the condensate begins to fall off with the result that the limiting heat transfer passes through a maximum before decreasing rapidly as $\alpha \rightarrow \pi/2$. This deteriorating behaviour was displayed earlier in Figures 3.33 and 3.37, and has been expressed by Nguyen-Chi and Groll (1982) in the form

$$\dot{Q}_{\max} \sim G(\alpha) = \left(\frac{\beta}{180^\circ} + (\sin 2\beta)^{1/2}\right)^{0.65},$$

comparable with that of Pioro given by eqn (3.58): $\beta = 90^\circ - \alpha$. With this empirical function, Nguyen-Chi and Groll found that their limiting heat transfer data for water could be correlated using the one-dimensional, two-phase counterflow model of Wallis (1969) and Sakhuja (1973). The comparison is given in Fig. 3.41 which also reveals the familiar improvement in performance as the saturation temperature is raised.

In very long tubes, the limiting heat transfer situation changes yet again when liquid in the upper reaches of the evaporator become highly superheated; this may convert the impulsive, and perhaps erratic, behaviour of large vapour slugs into convulsive behaviour. Vigorous boiling may thus become pulse boiling, eventually expressed in the geyser phenomenon. Tilt has a marked influence on all of these because it alters the opposing and restoring effects of gravity. Thus, for example, the nearer the horizontal the more likely that pulsing will occur (Negishi and Sawada 1983; Feldman and Srinivasan 1984). Furthermore, pulsing may also occur during startup, as in the vertical

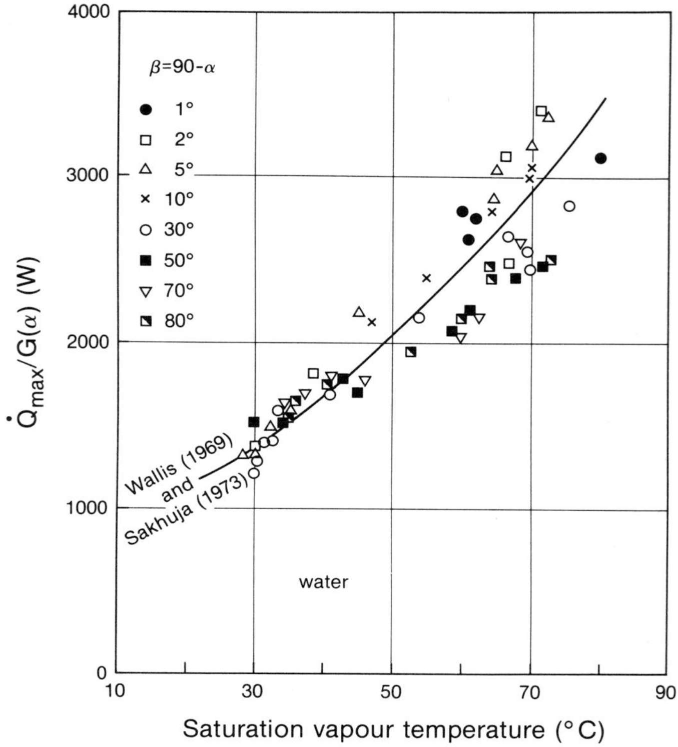

Fig. 3.41. Comparison of measured maximum heat transfer with an empirical correlation (after Nguyen-Chi and Groll 1982).

thermosyphon (Groll and Spendel 1984). In particular, small charges may again prevent normal steady conditions from being reached.

3.4.3 *Pathway modifications*

In closing this chapter, it is worth making brief mention of various modifications to the pathways which are followed by the vapour and liquid as they circulate within the thermosyphon tube. For the most part, these modifications are deliberately introduced in order to improve the performance of the device or make it better suited to the particulars of a given application. The discussion will be divided into two parts: modifications created through internal alterations, notably through plumbing; and those caused by alterations to the tube configuration, notably by bending it.

Strictly speaking, internal alterations should also include modifications caused by the introduction of incondensable gases, even though these were touched upon in the previous section. By and large, the vapour pathway shortening caused by a gas in the condensing section is not greatly altered by tilting, except through the natural stratification process attributable to any density difference between the gas and the vapour. It will be recalled that the incondensable gas introduces a dead space at the closed end of the condenser, thus reducing overall thermal effectiveness. This is clearly a deleterious effect when the gas finds its way into the tube through an accidental leakage. On the other hand, it makes possible a design strategy in which the condenser pathway length, shortened by the dead space, automatically adjusts to the vapour saturation pressure and temperature. The thermosyphon may then be adapted for a variety of responses, depending upon the desired variation of condenser path length with temperature. A proportionate increase of path length with saturation temperature may thus be used to maintain the evaporator wall temperature at a fairly constant operating level, while a sudden increase in effective path length at a specific saturation temperature permits the sudden increase of heat load that is characteristic of an emergency situation. It is only necessary to add that the designer may also choose to incorporate auxiliary pathways external to the thermosyphon proper, but these are beyond the scope of this text.

Internal plumbing arrangements may also change the flow pathways to such an extent that the device is not considered a tubular thermosyphon but as some form of related device. The discussion here thus serves partly as a prelude to a more general discussion of related devices in Chapter 5, notably in Sections 5.3 and 5.4, but its main purpose is to introduce the important idea that even minor plumbing arrangements may lead to significant performance improvements, and are thus important in themselves. In essence, internal plumbing falls into two categories: in one of these the function and purpose is distribution or redistribution of condensate in the evaporator; in the other, the intention is to separate the counterflows of liquid and vapour in the condenser. Each thus represents a strategy to avoid the limiting behaviour found in bare tubes; notably, the dryout and flooding limits.

The first category is best illustrated by the *artery*, which is employed to reroute condensate to or through the evaporator. Figure 3.42(a) provides a schematic representation. The work of Fukuda *et al.* (1984) provides an interesting example in which liquid leaving the condenser may follow either the normal (annular) route along the evaporator wall or flow into a central drain pipe leading directly to the evaporator base. Alternatively, this central drain or artery may be perforated to distribute condensate uniformly over the evaporator wall by means of a series of radial jets. Arteries may also be used in anti-gravity thermosyphons. In the second category, improvement is effected through a *flow divider* shown schematically in Fig. 3.42(b). Ideally, this occupies the original common surface of the liquid–vapour interface in the vicinity of the condenser entry; the increase in viscous drag is therefore

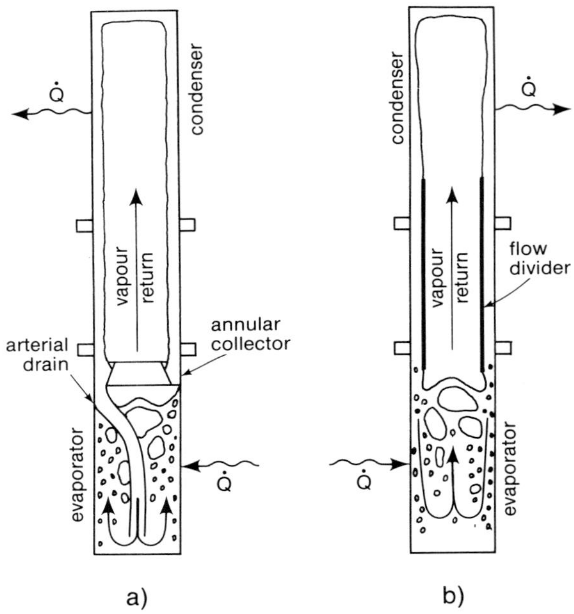

Fig. 3.42. Schematic representation of arterial return and flow divider in a vertical thermosyphon.

minimal. The divider may also be perforated to reduce viscous effects further while retaining the ability to dampen interfacial waves. The prospects for this device were explored by Tong *et al.* (1984) who used a perforated concentric tube in the evaporator and a holed concentric tube in the condenser where it appears to have been combined with a series of rings designed to interrupt the falling film, preventing it from thickening. Both of these measures generated improvements, the evaporator divider having the greater effect. Groll and Spendel (1984) also employed a concentric divider and found that it made a significant improvement under the axisymmetric (near-vertical) conditions for which they designed it but reduced performance when the tube was highly tilted, no doubt because of the attendant stratification of the flow. In general, the flow divider must be designed for a given tilt. Implicit in both types of plumbing is an attempt to control the overall circulation pattern. With the flow divider, the ideal vertical pattern is a single loop; vapour flows in a single central core surrounded by an annular condensate return. With the central artery, the ideal pattern is a figure of eight; the condenser and evaporator sections both have a central core flowing towards the nearest closed end and a surrounding annular return. The attainment of this second pattern is readily facilitated with the flow eversion device discussed previously (Maezawa and Ishizaki 1988).

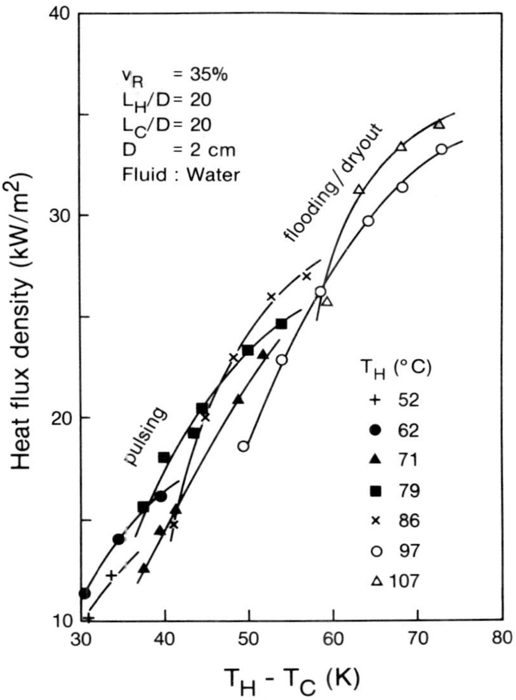

Fig. 3.43. Heat transfer characteristics of an evaporative elbow thermosyphon with a horizontal condenser (after Lock and Fu 1992*b*).

Turning now to thermosyphon tubes which are not straight, it will be recalled from Chapter 2 that the number of possible forms which non-linear tubes may take is very large. Accordingly, the discussion here can only provide a brief introduction to tubular arrangements in which the evaporator and the condenser do not have the same tilt or alignment. While the range of potential applications is large, the number of published studies is very small. As in Chapter 2, we limit ourselves to the elbow and cranked forms.

The elbow device has found service in the evaporative mode; see, for example, Figs. 1.12, 1.22, and 1.23. With the condenser standing upright and the evaporator lying horizontal it has been used to maintain the ground in a frozen condition. Alternatively, it may be employed as an anti-icing device with the condenser lying horizontal and the evaporator extending vertically beneath. It may be recalled from Section 2.4 that these two configurations showed identical heat transfer rates in the single-phase elbow thermosyphon, but this is unlikely to be true for the evaporative thermosyphon. The discussion given earlier in this section suggests that the performance of the

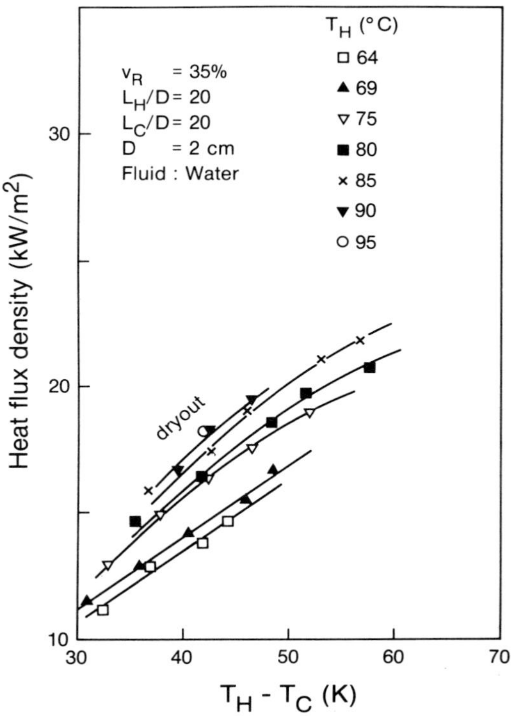

Fig. 3.44. Heat transfer characteristics of an evaporative elbow thermosyphon with an upright condenser (after Lock and Fu 1992*b*).

condensing section improves steadily as the tilt from the vertical increases, while that of the evaporative section grows steadily worse. It is thus to be expected that a horizontal condenser (with evaporator beneath) would perform better than a horizontal evaporator (with condenser above). The heat transfer characteristics shown in Figs 3.43 and 3.44 confirm this but reveal that the difference is not substantial. Lock and Fu (1992*b*) concluded that the higher heat transfer coefficients associated with a horizontal condenser were partially offset by poor drainage characteristics.

The elbow thermosyphons represented in Figs 3.43 and 3.44 had the same charge of water, and both were subjected to the same range of thermal conditions. However, their limiting behaviour was very different. With the condenser horizontal, failure was usually identified as a dryout condition corresponding to flooding and holdup in the condenser. Unlike the holdup condition in a vertical thermosyphon, this evidently reflects the limited drainage associated with indirect Archimedean buoyancy forces. When the condenser was upright, failure was attributable to simple dryout without any

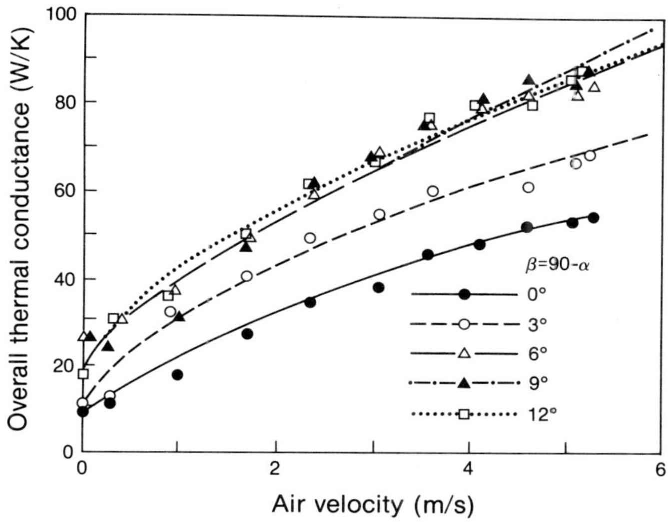

Fig. 3.45. Overall thermal conductance as a function of air velocity for a CO_2 elbow thermosyphon with upright condenser (after Haynes and Zarling 1988*b*)

holdup; that is, the returning condensate failed to reach the closed end of the horizontal evaporator. As in linear thermosyphons, oscillatory behaviour was not associated with the latter form of failure but was very much in evidence during the flooding of the horizontal condenser. In fact, as the vapour pressure and temperature were lowered it became increasingly difficult to distinguish between flooding oscillations and pulse boiling.

The experiments of Haynes and Zarling (1988*a*,*b*) provide additional data for the upright condenser configuration and also investigate the effect of tilting the evaporator separately. These data were obtained with the evaporator heated electrically while the condenser was placed in a cold air stream of variable velocity. Accordingly, the authors chose to plot the thermal effectiveness of the device against air speed, as shown in Fig. 3.45. This reveals that the overall thermal conductance of the device is a monotonically increasing function of air speed, as might be expected. Also evident is the effect of evaporator tilt, with β measured relative to the horizontal. Improvement is particularly rapid in the first 6°, but virtually absent thereafter. The explanation no doubt lies in the rapid growth of direct buoyancy and the attendant increase in circulation rate. To extend the dryout limit, the same authors have made use of an arterial return which supplies about 60% of the condensate to the closed end (Haynes *et al*. 1991). As in the vertical, linear tube study mentioned earlier, the artery proved to be beneficial.

In the cranked thermosyphon, the evaporator and condenser are parallel

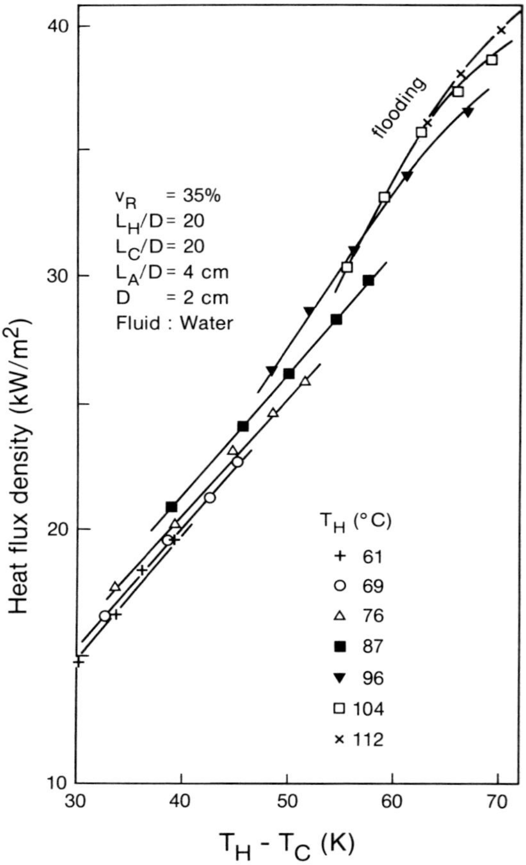

Fig. 3.46. Heat transfer characteristics of an evaporative, cranked thermosyphon standing upright (after Lock and Fu 1992*c*).

but their axes are offset from one another. The heat transfer characteristics of the device are displayed in Fig. 3.46 with the axes vertical (Lock and Fu 1992*c*). By comparison with Figs 3.43 and 3.44, it is evident that the cranked thermosyphon performs rather better than its elbow counterpart. For the data shown, the working fluid, tube diameter, heated length, and cooled length were identical in the two systems, thus indicating that the heat flux in the vertical, cranked thermosyphon is likely to be higher than in a comparable elbow thermosyphon exposed to the same temperature difference. Perhaps more important is the fact that the critical heat flux was higher; this was attributed to a flooding/dryout condition. With the tube axes aligned horizontally, the system failed completely in the presence of condenser holdup combined with evaporator dryout.

4
THE ROTATING, TUBULAR THERMOSYPHON

At first glance, it may seem that the replacement of gravitational acceleration by centrifugal acceleration is a minor change, and indeed it is often assumed that heat and mass transfer relations in a rotating field may be obtained from those in a uniform gravitational field by this simple substitution. As first-order approximations, the results are often useful but they cannot be expected to reflect with great accuracy the consequences of spatially varying centrifugal and Coriolis accelerations. In particular, it is often found that the Coriolis effect in a rotating thermosyphon plays an important role in the overall circulation pattern and heat transfer rate. In general, the vector addition of the centrifugal and gravitational accelerations produces a flow field which, at best, is steady but complex; at worst, it is transient, complex, and subject to instability.

4.1 Convection in rotating systems

Perhaps it is not surprising that the literature on rotating thermosyphons is not extensive. In recent years, some attention has been directed towards evaporative systems, the principal aim being the acquisition of empirical data for use in an engineering context. These experimental studies will be discussed in Sections 4.5 and 4.6. Very little has been written on single-phase behaviour despite the established literature on rotational fluid dynamics in a geophysical context (Greenspan 1969). Accordingly, the first three sections of this chapter have been designed to provide an introduction to single-phase, physico-mathematical models from which the performance of particular thermosyphons may be anticipated.

The equations governing the flow of a homogeneous, rotating fluid may be stated with reference to a coordinate system that is either fixed (inertial) or rotating (with respect to the inertial framework). In order to make comparisons with stationary thermosyphons, it is perhaps better to describe velocity and temperature fields in relation to the thermosyphon tube itself. This will be assumed to rotate at a steady speed Ω about an axis which, incidentally, may or may not coincide with its own. It is thus natural to undertake an analysis of flow and heat transfer in a coordinate system rotating at $\mathbf{\Omega}$, keeping in mind that

$$\mathbf{V}_{\text{in}} = \mathbf{V} + \mathbf{\Omega} \times \mathbf{R}$$

where $\mathbf{V}_{\text{in}}$ is the velocity in the inertial coordinate system. The appropriate equations[1] governing laminar flow of an incompressible fluid are as follows (Greenspan 1969).

Continuity:

$$\nabla \cdot \mathbf{V} = 0 \tag{4.1}$$

Motion:

$$\rho\left(\frac{\partial \mathbf{V}}{\partial t} + \frac{1}{2}\nabla(\mathbf{V}\cdot\mathbf{V}) + (\nabla \times \mathbf{V}) \times \mathbf{V} + 2\boldsymbol{\Omega} \times \mathbf{V} + \boldsymbol{\Omega} \times (\boldsymbol{\Omega} \times \mathbf{R})\right)$$
$$= -\nabla P + \rho \mathbf{g} - \mu[\nabla \times (\nabla \times V)] \tag{4.2}$$

Energy:

$$\frac{\partial \theta}{\partial t} + (\mathbf{V} \cdot \nabla)\theta = \kappa \nabla^2 \theta \tag{4.3}$$

where $\theta = T - T^0$ and T^0 is a reference temperature.

As noted earlier, the relative magnitudes of the gravitational and centrifugal accelerations are an important aspect of behaviour which is therefore easier to follow if the two terms are grouped together as

$$\mathbf{G} = \mathbf{g} - \boldsymbol{\Omega} \times (\boldsymbol{\Omega} \times \mathbf{R}).$$

This suggests that the equation of motion may be more conveniently stated in the form

$$\rho\left(\frac{\partial \mathbf{V}}{\partial t} + (\mathbf{V} \cdot \nabla)\mathbf{V} + 2\boldsymbol{\Omega} \times \mathbf{V}\right) = -\nabla P + \rho(1 - \beta\theta)\mathbf{G} + \mu \nabla^2 \mathbf{V} \tag{4.4}$$

in which the incompressible form of the continuity equation (4.1) has been invoked along with the Boussinesq approximation. It is worth noting that the Coriolis acceleration $2\boldsymbol{\Omega} \times V$ does not take the form appropriate to an inertial coordinate system.

In cylindrical coordinates *rotating steadily* about the Z axis, as ilustrated in Fig. 4.1, the momentum equation (4.4) may be expressed in the following component forms.

Axial momentum:

$$\frac{\partial W}{\partial t} + U\frac{\partial W}{\partial R} + \frac{V}{R}\frac{\partial W}{\partial \Phi} + W\frac{\partial W}{\partial Z}$$

[1] Certain simplifying assumptions are implicit in these equations and may be justified *a posteriori*. In particular, they include the neglect of viscous dissipation and compression work in the energy equation.

$$= -\frac{1}{\rho}\frac{\partial P'}{\partial Z} - \beta\theta G_Z + \nu\left(\frac{\partial^2 W}{\partial R^2} + \frac{1}{R}\frac{\partial W}{\partial R} + \frac{\partial^2 W}{\partial Z^2} + \frac{1}{R^2}\frac{\partial^2 W}{\partial \Phi^2}\right) \qquad (4.5a)$$

Radial momentum:

$$\frac{\partial U}{\partial t} + U\frac{\partial U}{\partial R} + \frac{V}{R}\frac{\partial U}{\partial \Phi} + W\frac{\partial U}{\partial Z} - 2\Omega V$$

$$= -\frac{1}{\rho}\frac{\partial P'}{\partial R} - \beta\theta G_R + \nu\left(\frac{\partial^2 U}{\partial R^2} + \frac{1}{R}\frac{\partial U}{\partial R} + \frac{\partial^2 U}{\partial Z^2} + \frac{1}{R^2}\frac{\partial^2 U}{\partial \Phi^2}\right) \qquad (4.5b)$$

Azimuthal momentum:

$$\frac{\partial V}{\partial t} + U\frac{\partial V}{\partial R} + \frac{V}{R}\frac{\partial V}{\partial \Phi} + W\frac{\partial V}{\partial Z} + 2\Omega U$$

$$= -\frac{1}{\rho R}\frac{\partial P'}{\partial \Phi} - \beta\theta G_\Phi + \nu\left(\frac{\partial^2 V}{\partial R^2} + \frac{1}{R}\frac{\partial V}{\partial R} + \frac{\partial^2 V}{\partial Z^2} + \frac{1}{R^2}\frac{\partial^2 V}{\partial \Phi^2}\right) \qquad (4.5c)$$

In each of these, $P' = P - P^0$ is the departure from the hydrostatic pressure P^0 given by $\nabla P^o = \rho^o \mathbf{G}$. The above, together with the corresponding forms of the continuity equation (4.1) and the energy equation (4.3), provide an adequate description of convection in a rotating tube when distances and velocities are measured with reference to the tube wall.

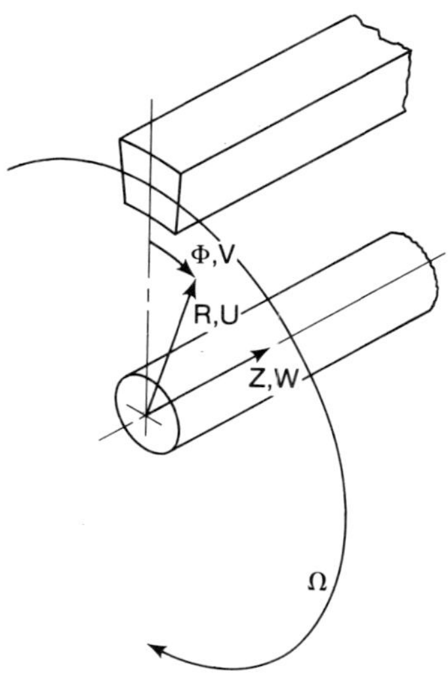

Fig. 4.1. Rotating coordinates and velocity components in a system revolving at Ω about a fixed axis

We have thus arrived at a useful starting point in the analysis of rotating thermosyphons, but eqns (4.1), (4.3), and (4.5) are not without limitations. Obviously, they refer to a homogeneous fluid and must therefore be appied separately to the liquid and vapour phases in evaporative systems, although this is not a problem in principle. Strictly, it is only necessary then to replace thermal buoyancy with Archimedean buoyancy but behaviour may vary considerably depending, for example, on whether or not one of the phases is dispersed throughout the other. In either phase, the equations describe laminar, incompressible flow. Such a description commonly applies to a liquid film when surface evaporation or condensation is taking place but it is clearly inadequate during boiling or turbulent conditions. In the vapour stream, the description is limited by the effects of compressibility and turbulence. Interfacial conditions add further to the complexity but otherwise pose no special difficulties provided that the interface does not become unstable.

Within the single-phase systems discussed immediately below, the description of behaviour is not much easier. It will be recalled from Chapter 2 that natural convection in cavities is typically complex. Beyond the primary flows which establish the longitudinal convective fluxes, the appearance of secondary and tertiary flows is common; and when the fluid is 'heated from below' bifurcation may occur as the result of a physico-mathematical ambiguity which is reflected in the heat transfer rate. Instability in the vicinity of bifurcated behaviour and, more generally, in the transition to turbulent flow is not easily understood or treated. Turbulence itself is not yet well understood in cavity flows.

Complexities in gravitational thermosyphons thus extend to rotating thermosyphons where the presence of a spatially varying force field directly reflecting the role of inertia is unlikely to introduce simplifications. Despite this fact, and indeed because of it, the scaling and ordering procedures developed earlier, notably in Chapter 2, emerge as valuable tools capable of yielding insight. At the very least, the natural scales thus uncovered lead to predictions of thermal performance which may be compared with scarce data; in the absence of such data, they may be the only means of making an assessment. It will become apparent in the next two sections, for example, that the single-phase heat transfer rate must be written, in general, as

$$Nu = f\left(Ra_D, Pr, Ek_D, \frac{\Omega^2 \mathfrak{R}}{g}, \frac{L}{D}, \frac{\mathfrak{R}}{D}\right)$$

where Ra_D and Pr are the familiar Rayleigh and Prandtl numbers, respectively; $\Omega^2 \mathfrak{R}/g$ is an acceleration ratio associated with rotation at a radius $\mathfrak{R}$; L/D and $\mathfrak{R}/D$ are geometrical ratios for a tube of length L and width D; and $Ek_D = \nu/\Omega D^2$ is the Ekman number which, in providing a measure of the relative importance of viscous and centrifugal (inertial) forces, may be regarded as an inverse Reynolds number. In what follows, basic particular forms of this equation will be explored in relation to the primary and secondary flow fields. Given the severe limits imposed by our knowledge of rotating thermosyphons, this exploration constitutes no more than an introduction.

4.2 The single-phase, radial thermosyphon

Various types of radial thermosyphon are illustrated in Fig. 4.2. In the open system, a purely radial tube, closed at its outer end, discharges heated fluid radially inwards from its open mouth which opens into a reservoir of cooler fluid. At first glance, this behaviour resembles that of the open, vertical thermosyphon discussed in Section 2.2; gravitational acceleration is replaced by centrifugal acceleration. Likewise, the closed system shown in Fig. 4.2 bears a resemblance to the closed, vertical thermosyphon treated in Section 2.3. To accommodate the lateral effect of the Coriolis acceleration, however, a more precise description requires a comparison with the behaviour of the tilted thermosyphons discussed in Section 2.4.

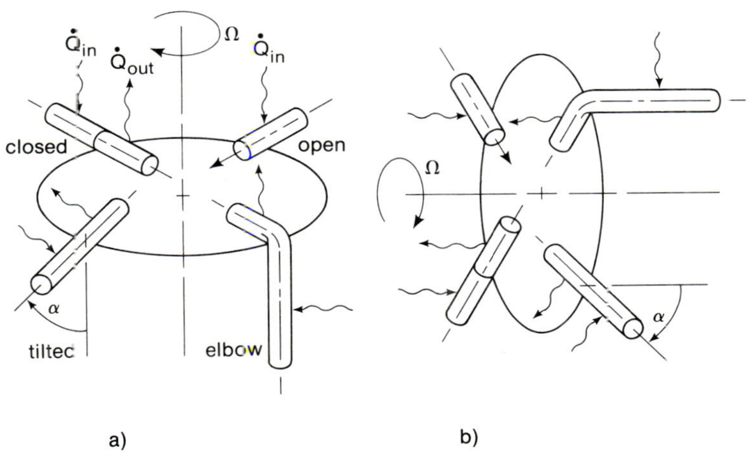

Fig. 4.2. Various configurations of radial, tubular thermosyphons: (a) vertical axis; (b) horizontal axis.

The two other devices depicted in Fig. 4.2 are variations on the same theme; they may be open or closed, although both are shown in closed form. The tilted tube is set at an angle to the axis of rotation and thus has a behaviour spanning the range from the purely radial to the purely axial. The elbow tube, on the other hand, combines heated and cooled sections with different tilts; the tilt angles are generally independent of each other and obviously depend upon the particular application. The configuration shown, for example, combines an axial heated section (as might be found in an electric motor armature) with a radial cooled section. Tilted and bent tubes were also treated in Section 2.4; they will not be discussed further here, except in passing.

This section will be limited to the simple, radial thermosyphon depicted in Fig. 4.3 which shows the components of the body force field when the device rotates about a vertical axis. Also indicated is a coordinate system appropriate to a *square* section tube of width D and length $L \ll \mathfrak{R}$, the radial distance from the axis of rotation to the radially innermost end of the tube. With this arrangement, the natural cylindrical polar coordinate system used earlier may be replaced by the more convenient, though approximate, representation of a rectangular coordinate system. Thus, the radial coordinate is replaced by X, measured outwards from the inner end of the tube, while the azimuthal coordinate is converted into the tangential coordinate $Y = \mathfrak{R}\Phi$ as shown; $Y = 0$ defines the trailing face of the tube and $Y = D$ defines the leading face. The elevation Z completes the right-handed set of Cartesian coordinates within which the more general forms of eqns (4.1), (4.3), and (4.5) reduce to the forms given below provided that the rotational speed Ω is fixed along with the imposed boundary conditions.

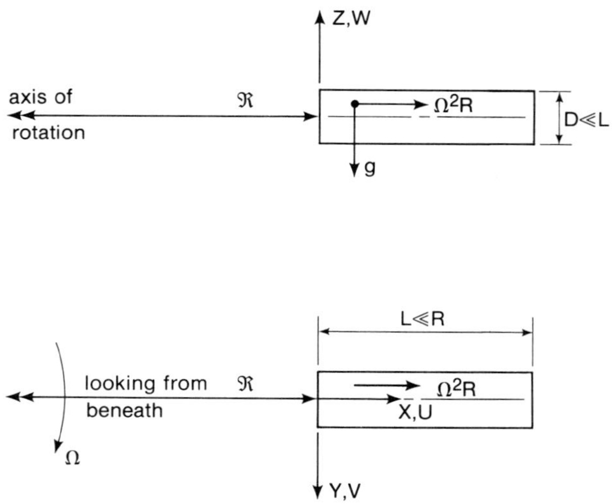

Fig. 4.3. Components of the body force field in a radial thermosyphon rotating about a vertical axis.

Continuity:

$$\frac{\partial U}{\partial X}+\frac{\partial V}{\partial Y}+\frac{\partial W}{\partial Z}=0 \tag{4.6a}$$

Axial momentum:

$$U\frac{\partial W}{\partial X}+V\frac{\partial W}{\partial Y}+W\frac{\partial W}{\partial Z}$$

$$=-\frac{1}{\rho}\frac{\partial P'}{\partial Z}+\beta g\theta+\nu\left(\frac{\partial^2 W}{\partial X^2}+\frac{\partial^2 W}{\partial Y^2}+\frac{\partial^2 W}{\partial Z^2}\right) \tag{4.6b}$$

Radial momentum:

$$U\frac{\partial U}{\partial X}+V\frac{\partial U}{\partial Y}+W\frac{\partial U}{\partial Z}-2\Omega V$$

$$=-\frac{1}{\rho}\frac{\partial P'}{\partial X}-\beta\Omega^2 R\theta+\nu\left(\frac{\partial^2 U}{\partial X^2}+\frac{\partial^2 U}{\partial Y^2}+\frac{\partial^2 U}{\partial Z^2}\right) \tag{4.6c}$$

Tangential momentum:

$$U\frac{\partial V}{\partial X}+V\frac{\partial V}{\partial Y}+W\frac{\partial V}{\partial Z}+2\Omega U$$

$$=-\frac{1}{\rho}\frac{\partial P'}{\partial Y}+\nu\left(\frac{\partial^2 V}{\partial X^2}+\frac{\partial^2 V}{\partial Y^2}+\frac{\partial^2 V}{\partial Z^2}\right) \tag{4.6d}$$

Energy:

$$U\frac{\partial \theta}{\partial X}+V\frac{\partial \theta}{\partial Y}+W\frac{\partial \theta}{\partial Z}=\kappa\left(\frac{\partial^2 \theta}{\partial X^2}+\frac{\partial^2 \theta}{\partial Y^2}+\frac{\partial^2 \theta}{\partial Z^2}\right). \tag{4.6e}$$

It is understood that $R=\mathfrak{R}+X$ in eqn (4.6c). These equations imply that radial planes are essentially parallel to each other over $0<X<L$, and curvature in the azimuthal coordinate is negligible; the forms are therefore accurate if $L \ll \mathfrak{R}$.

Under these circumstances it is to be expected that the primary circulation would be attributable to thermal buoyancy, centrifugally driven. This circulation creates the characteristically high convective flux, which may be estimated from the scaling analysis developed in Chapter 2. Before doing this, however, it is important to anticipate the general nature of the Coriolis effect which, as we shall see, can alter the heat transfer rate significantly. The most obvious effect is introduced through the vortical pattern shown schematically in Fig. 4.4. This is reminiscent of the secondary flow in a tilted tube situated in a gravitational field, but is fundamentally different because it is caused by the Coriolis acceleration, not by thermal buoyancy; that is, it is influenced by the temperature field only because the latter establishes the primary (radial) flow. Keeping this distinction in mind, the vigour and effect of the secondary flow may be estimated through the 'Coriolis tilt' α_c. By analogy with the tilted tube, this is defined as

$$\alpha_c = \frac{2\Omega U^c}{\beta\Omega^2\mathfrak{R}\theta^c} \tag{4.7}$$

where the numerator and denominator are taken to be the principal driving accelerations in eqns (4.6): $2\Omega U^c$, where U^c is the longitudinal velocity scale, thus represents the tangential Coriolis acceleration; $\beta\Omega^2\mathfrak{R}\theta^c$, where θ^c is the temperature (difference) scale, represents the centrifugally driven thermal buoyancy. Recalling the effect of tilt in gravitational systems, it is not difficult to see that the magnitude of α_c will provide a significant indication of the behaviour and performance of the rotating system.

Firstly, consider behaviour when the Coriolis effect is very weak. In essence, the flow then consists of a strong, annular, refluent circulation on which is superimposed a trace of the secondary pattern indicated in Fig. 4.4. Using the scaling technique developed in Chapter 2, it is found that under boundary layer conditions the thickness and velocity scales of the annular return flow are given by

$$\delta = \frac{L}{Ra_L^{1/4}}$$

$$U^c = \frac{\kappa}{L} Ra_L^{1/2} \tag{4.8}$$

where $Ra_L = \beta\Omega^2\mathfrak{R}\theta^c L^3/\nu\kappa$ and θ^c is the overall temperature difference. It is at once apparent that these forms are completely analogous to those obtained for a vertical tube where $Ra_L = \beta g\theta^c L^3/\nu\kappa$. The heat transfer rate may thus be estimated from

$$\bar{\dot{q}} = \frac{\rho c_p \theta^c U^c \delta}{L} \tag{4.9}$$

Hence, for an open, radial thermosyphon for which $\theta^c = T_H - T_0$, the difference between the wall temperature and the fluid entry temperature,

$$\bar{N}u_D = \frac{\bar{h}D}{k} = C_1\left(\frac{D}{L}Ra_D\right)^{1/4} \tag{4.10}$$

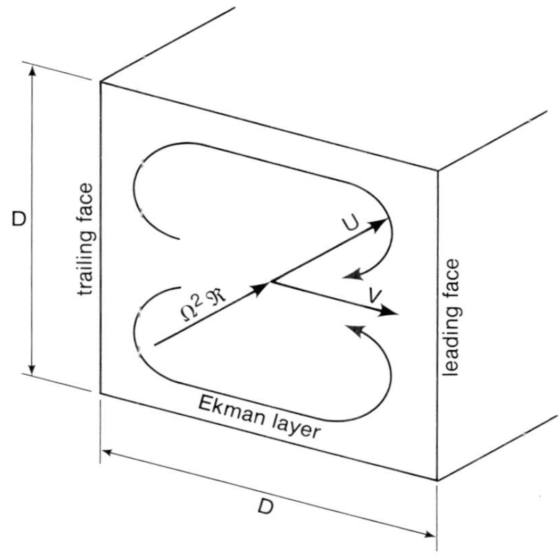

Fig. 4.4. Schematic representation of secondary circulation in a radial thermosyphon with a wall temperature T_H greater than the fluid core temperature.

where $Ra_D = \beta(T_H - T_0)\Omega^2\mathfrak{R}D^3/\nu\kappa$.

To complete the analogy under laminar conditions, it is necessary to determine the heat transfer relation when, according to the first of eqns (4.8), the Rayleigh number is low enough for the boundary layer to fill the tube; that is, $\delta = O(D/2)$. The velocity scale is then given by

$$U^c = \frac{\beta\Omega^2\mathfrak{R}\theta^c D^2}{4\nu}, \tag{4.11}$$

eqn (4.9) then yielding

$$\bar{N}u_D = C_2 \frac{D}{L} Ra_D. \tag{4.12}$$

This describes the impeded flow régime. The arguments developed in Section 2.2 suggest that transition between the two régimes would occur when $Ra_D D/L = O(10^5)$.

The laminar heat transfer predictions expressed in eqns (4.10) and (4.12) are based on the assumption that the Coriolis effect is very weak. Equation (4.7) provides a measure of its strength. For the boundary layer régime, in which U^c is given by eqn (4.8), it is found that the Coriolis tilt is given by

$$\alpha_c = \frac{2}{Ek_D t_D^{1/2}} \tag{4.13}$$

where $Ek_D = \nu/\Omega D^2$ is the Ekman number and $t_D = Ra_D D/L$. Under impeded conditions, on the other hand, when U^c is given by eqn (4.11),

$$\alpha_c = \frac{1}{2Ek_D}. \tag{4.14}$$

Since $t_D \gg 1$, typically, the severest of these conditions is the second, which indicates that unless $Ek_D \gg 1$ the Coriolis effect may not be neglected in the impeded flow régime. For a tube of 1 cm diameter containing a liquid with a representative viscosity of 10^{-6} m²/s, the speed at which $Ek_D = 1$ is found to be 10^{-2} rad/s, very much lower than the speed of most rotating machines, at least in a terrestrial context. In the boundary layer régime, the lowest Ekman number for which the Coriolis effect may be neglected (i.e. $\alpha_c \ll 1$) can be estimated from eqn (4.13) by imposing the requirement $Ek_D t_D^{1/2} > 20$, or $Ek_D^2 > 400/t_D$; for the tube mentioned above, this suggests that in the low-speed range, that is 10^{-1} rad/s $< \Omega < 10$ rad/s, t_D falls in the range $4 \times 10^4 < t_D < 4 \times 10^8$, and the lowest 'Coriolis-free' Ekman numbers fall in the range $10^{-3} < Ek_D < 10^{-1}$. The laminar–boundary layer régime is evidently not too susceptible to the Coriolis effect in this low-speed range, or indeed for even lower speeds. It is equally apparent that high-speed behaviour will certainly be influenced by Coriolis forces, even under boundary layer conditions. Furthermore, under laminar–impeded conditions, this influence extends down into the low-speed range and lower, thus implying that the above analysis of impeded flow is incorrect because it does not incorporate the Coriolis effect; in particular, eqns (4.12) and (4.14) are then invalid.

Returning to eqns (4.6), the most obvious influence of the Coriolis acceleration is seen in the tangential equation of motion. From this equation, the Ekman layer on both of the side faces shown in Fig. 4.4 may be defined by

$$\frac{\nu V^c}{(Z_E^c)^2} = O(2\Omega U^c) \tag{4.15}$$

where Z_E^C is the Ekman layer thickness scale. This scale is unknown *a priori* but cannot be greater than the thickness Z^c of the main return flow which drives it. Consider the situation when they are equal. If the heat conducted through the Ekman layers is removed convectively, then from the energy equation (4.6e)

$$V\frac{\partial\theta}{\partial Y} = O\left(\kappa\frac{\partial^2\theta}{\partial Z^2}\right)$$

or

$$\frac{V^c (Z_E^c)^2}{\kappa Y^c} = O(1) \tag{4.16}$$

in which $Y^c = O(D)$, the tube width. At the same time, a balance between the thermal buoyancy and principal viscous forces in the longitudinal (radial) equation of motion yields

$$\beta\Omega^2\mathfrak{R}\theta = O\left(\nu\frac{\partial^2 U}{\partial Z^2}\right)$$

or

$$\frac{\beta\Omega^2\mathfrak{R}\theta^c(Z_E^c)^2}{\nu U^c} = O(1). \qquad (4.17)$$

Equations (4.15), (5.16), and (4.17) permit the thickness and velocity scales to be determined. Thus,

$$Z_E^c = D\left(\frac{Ek_D}{2Ra_D}\right)^{1/6}$$

$$U^c = \frac{\kappa}{D}\left(\frac{Ra_D^2 Ek_D}{2}\right)^{1/3}$$

$$V^c = \frac{\kappa}{D}\left(\frac{2Ra_D}{Ek_D}\right)^{1/3} \qquad (4.18)$$

from which it follows that the heat transfer rate *through the Ekman layers*,

$$\bar{q} = \rho c_p \theta^c V^c Z_E^c / D$$

may be rewritten as

$$\bar{N}u_D = C_1\left(\frac{Ra_D}{Ek_D}\right)^{1/6}. \qquad (4.19)$$

As the rotational speed increases, eqn (4.19) indicates that the heat transfer rate also increases. Secondary vortices driven by the Ekman layers then enhance the convective ability of the thermosyphon, much as small tilts from the vertical improve the performance of the gravitational system. Larger tilts from the vertical, however, introduce a deteriorating performance attributable to the spread of a more sluggish bifilamental flow. The same trend in the radial thermosyphon will eventually lead to $Z^c = O(D/2) \gg Z_E^c$. This condition requires a change in eqn (4.17) but leaves eqns (4.15) and (4.16) unchanged. From the new set of equations, it is found that

$$\bar{N}u_D = C_1'\left(\frac{Ra_D}{Ek_D}\right)^{1/4}. \qquad (4.20)$$

By comparison with eqn (4.19), this indicates a steeper, and steepening, dependency on speed under these conditions; namely, when the Ekman layers exert a controlling influence on behaviour as the speed is increased. Both equations immediately above reveal that heat transfer is an inverse function of the Ekman number in what might be described as a Coriolis-enhanced régime.

The primary flow field in this régime is illustrated in Fig. 4.5 which refers to a short, closed thermosyphon filled with a viscous liquid. Near the closed end of the tube, an annular refluent flow is retained, but this gradually changes to a bifilamental form before reaching the mid-length region. This behaviour is reminiscent of near-vertical behaviour in the gravitational thermosyphon. A similar comment applies to the secondary flow field illustrated in Fig. 4.6. The latter shows that the end stagnation point is not located on the tube axis, and reveals the development of two symmetric pairs of vortices which eventually become equal and opposite at the mid-length plane.

It was noted earlier that $Ek_D \ll 1$ and $Ra_D \gg 1$ in practical engineering situations. From the first of eqns (4.18), this implies that $Z_E^c/D \ll 1$. The second of eqns (4.18) enables the Coriolis tilt to be determined as

$$\alpha_c = \left(\frac{4}{Ra_D Ek_D^2}\right)^{1/3}. \tag{4.21}$$

Thus, for example, if $Ra_D \gtrsim 10^4$ and $Ek_D \gtrsim 10^{-2}$, the Coriolis tilt will be small but not negligible. It is interesting to note that the second and third of eqns (4.18) combine to reveal that

$$\frac{V^c}{U^c} = \left(\frac{4}{Ra_D Ek_D^2}\right)^{1/3}.$$

Hence when $10^{-2} < Ek_D < 1$, the lateral velocity in the Ekman layer is usually less than the longitudinal velocity, their ratio being equal to the Coriolis tilt, as given by eqn (4.21).

When the Ekman number Ek_D is decreased beneath 10^{-2}, the Coriolis tilt ceases to be small. Equations (4.18) then suggest a thinner but faster Ekman layer accompanied by a weakening longitudinal flow, as implied in the derivation of eqn (4.20). This signals a fundamental alteration in behaviour as the Coriolis tit undergoes a change from the 'near vertical' ($\alpha_c < 1$) to the 'near horizontal' ($\alpha_c > 1$). Borrowing experience from the horizontal thermosyphon, this alteration may be attributed to the rising importance of the pressure field. From the tangential equation of motion, eqn (4.6d), it may be seen that the lateral pressure gradient is then caused principally by the Coriolis acceleration which thus acts to create a *longitudinal* pressure gradient, and thereby influences longitudinal motion in what might be termed a 'geostrophic' flow. Eliminating the pressure from the radial and tangential equations of motion, the principal terms strike a balance in which

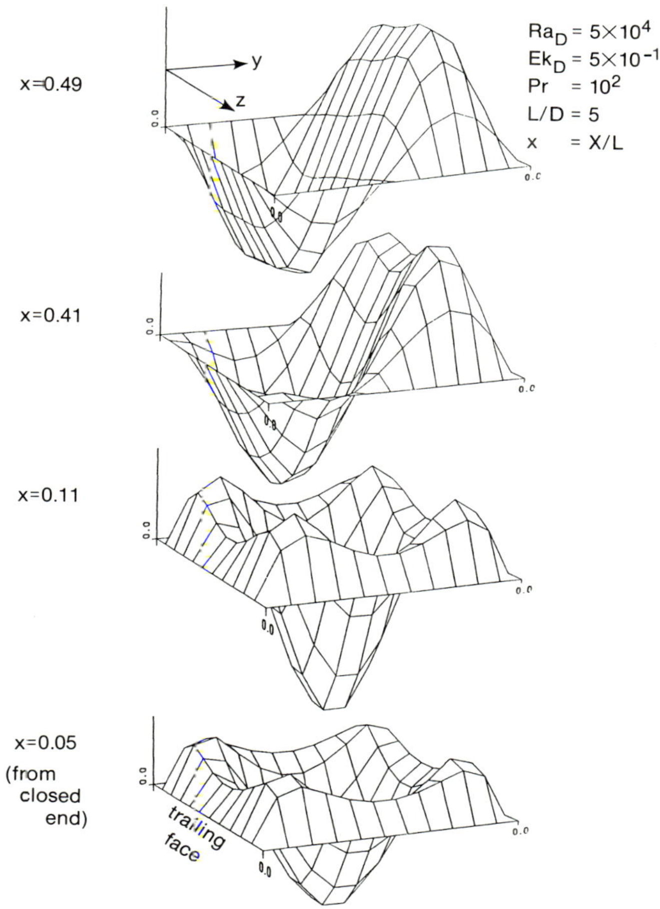

Fig. 4.5. Development of the primary flow in the heated section of a radial thermosyphon under Coriolis-enhanced conditions (after Lock and Zhao 1991*b*).

$$-2\Omega\frac{\partial U}{\partial X}\simeq-\beta\Omega^2\mathfrak{R}\frac{\partial\theta}{\partial Y}+\nu\frac{\partial^3 U}{\partial Y^3}.$$

As the Ekman number is lowered, the buoyancy and viscous terms on the right-hand side gradually fall in relation to the left-hand side. Eventually, this results in a weaker longitudinal flow in which

$$2\frac{\Omega U^c}{X^c}=O\left(\frac{\beta\Omega^2\mathfrak{R}\theta^c}{Y^c}\right)=O\left(\frac{\nu U^c}{(Y^c)^3}\right).$$

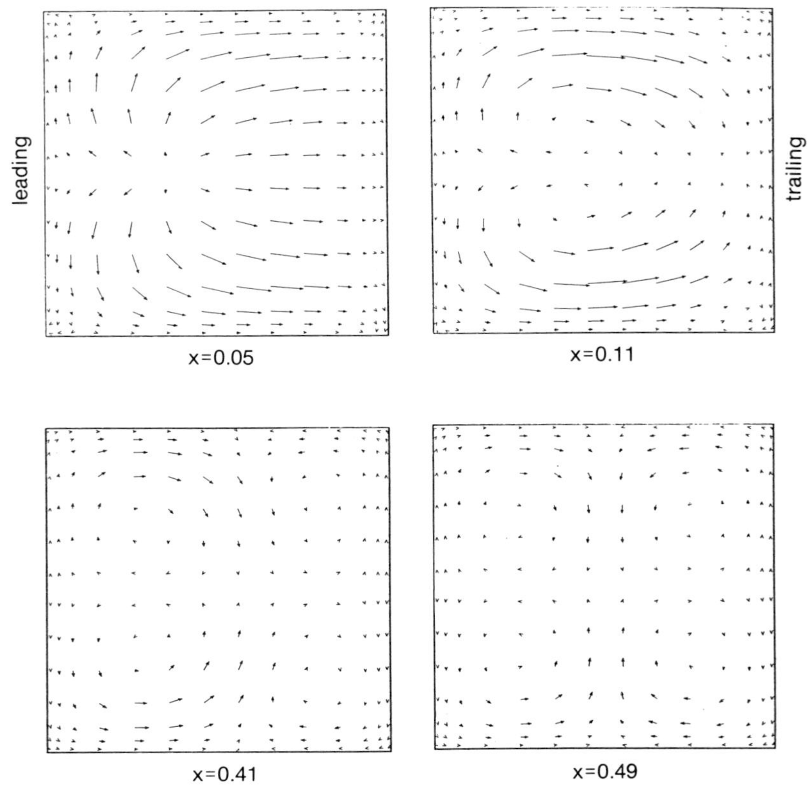

Fig. 4.6. Secondary flow patterns in a radial thermosyphon under Coriolis-enhanced conditions (see Fig. 4.5) (after Lock and Zhao 1991*b*).

This flow is flanked by much thinner Ekman layers and thus lies closer to a two-dimensional representation for which the unknown length and velocity scales may be determined from the above to be

$$Y^c = D\left(\frac{Ek_D}{2D/L}\right)^{1/3}$$

$$U^c = \frac{\kappa}{D}\left(\frac{L}{2D}\right)^{2/3} Ra_D Ek_D^{2/3}. \qquad (4.22)$$

The heat transfer attributable to the *primary* flow rate is then given by

$$\bar{N}u_D = C_2 Ra_D Ek_D. \qquad (4.23)$$

Figure 4.7 shows the development of the axial velocity profile in the same closed thermosyphon used for Fig. 4.5 when operating under these Coriolis-impeded conditions. In the mid-length region, a bifilamental flow is again evident. Nearer the closed end, a central core moves radially outwards while return flows move inwards over the trailing and leading faces. This end flow does not take the three-dimensional annular form observed in Fig. 4.5. Along with the bifilamental flow, it is seen to be almost two-dimensional in the central plane of rotation ($z=0.5$). This behaviour contrasts strongly with that in the Coriolis-enhanced régime.

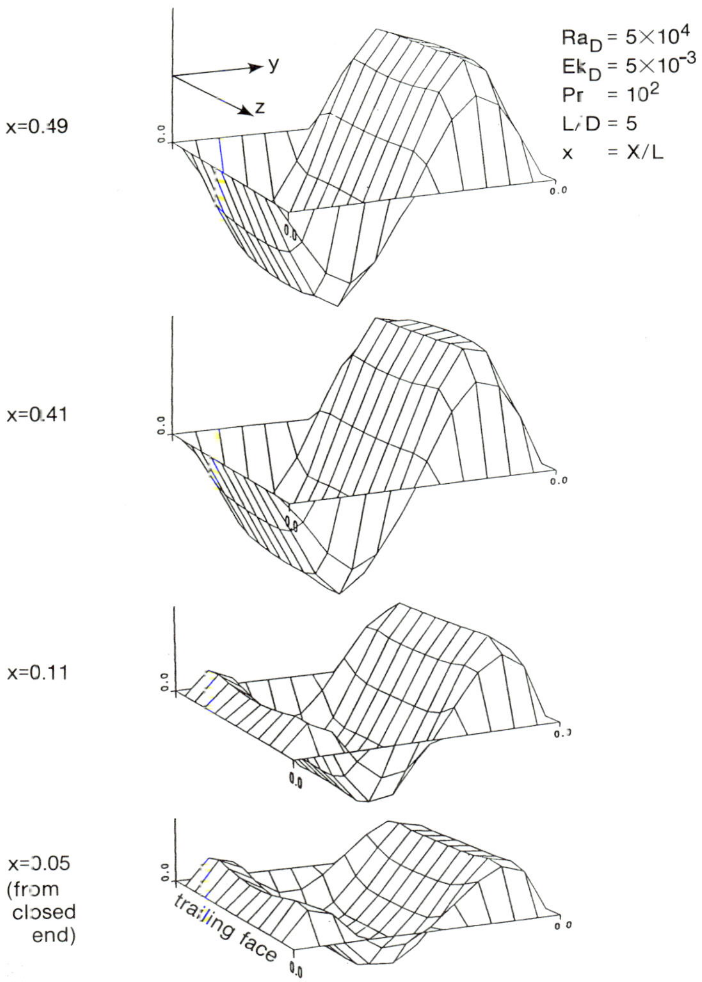

Fig. 4.7. Development of the primary flow in the heated section of a radial thermosyphon under Coriolis-impeded conditions (after Lock and Zhao 1991*b*).

Comparing these results with those obtained earlier, a number of conclusions may be drawn. Equations (4.10) and (4.19), for example, refer to boundary layer types of régimes, the former reflecting an annular type of flow while the latter emphasizes the Ekman layers on the side walls. Equation (4.23), on the other hand, typifies what may be described as an impeded flow régime, by analogy with the gravitational thermosyphon in which decreases in Rayleigh number alone create the condition. Such differences and similarities bring out the complexities of rotating thermosyphons. While they exhibit similar flow régimes to tilted gravitational thermosyphons, the driving mechanisms are clearly not identical. The role of the pressure field in a stationary tube approaching the horizontal, for example, is to replace the direct buoyancy force, originally acting longitudinally, with a *supportive* indirect buoyancy force which maintains the circulation; in a radial tube under low Ekman number conditions, the pressure field produces an *opposing* force which slows the circulation. Furthermore, the use of the Rayleigh number as the sole determinant of régime behaviour must be abandoned. It is clear that the Ekman number also plays an important role. This is most easily seen through the Coriolis tilt. For the Coriolis-impeded flow described immediately above, it is found that

$$\alpha_c = \left(\frac{2(L/D)^2}{Ek_D}\right)^{1/3} \tag{4.24}$$

while the corresponding expression for the Coriolis-enhanced flow was given by eqn (4.21). When $10^{-2} < Ek_D < 1$, the latter reveals a small tilt effect, but when $10^{-4} < Ek_D < 10^{-2}$, eqn (4.24) suggests a large tilt effect. Given the established near-vertical and near-horizontal behaviour in gravitational thermosyphons it is important to recognize the corresponding Coriolis-enhanced and Coriolis-impeded régimes, respectively, of the rotating thermosyphon. It is also important to locate the transition between these régimes.

Equations (4.19) and (4.23) provide two very specific measures of the Coriolis effect on heat transfer; namely, over the two pairs of opposite long surfaces under differing conditions. To the extent that these differing effects reflect changes in the overall interdependence of the centrifugal and Coriolis accelerations, they will also provide a representative indication of overall behaviour in the two identified régimes. In the Coriolis-impeded régime, an increase in heat transfer rate attributable to an increasing Ekman number is indicated by eqn (4.23); a subsequent decrease in the Coriolis-enhanced régime is indicated by eqn (4.19). This implies an optimal value of Ek_D for a given Rayleigh number. Figure 4.8 shows some results obtained for the same thermosyphon used above. Maximal behaviour is confirmed. The asymptotes, based on eqns (4.19) and (4.23), suggest that the transition from low-tilt, Coriolis-enhanced behaviour to high-tilt, Coriolis-impeded behaviour occurs

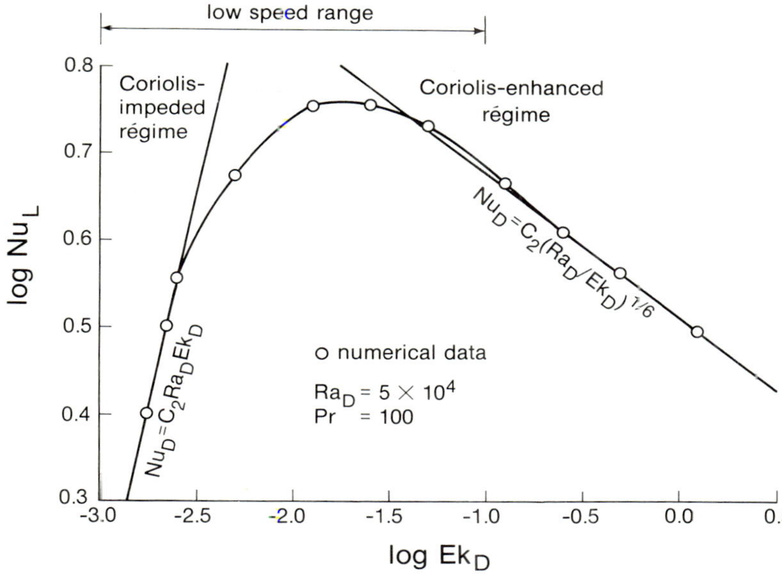

Fig. 4.8. The effect of Ekman number on laminar heat transfer in a radial thermosyphon.

over an Ekman number range of about an order of magnitude. As Ek_D decreases from 10^{-1} to 10^{-2}, the slight steepening of the curve, anticipated in eqn (4.20), heralds the onset of the Coriolis-impeded régime with increasing rotational speed. The low-speed range, arbitrarily defined by $10^{-3} < Ek_D < 10^{-1}$, evidently spans the transition for this particular Rayleigh number. According to arguments advanced earlier on the effect of Coriolis forces, it is to be expected that the optimal heat transfer rate would shift to a lower Ekman number if the Rayleigh number were raised. This suggests that the range of the rough analogue with tilted gravitational systems, being based on the existence of maximal behaviour, would also shift with Rayleigh number.

The behaviour for much lower Ekman numbers is even more important from an engineering standpoint but is currently unknown. There is some evidence (Zhao 1991) to suggest that further decreases in the Ekman number may lead to thermal instability which creates a multi-cellular flow pattern reminiscent of the 'cat's eyes' found in long cavities heated from the side, and exemplified earlier in Fig. 2.6. Such a régime change, again influenced by the Ekman number, would create a rapid and substantial increase in heat transfer rate, thereby implying a minimum to the left of the data points shown in Fig. 4.8. However, this is merely speculation and, in any event, the flow pattern may be completely altered by the onset of turbulence, particularly for higher Rayleigh numbers.

The governing equations (4.6) indicate that the orientation of the axis of rotation is unimportant provided that the acceleration ratio $\eta_R = \Omega^2\mathfrak{R}/g$ is very large. Clearly, this is not true if $\eta_R = O(1)$, when thermosyphons rotating about a vertical axis behave very differently from those rotating about a horizontal axis. In the former situation, a decrease in η_R effected through a speed reduction has no influence on α_c in the boundary layer régimes, according to eqns (4.13) and (4.21), but decreases α_c in the Coriolis-impeded régime, according to eqn (4.24). Thus, while a speed reduction produces a decrease in the Rayleigh number, and hence a less vigorous primary flow, it will not increase the Coriolis tilt under laminar conditions.

At the same time, the Coriolis acceleration, acting parallel to the planes of rotation, is accompanied by a more significant gravitational acceleration acting normal to that plane. The resultant acceleration is thus a vector which, at any point in the thermosyphon, is inclined to the vertical. As noted above, the Coriolis acceleration creates a vortical secondary circulation symmetric about the central (horizontal) plane of rotation and, if the Coriolis tilt is high enough, influences the bifilamental primary circulation in the same plane. The gravitational acceleration tends to act in a similar manner but with a vertical plane of symmetry. Lowering η_R thus resembles the effect produced by rotating a horizontal tube through intervals about its longitudinal axis. Figure 4.9 shows the combined influence of the two lateral acceleration vectors on the primary flow in a closed, square section thermosyphon; the flow is seen to be skewed.

When the axis of rotation is horizontal, steady behaviour is gradually altered by growing periodicity as η_R is decreased from a high value. With every revolution, the longitudinal acceleration must vary in the range $\Omega^2\mathfrak{R}(1 \pm 1/\eta_R)$ and hence for all but the very lowest speeds (when $\eta_R \ll 1$ and conditions vary quasi-steadily) a sinusoidal gravitational perturbation is superimposed on a steady centrifugal acceleration. This periodicity will become important when the time for one revolution t^c satisfies the relation

$$t^c = \frac{2\pi}{\Omega} \geq \frac{L}{U^c}.$$

The conditions for which this is true may be found from eqn (4.18), for the Coriolis-enhanced régime, and from eqn (4.22), for the Coriolis-impeded régime. The expressions thus obtained reveal that periodicity will be less prevalent in viscous liquids than in liquid metals.

4.3 The single-phase, axial thermosyphon

Various configurations of the axial, tubular thermosyphon rotating steadily with speed Ω are shown in Fig. 4.10. When the axis is vertical, Fig. 4.10(a), the

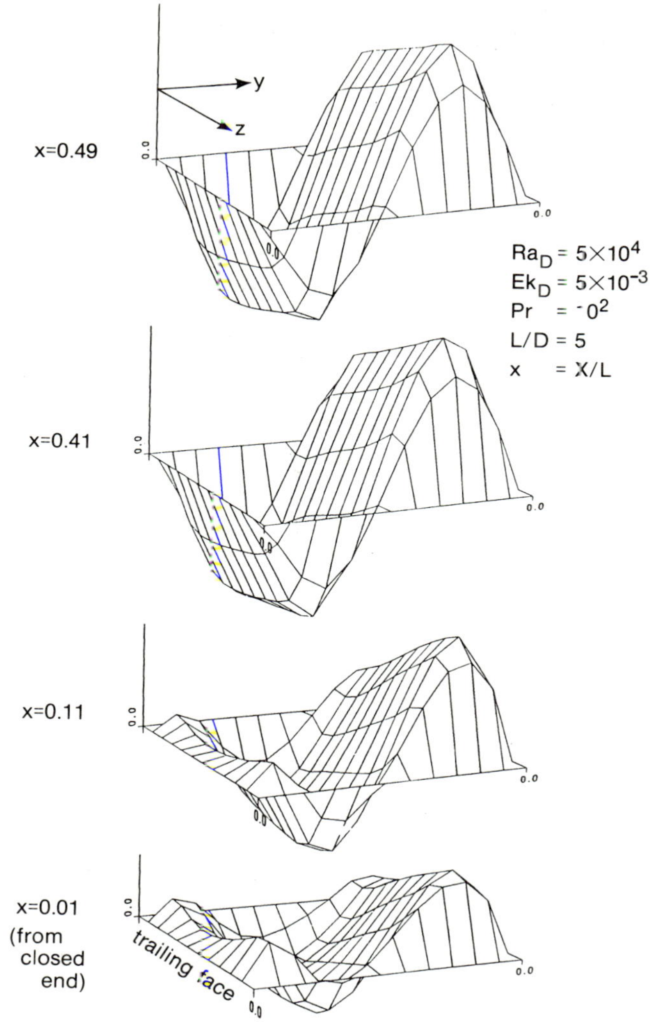

Fig. 4.9. Development of the primary flow in the heated section of a radial thermosyphon with $\eta_R = 1$ (after Lock and Zhao 1991*b*).

centrifugal and gravitational accelerations are normal to each other, but since neither is a function of time it is natural to expect the system to behave steadily providing the imposed wall temperatures are also steady. The discussion in this section will deal principally with such behaviour, when (numerical) solutions to the governing differential equations are not only steady and stable but unique; that is, they are independent of initial conditions. However, in light of earlier comments, particularly in Chapter 2, it must always be borne in mind that a stable solution may not be unique; and should instability occur, the idea of steadiness must be abandoned entirely. When the tube axis is horizontal, as shown in Fig. 4.10(b), the problem is intrinsically transient except for a high acceleration ratio (i.e. $\Omega \gg (g/\mathfrak{R})^{1/2}$) when the periodic effect of gravity may be ignored.

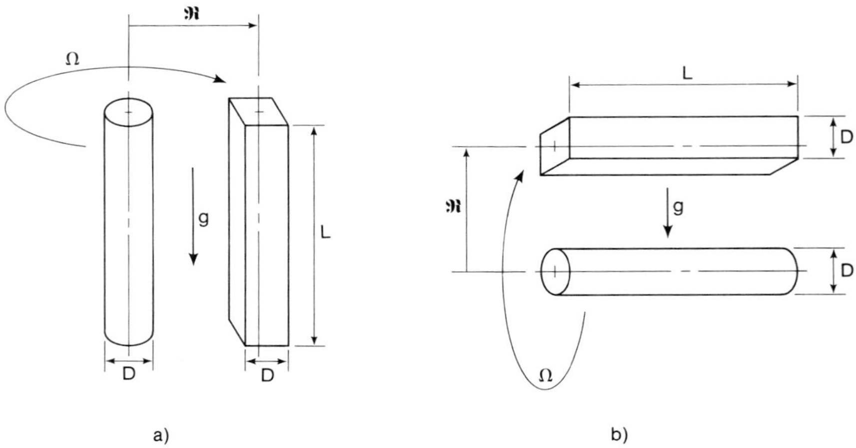

Fig. 4.10. Various configurations of axial, tubular thermosyphons: (a) vertical; (b) horizontal.

With a high acceleration ratio, the flow pattern in the concentric tube is essentially axisymmetric, regardless of the orientation of the tube axis; a core of hot fluid moving towards the cold end is surrounded by an annulus of cold fluid moving in the opposite direction. By way of contrast, the corresponding high η_R pattern in the eccentric tube is not axisymmetric though it does exhibit symmetry about the plane containing the axis of rotation and that of the tube; by analogy with the horizontal tube, the pattern is essentially bifilamental. For very low acceleration ratios, and especially as $\Omega \rightarrow 0$, these flow patterns undergo radical change, depending upon orientation. Thus, in the limit, the vertically oriented tubes recover the axisymmetric flow found in the vertical, gravitational thermosyphon while horizontally oriented tubes adopt the pattern of the horizontal, gravitational thermosyphon.

Figure 4.11 shows the limiting flow patterns for all four configurations. When the axis is vertical, Fig. 4.11(a), the high acceleration ratio pattern in the concentric tube resembles that in the cold section of the static tube. This suggests that the effect of a speed reduction would be to create in the heated end region an opposite circulation which grows eventually to fill the lower half of the tube. For the vertical, eccentric tube, a comparison of the limiting patterns suggests that the intermediate pattern wil be similar to that in a gravitational thermosyphon tilted at an angle $\Omega^2\mathfrak{R}/g$ from the vertical; that is, as the acceleration ratio falls, the end region flows will convert from a bifilamental pattern into an annular refluent pattern which gradually grows to fill the tube. In both of the above (vertical) situations, gravity acts consistently on the *primary* flow.

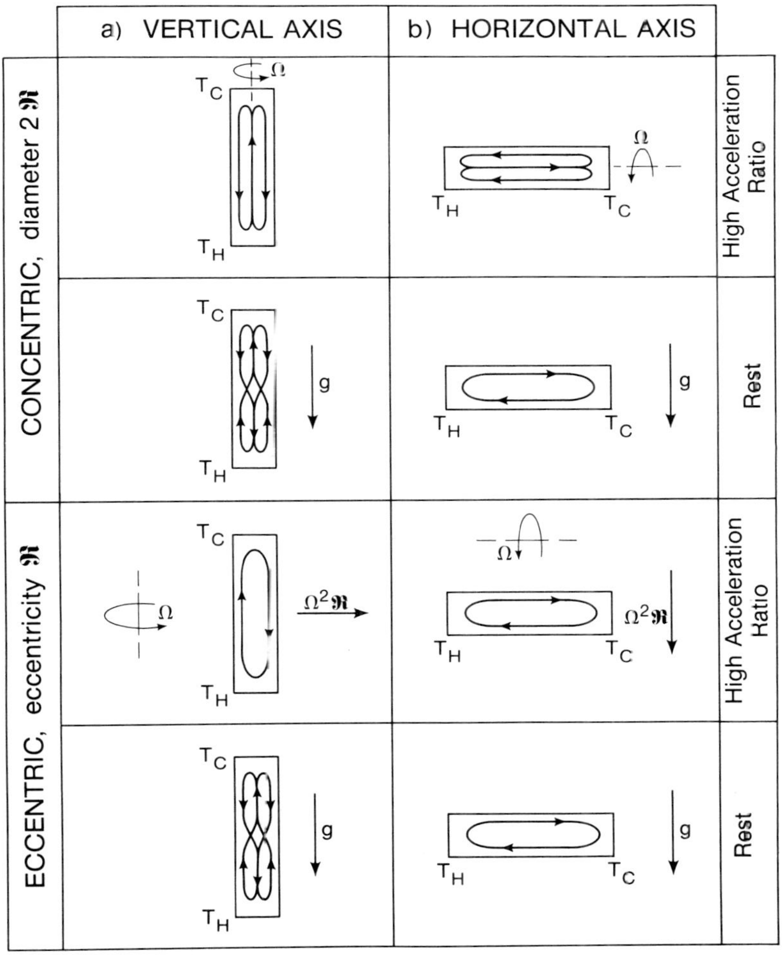

Fig. 4.11. Primary flow patterns in axial thermosyphons for $\eta_R \gg 1$ and $\eta_R = 0$.

While no published experimental data are available, it is believed that if the speed is varied slowly these transitions in vertical tubes will not be accompanied by instabilities and will therefore represent steady, though variable, behaviour; that is, the transformation from one flow pattern to the other is likely to be gradual. The same is not true when the tube is horizontal, Fig. 4.11(b). At the higher speed, the effect of gravity is to introduce a small *periodic* disturbance acting on both primary and secondary flows; but as the speed falls, this periodicity gradually lengthens and subsides with the eventual establishment of a gravitationally induced primary flow pattern. The details of flow behaviour will be especially complex in the intermediate-speed range (i.e. $\Omega = O(g/\mathfrak{R})^{1/2}$) when the periodic effect of the gravitational acceleration on a shifting flow pattern makes the problem particularly difficult. Very little, if any, discussion of these difficulties has appeared in the open literature.

4.3.1 *The eccentric tube*

Consider the vertical, eccentric thermosyphon shown schematically in Fig. 4.12. If the eccentricity is large, that is $\mathfrak{R} \gg D$, where D is the tube width, the gravitational and centrifugal accelerations may be taken as constant throughout the system. Under this condition, the natural, cylindrical polar coordinate system may again be approximated by the rectangular, Cartesian coordinates shown. As in the previous section, X replaces the radial coordinate while the tangential coordinate $Y = \mathfrak{R}\Phi$ replaces the azimuthal coordinate; Z remains as the longitudinal coordinate, or elevation. The governing differential equations for steady, laminar conditions once more take the forms given by eqns (4.6).

The primary, bifilamental flow under these conditions is driven by the indirect effect of thermal buoyancy. Eliminating pressure between the longitudinal and radial equations of motion, eqns (4.6b) and (4.6c), respectively, a balance between the indirect buoyancy forces and the principal viscous forces acting longitudinally suggests that

$$\frac{\partial}{\partial Z}(\beta\Omega^2\mathfrak{R}\theta) = O\left(\nu\frac{\partial^3 W}{\partial X^3}\right)$$

or

$$\frac{\beta\theta^{\mathrm{c}}\Omega^2\mathfrak{R}(X^{\mathrm{c}})^3}{\nu L W^{\mathrm{c}}} = O(1). \tag{4.25}$$

Likewise, a balance between lateral conduction and longitudinal convection in the energy equation indicates that

$$W\frac{\partial\theta}{\partial Z} = O\left(\kappa\frac{\partial^2\theta}{\partial X^2}\right)$$

or

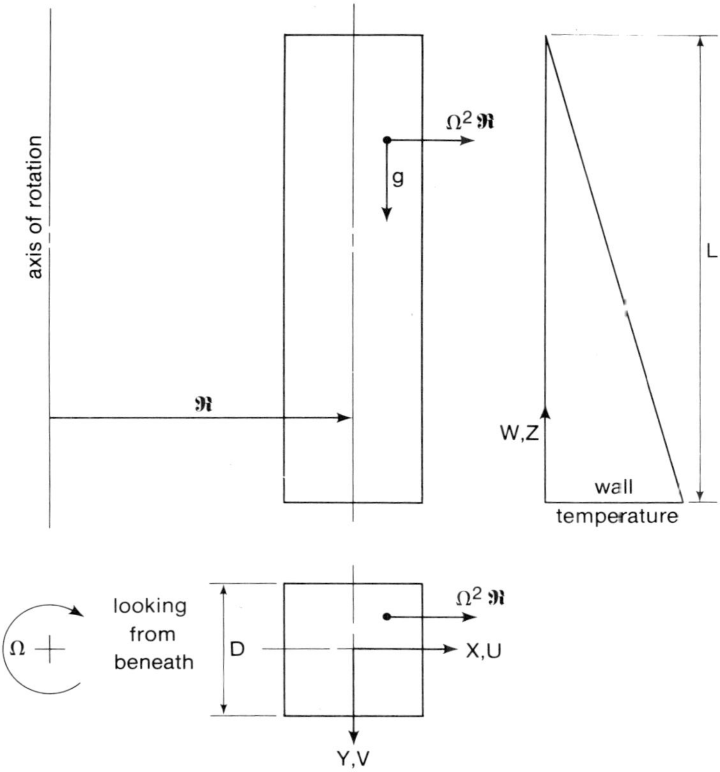

Fig. 4.12. Components of the body force field in a vertical, eccentric thermosyphon with a square cross section.

$$\frac{W^c(X^c)^2}{\kappa L} = O(1) \tag{4.26}$$

in which $Z^c = L$. Equations (4.25) and (4.26) now yield the thickness and velocity scales as

$$X^c = \frac{L}{Ra_L^{1/5}}$$

$$W^c = \frac{\kappa}{L} Ra_L^{2/5} \tag{4.27}$$

where $Ra_L = \beta\theta^c\Omega^2\mathfrak{R}L^3/\nu\kappa$.

These scales correspond to boundary layer conditions under which $X^c \ll D$. The corresponding heat transfer rate may be estimated from

$$\bar{\dot{q}} = \frac{\rho c_p \theta^c W^c X^c}{L} \tag{4.28}$$

which, for an open thermosyphon with $\theta^c = T_H - T_0$, where T_0 is the fluid entry temperature, yields

$$\bar{N}u_D = \frac{\bar{h}D}{k} = C_1\left[\left(\frac{D}{L}\right)^2 Ra_D\right]^{1/5}. \tag{4.29}$$

When the Rayleigh number is low enough for the thickness scale in eqns (4.27) to be comparable with the tube half-width, the requirement of eqn (4.26) must be relaxed. The longitudinal velocity scale may then be determined directly from eqn (4.25) with $X^c = O(D/2)$. Thus, under such impeded flow conditions,

$$W^c = \frac{\beta\theta^c\Omega^2\mathfrak{R}D^3}{8\nu L}.$$

Equation (4.28) then yields

$$\bar{N}u_D = C_2\left(\frac{D}{L}\right)^2 Ra_D. \tag{4.30}$$

Equations (4.29) and (4.30) are analogous to the expressions obtained earlier for a horizontal, gravitational thermosyphon, as would be expected from estimates based on the primary flow. This analogy also suggests that $C_1 = O(1)$ whereas $C_2 \ll O(1)$; however, no experimental data appear to be available for comparison.

Laminar behaviour in a closed, eccentric thermosyphon with a vertical axis has been explored by Zhao and Lock (1992*a*). Some of their numerical results are shown in Figs 4.13, 4.14, and 4.15. The velocity field on a radial plane containing the tube axis is shown in Fig. 4.13(a). This represents a Coriolis-impeded flow régime and clearly confirms a bifilamental structure, with the centrifugal acceleration acting from left to right. The corresponding velocity field in a tangential plane containing the tube axis is displayed in Fig. 4.13(b) from which it is evident that the bifilamental pattern is not two-dimensional, This anti-symmetry is caused by the Coriolis effect which will be discussed in more detail below. At this point, it is sufficient to note that the Coriolis acceleration produces a 'Coriolis twist' defined by

$$\gamma_c = \frac{2\Omega U^c}{\beta\theta^c\Omega^2\mathfrak{R}}. \tag{4.31}$$

Unlike the Coriolis tilt defined in the previous section, however, this expression cannot be used to measure the Coriolis effect directly because the radial velocity scale is not available from the primary flow; in the mid-length region, U^c may be estimated using the continuity equation, that is

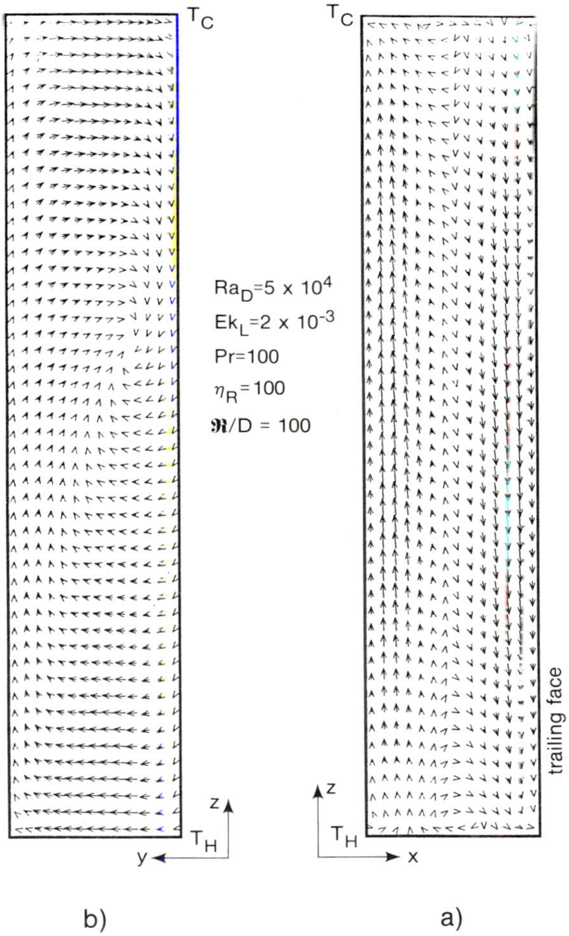

Fig. 4.13. Velocity field on central, longitudinal planes in an eccentric thermosyphon (after Zhao and Lock 1992*a*): (a) central radial plane; (b) central tangential plane.

$U^c = O(DW^c/L)$, but this scale is inappropriate near the closed ends where the flow must double back under the influence of buoyancy and Coriolis forces. However, if it is assumed that the radial velocity scale near the ends is comparable with the primary (longitudinal) velocity scale, the twist may be estimated roughly from eqn (4.31) as

$$\gamma_c = \frac{2(L/D)^{1/5}}{Ek_D Ra_D^{3/5}}$$

for a boundary layer flow, or

$$\gamma_c = \frac{D}{4LEk_D}$$

for an impeded flow. The conditions prescribed in Figs 4.13, 4.14, and 4.15 do not correspond to a boundary layer flow, thus suggesting that the patterns refer to a Coriolis-impeded régime, that is $\gamma_c \gg 1$.

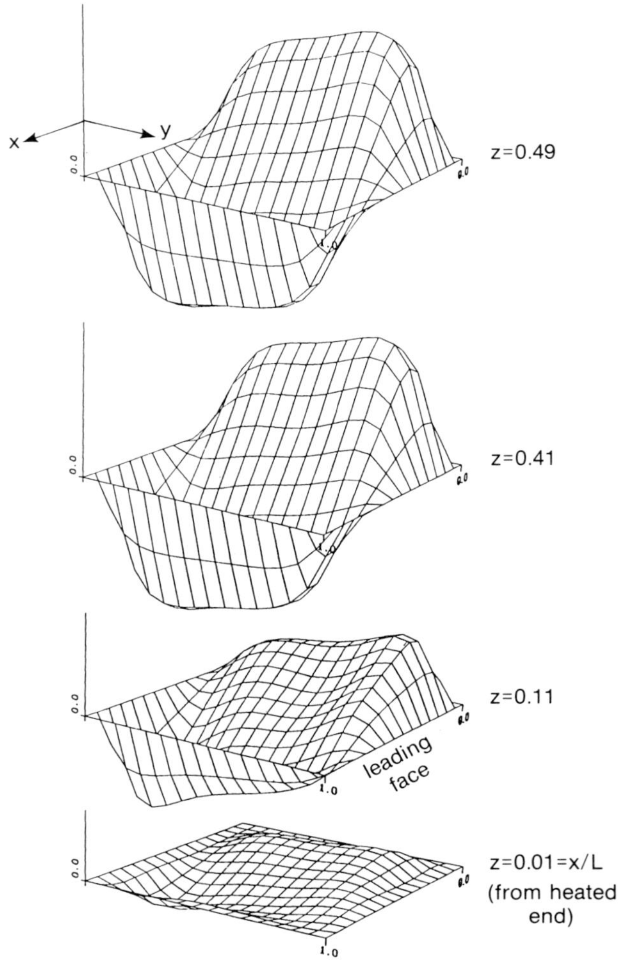

Fig. 4.14. Development of the primary flow in an eccentric thermosyphon for the conditions given in Fig. 4.13 (after Zhao and Lock 1992*a*).

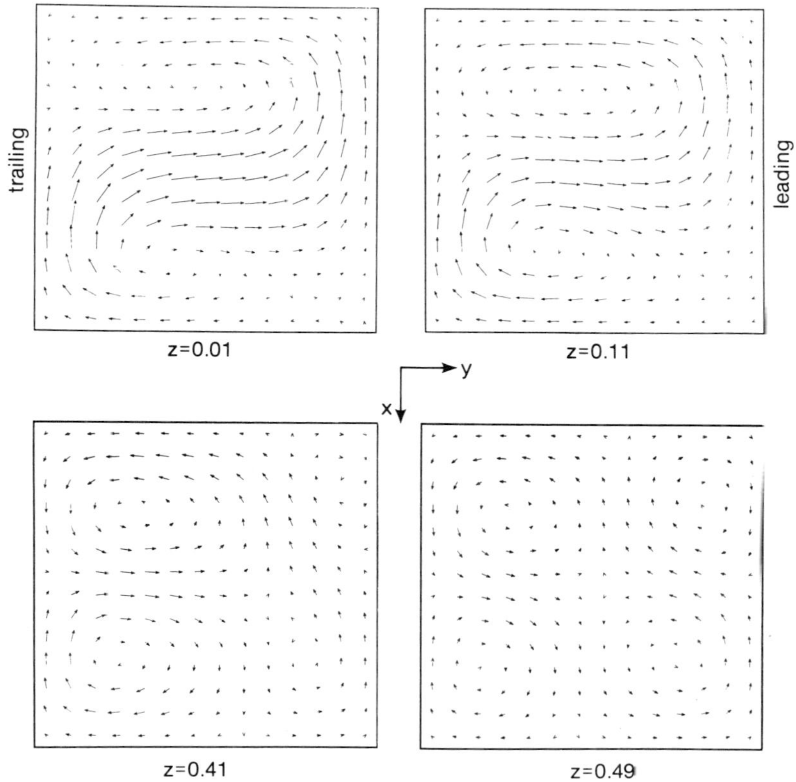

Fig. 4.15. Secondary flow patterns in an eccentric thermosyphon for the conditions given in Fig. 4.13 (after Zhao and Lock 1992*a*).

Figure 4.14 illustrates the development of the primary flow for the same conditions as in Fig. 4.13. The asymmetry noted above is again displayed, the entire flow field being reminiscent of that found in the horizontal thermosyphon or a radial thermosyphon under the condition of a large Coriolis tilt; that is, when the bifilamental structure dominates. The velocity profile is noticeably flat, a reflection of the Coriolis-impeded conditions.

The corresponding secondary flow pattern is illustrated in Fig. 4.15 which reveals how the Coriolis and direct buoyancy effects are combined. At the mid-length plane, the flow field consists essentially of two opposite pairs of vortices. Lateral conduction through the side walls, the leading and trailing faces in particular, thus promotes a flow pattern similar to that seen earlier in the horizontal and radial thermosyphons. Coriolis forces distort this pattern more and more as distance from the mid-length plane increases, the main effect apparently being the reinforcement of tangential motion in the same direction.

From Fig. 4.15, it is evident that this tendency encourages the enlargement of the vortices near the trailing face in the heated section; an anti-symmetric development takes place near the leading face in the cooled section. In the end regions, the resulting emergence of a single pair of vortices, one above the other, is reconciled with an S-shaped flow created by the combination of two comparable effects acting at right angles to each other: a vortex pair attributable to Coriolis acceleration acting alone; and a uniform radial flow attributable to thermal buoyancy acting alone. It is to be expected that the latter would gradually dominate the former if Ek_D were increased. The Coriolis-impeded régime would then be succeeded, not by a Coriolis-enhanced régime, but by one in which the Coriolis effect is negligible. However, this would correspond to very low speeds. For very low Ekman numbers, on the other hand, it is quite possible that the Coriolis-impeded régime will be succeeded by a new flow régime but confirmation, or denial, of this must await further study.

4.3.2 *The concentric tube*

Since there are few, if any, studies of the single-phase, concentric thermosyphon in the open literature, this subsection will be limited to a brief introduction to the device: its flow behaviour and heat transfer performance in particular. At the outset, it is important to recognize that the smallness of the effective radius in a concentric thermosyphon implies a weaker flow than in the eccentric thermosyphon which, in turn, has a weaker flow than the radial thermosyphon driven by direct thermal buoyancy. This suggests that the concentric device will not likely achieve particularly high heat transfer rates without exceptionally high rotational speeds; that is, practical engineering applications may be limited to extremely small Ekman numbers.

A vertically oriented, concentric thermosyphon is shown schematically in Fig. 4.16 which also displays a cylindrical polar coordinate system. Since the latter is more natural to the concentric tube than the rectangular system used earlier, it provides a more convenient framework for the statement of the governing differential equations. The laminar equations of motion are thus represented by eqns (4.5) after the transient terms are removed and appropriate substitutions are made for the driving accelerations; namely,

$$G_Z = -g, \quad G_R = \Omega^2 R, \quad \text{and} \quad G_\Phi = 0. \tag{4.32}$$

The continuity and energy equations respectively take the forms

$$\frac{\partial U}{\partial R} + \frac{U}{R} + \frac{\partial W}{\partial Z} = 0$$

$$U\frac{\partial \theta}{\partial R} + W\frac{\partial \theta}{\partial Z} = \kappa\left(\frac{\partial^2 \theta}{\partial R^2} + \frac{1\partial \theta}{R\partial R} + \frac{\partial^2 \theta}{\partial Z^2}\right) \tag{4.33}$$

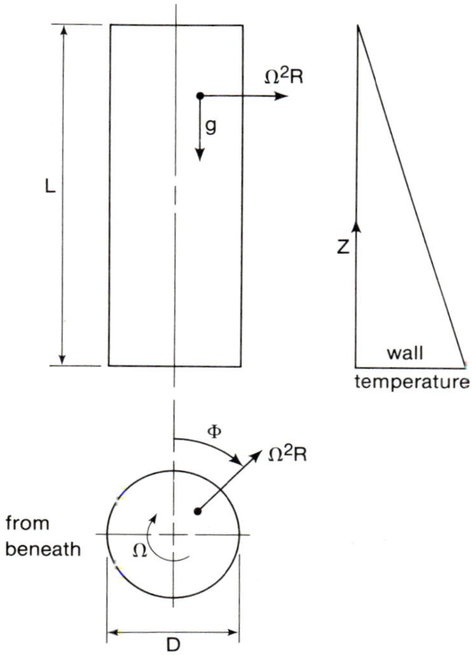

Fig. 4.16. Components of the body force field in a vertical, concentric thermosyphon with a circular cross section.

assuming the flow to be axisymmetric, as suggested by eqns (4.32).

It was noted earlier that the primary flow consists of an annulus of cold fluid moving towards the heated end while a central core of hot fluid moves away from it. This flow, like its counterpart in the eccentric thermosyphon, is driven principally by indirect buoyancy forces. Proceeding similarly, the elimination of pressure between the radial and longitudinal equations of motion permits the balance of the longitudinal buoyancy and viscous terms to be written as

$$\frac{\partial}{\partial Z}(\beta\theta\Omega^2 R) = O\left(\nu\frac{\partial^3 W}{\partial R^3}\right)$$

or

$$\frac{\beta\theta^{c}\Omega^2 D(R^{c})^3}{2\nu L W^{c}} = O(1) \tag{4.34}$$

in a thin, annular boundary layer. The balance between lateral conduction and longitudinal convection in the energy equation reuires that

$$W\frac{\partial\theta}{\partial Z} = O\left(\kappa\frac{\partial^2\theta}{\partial R^2}\right)$$

or

$$\frac{W^{c}(R^{c})^{2}}{\kappa L} = O(1) \tag{4.35}$$

which, taken together with eqn (4.34), yields the thickness and velocity scales as

$$R^{c} = \frac{L}{Ra_{L}^{1/5}}$$

$$W^{c} = \frac{\kappa}{L} Ra_{L}^{2/5} \tag{4.36}$$

where $Ra_{L} = \beta\theta^{c}\Omega^{2}DL^{3}/2\nu\kappa$. These are analogous to eqns (4.27).

Both of these scales, and the equations on which they are based, apply to boundary layer conditions when $R^{c} \ll D/2$. For low enough Rayleigh numbers, the first of eqns (4.36) reveals that the boundary layer will extend across the tube in which case the requirement leading to eqn (4.35) may be relaxed. The longitudinal velocity scale may then be determined directly from eqn (4.34); that is,

$$W^{c} = \frac{\beta\theta^{c}\Omega^{2}D^{4}}{16\nu L} \tag{4.37}$$

in the impeded régime where $R^{c} = O(D/2)$.

The heat transfer rate in a concentric thermosyphon may be estimated from

$$\bar{\dot{q}} = \frac{\dot{m}c_{p}\theta^{c}}{\pi DL} \tag{4.38}$$

where $\dot{m}$ is the axial mass flow rate in either the annulus or the core. Under boundary layer conditions,

$$\dot{m} \simeq \rho\pi DR^{c}W^{c},$$

and hence

$$\overline{Nu}_{D} = C_{1}\left[\left(\frac{D}{L}\right)^{2} Ra_{D}\right]^{1/5} \tag{4.39}$$

where $Ra_{D} = \beta\theta^{c}\Omega^{2}D^{4}/2\nu\kappa$. Similarly, under impeded conditions

$$\dot{m} \simeq \frac{\rho\pi D^{2}W^{c}}{8},$$

and hence

$$\overline{Nu}_{D} = C_{2}\left(\frac{D}{L}\right)^{2} Ra_{D}, \tag{4.40}$$

analogous to eqn (4.30). As noted previously in this chapter, $C_1 = O(1)$ while $C_2 \ll O(1)$.

Equations (4.39) and (4.40) are estimates based on the primary flow and thus fail to take into account the effect of the Coriolis acceleration, especially near the closed ends. In a concentric thermosyphon, the Coriolis effect may be measured through the 'Coriolis swirl' defined by

$$\varepsilon_c = \frac{4\Omega U^c}{\beta\theta^c\Omega^2 D}. \tag{4.41a}$$

This is analogous to the earlier definitions for Coriolis tilt and twist; like the latter, given by eqn (4.31), it suffers from the disadvantage that it cannot be used directly because the radial velocity scale near the ends is again not available from the primary flow. But again assuming the radial velocity scale near the ends to be comparable with the primary (longitudinal) velocity scale, eqn (4.41a) suggests that

$$\varepsilon_c = \frac{4(L/D)^{1/5}}{Ek_D Ra_D^{3/5}} \tag{4.41b}$$

in the boundary layer régime and

$$\varepsilon_c = \frac{D}{16LEk_D} \tag{4.41c}$$

in the impeded régime, thus providing rough estimates of the Coriolis effect.

The Coriolis effect in a concentric thermosyphon is relatively easy to understand. Near the heated end, direct thermal buoyancy drives the fluid radially inwards, thus aiding the primary flow. At the same time, the Coriolis acceleration causes the fluid to move tangentially in the direction of tube rotation, thus creating a swirl with the same direction. The reverse effect occurs near the cold end, again reinforcing the primary flow but producing a swirl in the opposite direction. The strength of the tangential velocity field may thus be expected to increase from zero on the mid-length plane to a maximum near the edge of the Ekman layers. The overall effect on heat transfer may appear to be beneficial but, as in the other thermosyphons considered earlier, it will create a deteriorating performance under Coriolis-impeded conditions. Furthermore, the reconciliation between Coriolis-impeded behaviour and its successor at very low Ekman numbers has yet to be determined.

4.4 Evaporation and condensation in rotating systems

Studies of film condensation in rotating systems, although not numerous, have generally established that the process is very similar to that described earlier for gravitational systems. Work on the rotating disk (Sparrow and Gregg

1960; Espig and Hoyle 1968; Wang and Greif 1981), while not directly applicable to tubes, does indicate that vapour shear is not often important but the influence of surface waves may be very significant. More specifically, Marto and Wanniarachchi (1987) have studied the effect of rotation on condensation inside a horizontal tube. They report that surface ripples and an early transition to turbulent conditions both exert a strong positive influence on heat transfer (and condensation) rates. This suggests that the transitional and turbulent régimes in a concentric tube may be shifted down the Reynolds number scale, with corresponding alterations in the condensation heat transfer coefficients. Should the tube be eccentric, similar arguments apply although the flow details are considerably different.

The first theoretical analysis of film condensation in a concentric tube rotating at high speed appears to be that of Ballback (1969) who applied Nusselt's film theory to a tapered surface and obtained the result

$$\bar{N}u = C(\alpha)\left(\frac{\Omega L D_o}{2\nu}\right)^{1/2}\left(\frac{Pr}{Ja}\right)^{1/4} \tag{4.42}$$

where D_o is the smallest diameter, L is the tube length, and $C(\alpha)$ is a function of tube geometry and the taper angle α. This form has since been discussed by Dhir and Lienhard (1971), Marto (1973), and Daniels and Al-Jumaily (1975). The experimental results of Marto (1976) are in satisfactory agreement with this description. Figure 4.17 gives a comparison for three different fluids.

Alternatively, the problem may be analysed using the pressure and velocity scales implicit in the equations governing motion and continuity in the liquid film. Using primitive variables, these equations are:

$$\frac{\partial P}{\partial Z} = \mu\left(\frac{\partial^2 W}{\partial R^2} + \frac{1}{R}\frac{\partial W}{\partial R}\right), \tag{4.43}$$

for longitudinal momentum,

$$\frac{\partial P}{\partial R} = \Omega^2 R \Delta\rho, \tag{4.44}$$

for lateral momentum, with $\Delta\rho = \rho_L - \rho_V$, and

$$O = \frac{\partial W}{\partial Z} + \frac{U}{R} + \frac{\partial U}{\partial R}, \tag{4.45}$$

for continuity; axisymmetry is assumed to exist. From eqn (4.44), the pressure scale is seen to be

$$P^c = O\left(\frac{\Omega^2 D \delta \Delta\rho}{2}\right) \tag{4.46}$$

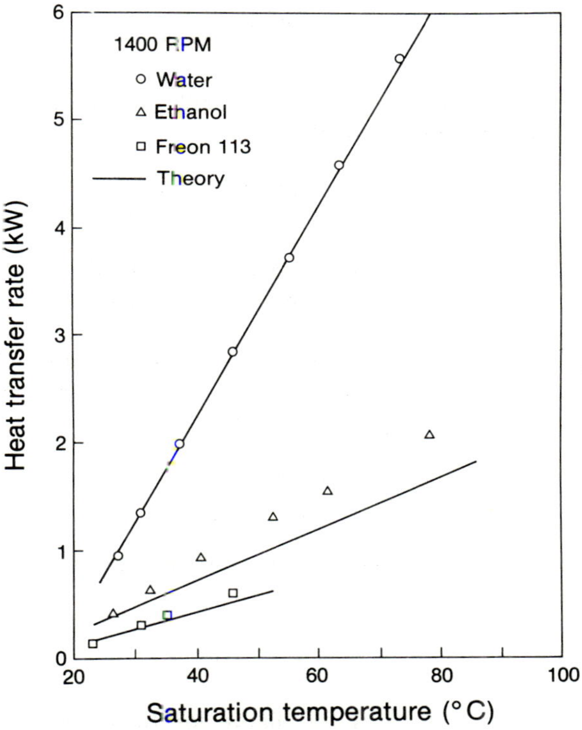

Fig. 4.17. Comparison of theory and experiment for laminar condensation in a concentrically rotating tube (after Marto 1976).

where D is the tube diameter and $\delta \ll D/2$ is the film thickness. It thus follows from eqn (4.43) that the axial velocity scale is

$$W^c = O\left(\frac{\Omega^2 D \delta^3 \Delta\rho}{\mu L}\right) \tag{4.47}$$

where $Z^c = L$ is the length of the (condenser) tube, while eqn (4.45) dictates that

$$W^c = O\left(\frac{U^c L}{\delta}\right) = O\left(\frac{\dot{q} L}{\rho \lambda \delta}\right) \tag{4.48}$$

because the condensation rate ρU^c is directly related to the latent heat transfer rate at the interface. Using eqns (4.47) and (4.48), it is evident that the film thickness is given by

$$\delta = O\left(\frac{\mu \dot{q} L^2}{\Omega^2 D \lambda \rho \Delta\rho}\right)^{1/4} \tag{4.49}$$

and therefore the heat transfer rate through the film, $\dot{q} = k\theta^c/\delta$, where $\theta^c = T_V - T_C$, may be found explicitly. Thus

$$\bar{N}u_L = \frac{\dot{q}L}{\theta^c k} = O\left(\frac{\Omega^2 D L^3 \lambda \rho \Delta\rho}{\mu k \theta^c}\right)^{1/5} \tag{4.50}$$

which reveals the same structure reported independently by several research workers, for example Nimmo and Lepert (1970) and Vasiliev and Khrolenok (1983).

This simple result, although limited to concentric laminar films rotating rapidly, may be written in a number of revealing ways. In terms of the heat transfer coefficient, for example, it becomes

$$h = O\left(\frac{\lambda k^4 \rho \Delta\rho}{\mu} \cdot \frac{\Omega^2 D}{\theta^c L^2}\right)^{1/5}$$

which indicates that the effect of thermophysical properties is essentially determined by a figure of merit

$$Fm = \frac{\rho^2 \lambda k^4}{\mu}$$

if $\Delta\rho \simeq \rho_L$. Table 4.1 compares this figure for three representative fluids suggesting the merit of water. Alternatively, the right-hand side of eqn (4.50) may be rewritten using the dimensionless groups encountered in the previous chapter; thus,

$$\bar{N}u_L = O\left(\frac{ArPr}{Ja}\right)^{1/5} \tag{4.51}$$

in which $ArPr = \Omega^2 D L^3 \Delta\rho/\rho\nu\kappa$ is the Archimedean equivalent of the Rayleigh number. This expression is analogous to eqn (4.39) which applies to boundary layer natural convection in a concentrically rotating tube; it clearly reveals the substantial increase in heat transfer rate (e.g. 4–10 times) which may be attributed to Archimedean buoyancy under typical phase-change conditions, that is, when $Ja \ll 1$. In general, the performance ratio is $(\lambda\Delta\rho/\beta\rho c_p\theta^2)^{1/5}$.

Evaporation in rotating systems has received much more attention (Dakin 1978; Sundell *et al.* 1980; El-Masri and Louis 1978; Verkin *et al.* 1978; Scurlock and Thornton 1977; Kirichenko 1976, 1980, 1982; Mudawwar *et al.* 1985). The work on pool boiling is particularly instructive. Figure 4.18 is a comparison of the temperature and pressure profiles in a pool of liquid ('depth' δ) in stable equilibrium with its vapour at the interface where $P = P_I$ and $T = T_I$.

TABLE 4.1. *Figure of merit ratio of representative fluids (following Marto and Weigel 1982)*

Fluid	Figure of merit ratio $Fm/Fm(H_2O)$
Water	1.0
Ethyl alcohol	0.92×10^{-3}
R113	1.57×10^{-5}

When no body force field is operative, the pressure is uniform throughout at $P_{sat} = P_I = P_W$, as indicated in Fig. 4.18a. Correspondingly, the saturation temperature is uniform at $T_{sat} = T_I$. Now if T_V represents the minimum temperature at which a vapour bubble may exist in equilibrium, the occurrence of nucleate boiling requires the liquid temperature T_L to lie in the range $T_W > T_L > T_V$, where T_W is the wall temperature. If this relation is satisfied only in a region extending from the wall to a point less than δ distant from it, subcooled boiling occurs; otherwise nucleate boiling occurs throughout the pool. Figure 4.18(a) illustrates the temperature profiles (dashed) for both nucleate and subcooled boiling. A body force lying normal to the wall changes this situation by requiring an additional temperature rise before boiling may be initiated. In a gravitational field, for example, the liquid pressure varies linearly, as indicated in Fig. 4.18(b). This creates a linearly increasing saturation temperature curve, and an additional wall temperature rise is then necessary because of the hydrostatic head $P_W - P_I = \delta\rho g$.

In a centrifugal field, mechanical equilibrium of the liquid is described by eqn (4.44), and hence

$$P_W - P_I = \frac{\Omega^2}{2}(R_W^2 - R_I^2)\Delta\rho. \tag{4.52}$$

This curve is suggested in Fig. 4.18(c) along with the corresponding curves[2] for T_{sat} and T_V. Once again, the effect of the body force field is to increase the value of $T_W - T_I$ necessary for incipient boiling. Rearranging the above equation slightly,

$$P_W - P_I = \frac{\Delta\rho\Omega^2}{2}(R_W + R_I)\delta,$$

in accord with eqn (4.46). Together with the Clausius–Clapeyron equation,

[2] They have been drawn roughly parallel in Fig. 4.18 implying that $T_V - T_{sat}$ is fixed. In fact, it is a function of pressure (Verkin *et al.* 1978), but is never very large.

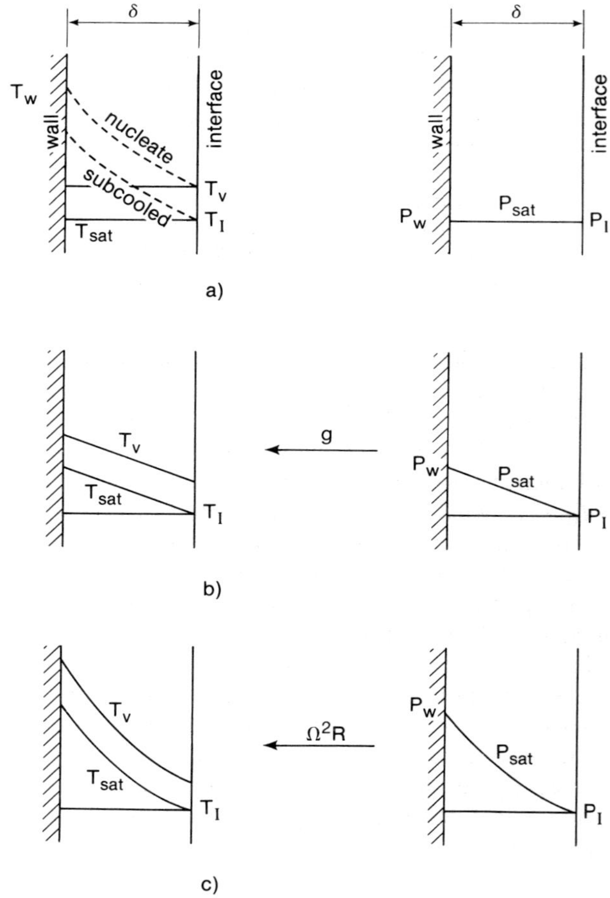

Fig. 4.18. Equilibrium temperature and pressure profiles in a liquid pool subject to force fields: (a) negligible force field; (b) gravitational field; (c) centrifugal field.

this reveals that the additional temperature rise required for incipient boiling in a centrifugal field varies directly with the radius of the tube, the 'depth' of the pool, and the square of the rotational speed. The work of Kirichenko and others demonstrates that the pool boiling curve in a rotating system has the same general shape as that found in gravitational systems, keeping in mind that the centrifugally induced pressure gradient also tends to promote subcooling. The observed growth of the heat transfer coefficient with speed may thus be attributed not to rotation itself but to the increase in saturation pressure which this induces in the pool (Verkin *et al.* 1978). Presumably for the same reason, the critical (burnout) heat flux is also influenced by the pressure field, as reflected in the speed; $\dot{q}_{cr}$ is found to increase monotonically with speed. Beckman and Merte (1965) have observed that while the mechanics of bubble growth is unchanged in a centrifugal field, two opposing influences are at work: rotationally induced pressure tends to suppress nucleation sites and decrease bubble size, while increased buoyancy tends to increase bubble frequency, bubble velocity, and detached growth rate. Broadly speaking, these opposing effects cancel out the direct influence of the acceleration ratio $\Omega^2\mathfrak{R}/g$, but leave a residual influence in the pressure field. For example, there is evidence (Eschweiler *et al.* 1967) to suggest that speed-induced subcooling near a heated surface permits an inverted form of subcooled boiling with ebullition being suppressed near the surface but not distant from it.

4.5 The evaporative, axial thermosyphon

The findings of Section 4.3 provide a natural point of departure in the discussion of the evaporative, axial thermosyphon. Although Archimedean buoyancy replaces thermal buoyancy, the general flow patterns are very similar. Thus, the concentric thermosyphon retains its annular structure with the more dense fluid flowing over the tube wall towards the heated end while less dense fluid returns along a central core. Similarly, the eccentric thermosyphon again consists essentially of a single, bifilamental loop in which the fluid is stably stratified, the more dense fluid flowing over the radially outermost surface towards the heated end. For all versions of the axial thermosyphon, the circulation is driven by the indirect buoyancy forces attributable to variations in the longitudinal pressure field.

Despite this similarity with natural convection, it is important to recall the discussion of Chapter 3 in which emphasis was placed upon the central role of the liquid–vapour interface: its shape, extent, and stability. Many of the heat transfer characteristics of the gravitational, evaporative thermosyphon are directly attributable to interfacial behaviour, and it is reasonable to expect that these characteristics would reappear in modified form in rotating thermosyphons. However, rotation introduces additional phenomena not observed in gravitational thermosyphons. The behaviour of liquid inside a horizontal tube rotating about its own axis provides a good illustration. When the liquid shares the total occupied volume with a small amount of gas, for example, it is known that the buoyant gas bubble thus formed does not behave in an obvious manner. As the tube begins to rotate, the bubble, initially at the top of the tube, is progressively displaced in the direction of rotation until a critical speed is reached when the bubble may divide in two (Benjamin and Pathak 1987). In other circumstances, a single bubble existing at higher speed breaks into smaller bubbles as the speed is progressively reduced (Preziosi and Joseph 1988). Should the gas phase occupy most of the volume, the situation changes yet again. Johnson (1988), for example, describes the sequence of events as the liquid, initially pooled at the bottom of the tube, rises up the wall in a number of different configurations as the speed increases until, eventually, the film coats the entire wall. It will become apparent later that a decreasing speed does not simply reverse the effect of an increasing speed; hysteresis enters the picture.

In thermodynamical terms, Chapter 3 again provides a useful guide. The basic model still consists of a liquid and its vapour in the vicinity of stable equilibrium where the local pressure, in determining the local saturation temperature, also determines the degree to which the local temperature distribution exhibits a metastable departure from saturated conditions, specifically in the superheated liquid. Thus, in rotating thermosyphons, vapour bubbles may again be treated as though they were in mechanical, but not thermal, equilibrium, the principal difference being caused by the replacement of the gravitational field with a centrifugal field which may be expected to affect the processes of nucleation, growth, and detachment. Ultimately, the performance of rotating thermosyphons, open or closed, will be limited by the same phenomena described in Chapter 3; namely, dryout, burnout, and flooding.

4.5.1 *The concentric tube*

It is useful to keep in mind the earlier discussions of phase change when considering the models of partially filled, axial thermosyphons shown in Fig. 4.19 and later in Fig. 4.26. In the first of these, three variations of the concentrically rotating system are presented. Figure 4.19(a) depicts the primary circulation in a tube of fixed diameter under conditions where the rotational speed is high enough to create a coating flow around the walls. Only at high speeds will the liquid cover the wall with an axisymmetric film. The film motion is driven by three effects. Firstly, Archimedean buoyancy acts at the cold end to propel the condensate radially outwards, thereby creating an axial inertial force when the film turns towards the axial direction. Secondly, the axial temperature gradient creates an axial pressure gradient, discussed in Section 4.1. Both of these effects are usually weak and are opposed by the axial shear stress exerted on the surface of the film by the returning vapour. Thirdly, and most importantly, the film gradually thins as it progresses towards the evaporative section which acts as a mass sink. As demonstrated earlier, it is this thinning which creates an axial gradient of 'hydrostatic' pressure and thereby provides the principal drive. Archimedean buoyancy acting near the closed, heated end may or may not influence the primary circulation indicated. While the liquid undoubtedly experiences a radially outward force, the radially inward flow of vapour bubbles may compensate; both will be influenced by Coriolis forces, though oppositely. A weak liquid vortex is suggested in the figure.

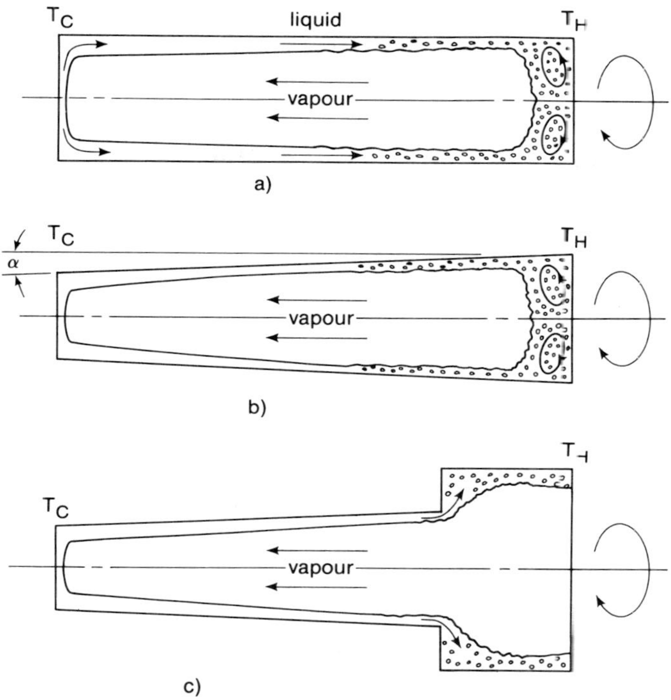

Fig. 4.19. Models of the concentrically rotating, evaporative thermosyphon: (a) parallel wall; (b) tapered wall; (c) tapered and stepped wall.

In Fig. 4.19(b), the same aguments apply but with one important change. As indicated, the tube is tapered from the hot end to the cold end, the effect being to produce an axial acceleration of $\Omega^2 R \sin \alpha$ where α is the taper angle. This contributes to the further thinning of the film and may significantly increase the rate of circulation, depending on the taper chosen. To this effect may be added a step change in the tube diameter, as indicated in Fig. 4.19(c). Such a measure increases the momentum of the film entering the evaporative section and is particularly useful when the heated length must be shorter than the cooled length. In the sketch given, the hot end wall is taken to be dry, thereby suggesting a means by which a slight superheat may be acquired by the vapour.

While film condensation is common in a concentric thermosyphon, it may be replaced by drop condensation if the condenser surface is first treated with a non-wetting agent. This alternative has been considered by Marto (1976) and Marto and Wagenseil (1979) who obtained experimental results using vacuum grease and n-octadecyl mercaptan dissolved in octanoic acid. The improvement in condensation rate was found to be greatest at low speeds, a fact attributable to a weaker centrifugal flattening effect: as the speed is increased, this flattening tends to promote drop coalescence, akin to wetting, and thus condensation reverts to a film mode. A comparative indication of the potential improvement is given in Fig. 4.20. If the tube were tapered, an increase in speed would presumably facilitate the removal of larger drops, thereby offsetting the flattening effect.

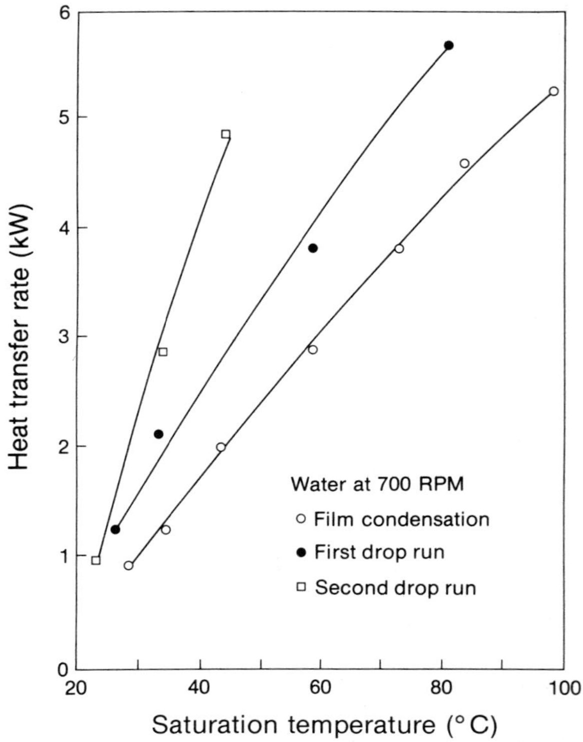

Fig. 4.20. Comparison of film and drop condensation in a concentrically rotating thermosyphon (after Marto 1976).

The performance of the concentric condenser tube may also be enhanced by using longitudinal grooves or fins on the inside (Vasiliev and Khrolenok 1983; Marto and Wagenseil 1979; Marto and Weigel 1982; Marto and Wanniarachchi 1987). At the very least, internal finning has the effect of increasing the available surface area and decreasing the mean film thickness, but it may also provide an impeller effect if the grooves are spiralled. Figure 4.21 shows the data of Vasiliev and Khrolenok (1983) which indicate that under laminar conditions the improvements resulting from six longitudinal grooves 1 mm wide (20% of the bare tube surface) may amount to several hundred per cent. This is consistent with the observations of Marto, but stands in sharp contrast to the conclusion of Nakayama *et al.* (1984b) who used a fine surface structure. More will be said about surface alterations during the discussion of heat pipes in Chapter 5.

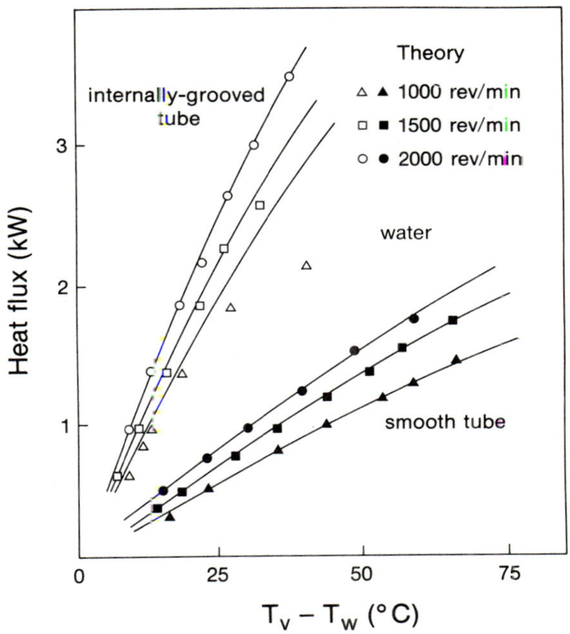

Fig. 4.21. The effect of longitudinal grooves on condenser performance in a concentrically rotating thermosyphon (after Vasiliev and Khrolenok 1983).

Improvements may also be effected in the evaporative section where the general models illustrated in Fig. 4.19 suggest a combination of surface evaporation, immediately downstream of the condenser, and boiling near the closed end. If a long adiabatic section separates the evaporative and condensing sections, the returning vapour, which may be superheated, will tend to heat the condensate towards its saturation temperature; upon entry to the evaporative section, this 'warm' condensate has a greater tendency to boil than if it had returned, subcooled, directly from the condensing section. On the other hand, the centrifugal field tends to induce local subcooling at the evaporator wall and this may reduce, and perhaps eliminate, any boiling.

When boiling is absent, surface evaporation is controlled by a buoyancy-driven circulation which, since $\Omega^2 D \gg 2g$, creates higher evaporation rates than are usually found in gravitational systems. Natural convection is commonly important in these circumstances during which

$$Nu = CRa^m \quad \text{or} \quad \dot{q} \propto \theta^{m+1} \tag{4.53}$$

where $0.25 \leq m \leq 0.35$ (Marto and Gray 1971; Körner 1970; Vasiliev and Khrolenok 1976). Surface evaporation data are included in Fig. 4.22 in the region where superheats are less than about 20 K and heat flux densities are typically under 4×10^4 W/m^2, that is for heat transfer coefficients up to 2 kW/m^2 K. The water and ethanol data in this range indicate that $m \simeq 0.67$, which is remarkably high for a natural convection flow and suggests that some other factor, such as wave interaction, may have been at work. The acetone data, on the other hand, reflect $m = 0$; that is, heat transfer through the film was by pure conduction. Such behaviour resembles classical film condensation in reverse. In general, it is likely that the circulation accompanying surface evaporation will be driven partly by thermal buoyancy and partly by Archimedean buoyancy, and may be modified by interfacial disturbances.

Boiling behaviour in Fig. 4.22 is revealed in nucleate form in the uppermost water data which evidently follow the established relation $\dot{q} \sim \theta^3$. Between the limits set by surface evaporation and nucleate boiling lies a modified boiling régime in which

$$\dot{q} \sim \theta^n$$

where $n > 3$. Vasiliev and Krohlenok (1990) have used the data of Fig. 4.22 to suggest a more general relation in the form

$$h = K\dot{q}^{\alpha_1} \eta_R^{\alpha_2} P^{\alpha_3}$$

which separates the thermal effect, discussed above, from the effects of pressure and acceleration ratio $\eta_R = \Omega^2 \mathfrak{R}/g$. Table 4.2 lists their empirical exponents which reveal the well-known positive effect of pressure increases mentioned earlier. Less obvious is the influence of the acceleration ratio. Surprisingly, this appears to have no effect on surface evaporation. On the other hand, an increase in acceleration ratio slightly reduces the boiling heat transfer rate, presumably by delaying the onset of nucleation at each site.

Before leaving the discussion of high-speed behaviour it is worthwhile noting that axisymmetry does not confer upon the rotational thermosyphon the identical characteristics of the vertical, gravitational system. While both possess an annular condensate film which moves towards the closed end of the evaporator, the behavioural details of the film may be expected to reflect the nature of the driving acceleration; the vertical system is driven by direct buoyancy but the concentric system is driven indirectly. There appear to be few, if any, experiments or analyses which provide comparisons, but a few

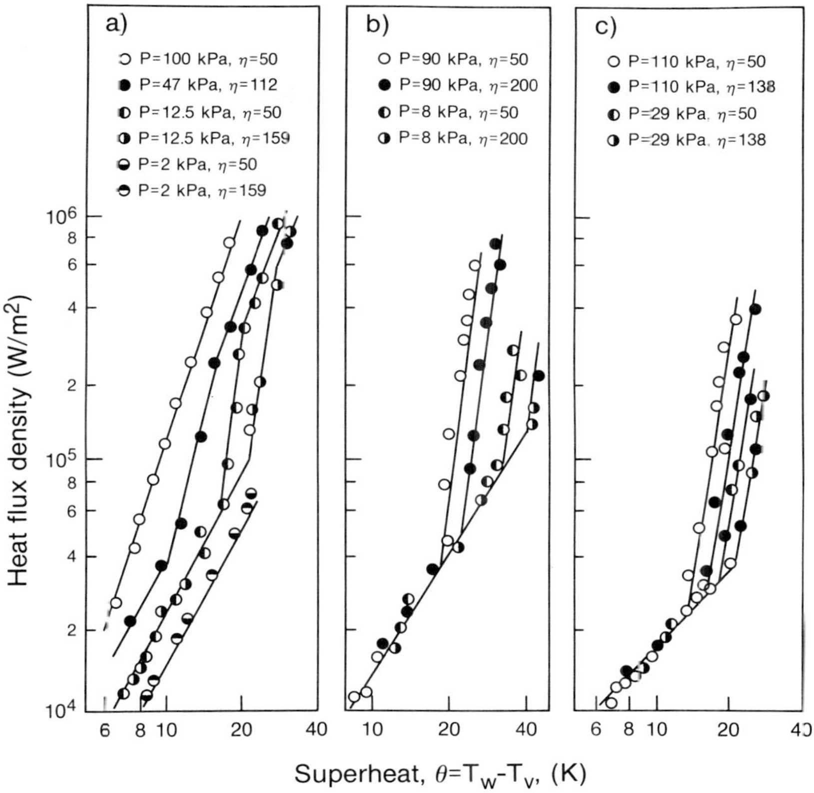

Fig. 4.22. Heat transfer in a high-speed, concentric thermosyphon (after Vasiliev and Krohlenok 1990): (a) water; (b) ethanol; (c) acetone.

TABLE 4.2. *Empirical exponents for evaporation in concentrically rotating films (following Vasiliev and Krohlenok 1990)*

Fluid	Evaporative mode								
	Surface evaporation			Modified boiling			Nucleate boiling		
	α_1	α_2	α_3	α_1	α_2	α_3	α_1	α_2	α_3
Water	0.4	0	0.16	0.78	−0.17	0.21	0.68	−0.14	0.15
Ethanol	0.4	0	0	0.88	−0.11	0.21	–	–	–
Acetone	0	0	0	0.84	−0.10	0.20	–	–	–

general remarks may be in order if only to point up the need for further study. Using the concentrically rotating, vertical thermosyphon as an illustration, it is clear that superimposed rotation would tend to redistribute the boiling pool up the evaporator wall, thereby influencing the dryout condition. Centrifugal acceleration may also act to suppress some nucleation sites along the evaporator wall while at the same time creating a higher void fraction near the tube axis; the first of these effects would lead to higher wall temperatures (or lower heat fluxes), while the second would promote vapour pocketing and plug flow, thereby influencing flooding behaviour. On the other hand, if the charge were smaller, the returning condensate film may not behave in the way described by Negishi and Sawada for the stationary tube, and illustrated in Fig. 3.34. Descending rivulets would presumably be flattened by centrifugal force and may exhibit entirely different surface wave patterns, given the larger Bond number. The centrifugally induced delay or suppression of nucleate boiling places a greater emphasis on surface evaporation, particularly at low heat fluxes, while centrifugal and Coriolis forces would modify the radially inward ejection of droplets coming from burst bubbles under high heat flux conditions. It is thus likely that rotation will introduce some revisions to the circulation and evaporation patterns familiar in the gravitational thermosyphon.

A more radical departure from gravitation-like behaviour occurs at lower speeds. The condition $\Omega^2 D \gg 2g$ does not then apply and the condensate film is no longer distributed uniformly or steadily over the tube surface. This situation has been considered by Nakayama *et al.* (1984*a*) who undertook a comprehensive study of behaviour in a horizontal tube rotating concentrically. Figure 4.23 shows the film distribution they observed in a visualization experiment. As indicated, an annular coating flow was not formed until a speed of 1100 rev/min had been reached. This corresponds to $\Omega^2 D = 38g$ ($\eta_R = \Omega^2 D/2g = 19$) but the threshold value was found to depend upon the amount of liquid occupying the tube, as might be expected. It was also discovered that 'rigid' rotation of the annulus did not occur suddenly but required a transitional régime which began at lower speed as a three-dimensional development near one end. Once formed, however, the annular configuration proved to be very stable with respect to further changes in speed. In particular, speed reductions *below* the point of initial formation did not produce an immediate reversion to the pooled structure. Evidently, gravity tends to promote an azimuthal perturbation in the coating thickness which is flattened by centrifugal force and surface tension until $\Omega = O(2g/D)^{1/2}$ and the liquid suddenly falls to the bottom of the tube. Behaviour is thus hysteretic. Figure 4.24 illustrates the effect of charge on this behaviour.

From these flow observations, Nakayama *et al.* were able to offer plausible explanations of their heat transfer data. In particular, they noted the existence of an optimum charge of 10–14% of the tube volume for low speeds (e.g.

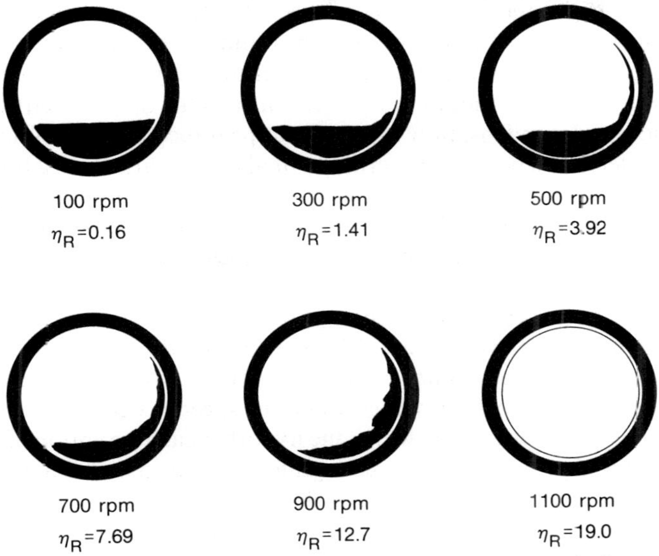

Fig. 4.23. Effect of speed increase on film distribution in a concentrically rotating tube: charge—20% of volume (following Nakayama *et al.* 1984*a*).

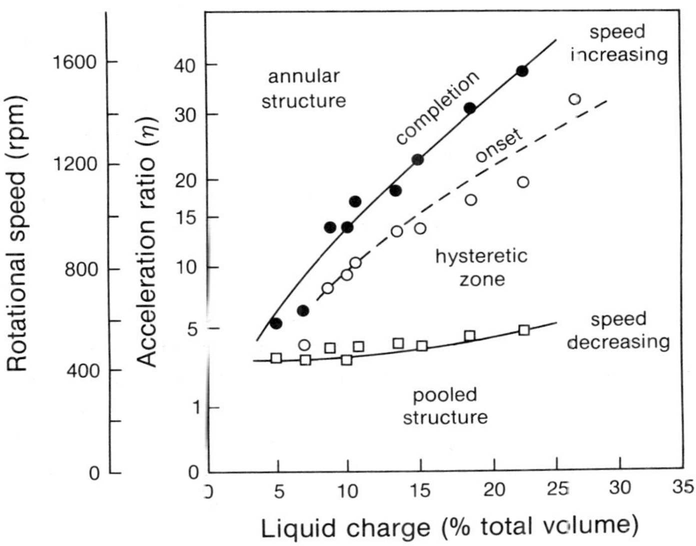

Fig. 4.24. Effect of charge on R113 film hysteresis in horizontal, concentrically rotating tube (after Nakayama *et al.* 1984*a*).

$\Omega^2 D < 13g$). Below this charge, the film evidently did not cover the entire evaporator wall, a shortcoming which rapidly became worse with the establishment of an annulus and the occurrence of dryout. For much greater charges, the volume of liquid created a thicker film thereby reducing both evaporation and condensation rates; the appearance of an annulus is then generally deleterious because it marks the final disappearance of a thinner film which, although spread over only a fraction of the tube circumference, yields higher local evaporation and condensation rates. This leads to the hypothesis that thermal performance may be maximized with respect to the charge in the low-speed range: a thin coating causes dryout with a suboptimal charge, while annular flow may occur with a superoptimal charge.

Similar conclusions were reached by Katsuta *et al.* (1984) who suggested empirical correlations to define the boundaries of the hysteretic zone in which the annular structure may only exist when the speed is decreasing. They also presented heat transfer data on which the hypothetical description of Fig. 4.25 is based. This figure reflects the changing structure and behaviour of the condensate as the speed is increased or decreased in the evaporator. Starting from rest, when the condensate is pooled in the bottom of the tube, the initial effect of rotation is to add a washing or sweeping process which evidently augments the evaporation heat transfer taking place on the surface of the pool. At the same time, the rising wall lifts a condensate film which, in spreading, adds further to the evaporation rate. Figure 4.23 suggests that this film is initially thin and spreads rapidly when $\eta_R = O(1)$. Beyond this speed range, however, the film thickens without spreading, thus causing a decrease in heat transfer rate while the pool takes on a crescent shape skewed to the vertical.

Further increases in speed have little effect until the pool begins its transition to an annulus. Over a considerable range of speed this transitional process continues with the effect that the overall thermal resistance, as dictated by the effective film thickness, gradually increases. The corresponding decrease in heat transfer rate continues until the pool liquid has been completely redistributed as an annulus. The speed at which the annulus is completed, and the corresponding heat transfer asymptote, both depend upon the charge, as suggested in Fig. 4.25. It will be recalled that a subsequent decrease in speed leaves the annular structure intact until, at a speed much lower than at the beginning of the initial transition, it collapses abruptly into a pooled structure. This is also suggested in Fig. 4.25. Similar arguments apply to condenser heat transfer except that maximal behaviour requires a slightly different explanation; washing again produces an enhancement as the speed increases from rest, but the development of the crescent-shaped pool now represents a reduction in the extent of the thin condensate film formed elsewhere on the tube surface, and thus reduces the heat transfer rate. Near the rest speed, the heat flux density through this film is usually greater than through the evaporator pool if surface evaporation alone takes place; the

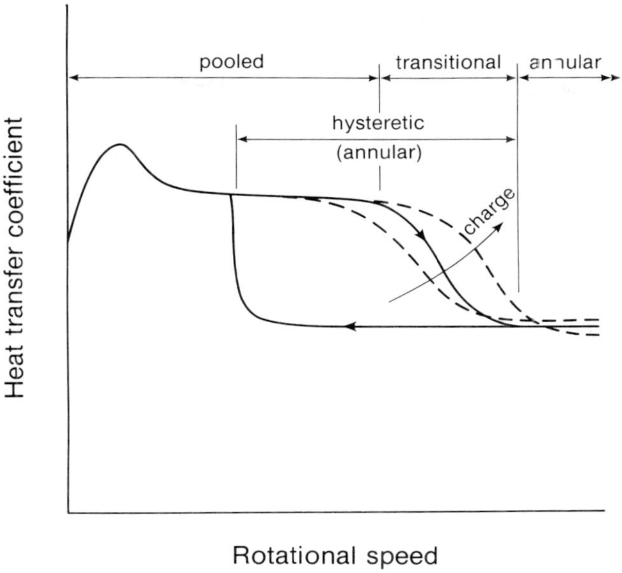

Fig. 4.25. Effect of speed on evaporator heat transfer coefficient in a horizontal, concentric thermosyphon.

condenser then outperforms the evaporator. However, if boiling takes place in the evaporator pool this difference in heat flux densities will be reduced.

4.5.2 *The eccentric tube*

The discussion of the pool brings us naturally to the other type of axial thermosyphon: namely, the eccentric variety. Here the tube rotates about an axis parallel to its own but displaced from it by an amount $\mathfrak{R}$, the eccentricity. When the rotational speed is high enough ($\Omega^2\mathfrak{R} \gg g$) and the eccentricity great enough ($2\mathfrak{R} \gg D$) the flow system resembles that in a horizontal tube, except that the driving acceleration is much greater. The resulting flow system is modelled in Fig. 4.26(a) which depicts a long liquid pool supplied by a film draining circumferentially in the condenser. The pool is driven by the same three mechanisms identified for the horizontal tube. At and near the cold end, Archimedean buoyancy (here modified by a Coriolis force) moves the condensate film outwards from the centre of rotation thereby supplying and driving the upstream end of the pool. This condensate flow joins with that accumulating from the side walls to create a 'hydrostatic' pressure gradient driving the pool towards the evaporator where its depth is further reduced by evaporation; Archimedean buoyancy thus acts indirectly over the length of the pool. In addition, the longitudinal temperature gradient in the pool generates

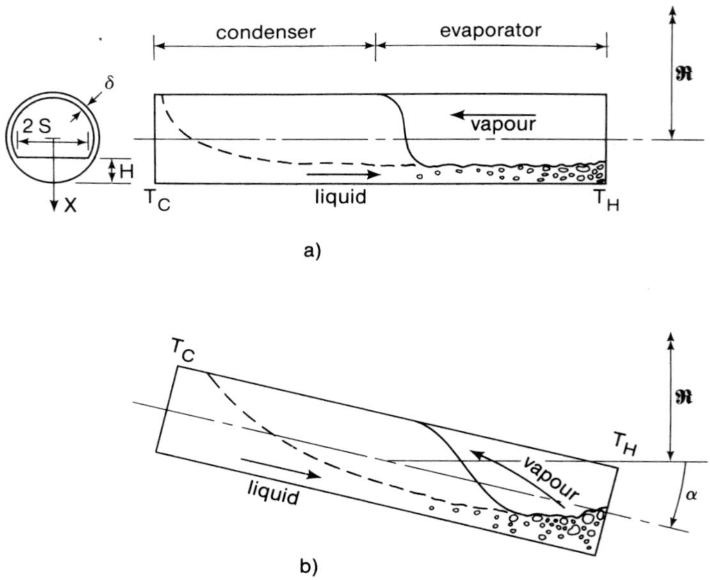

Fig. 4.26. Model of eccentrically rotating, evaporative thermosyphon: (a) parallel axis; (b) inclined axis.

a pressure gradient which assists the drive from the cold end to the hot end; that is, thermal buoyancy also acts indirectly.

In the condensing section, the heat transfer rate is largely governed by drop or film condensation on the curved sides of the tube; the pool thus plays a dependent role, although some vapour will condense on the free surface, the width of which is taken to be $2S$. The Nusselt film analysis developed in Chapter 3 provides an adequate and conservative description of behaviour under these circumstances for which the scales of the film thickness δ and circumferential velocity U^c are given, respectively, by

$$\delta = O\left(\frac{\pi D \mu \dot{q}}{2\Omega^2 \mathfrak{R} \lambda \rho \Delta\rho}\right)^{1/3} \tag{4.54}$$

and

$$U^c = O\left(\frac{\dot{q}\pi D}{2\rho\lambda}\right)^{2/3}\left(\frac{\Omega^2 \mathfrak{R} \Delta\rho}{\mu}\right)^{1/3}. \tag{4.55}$$

It is thus found that heat transfer is described by the relation

$$\bar{N}u_D = \frac{\dot{q}D}{k\theta^c} = C\left(Ar_D \frac{Pr}{Ja}\right)^{1/4}. \tag{4.56}$$

These expressions must be reconciled with the size and shape of the pool which carries away the condensate.

Conservation of mass and energy demand that the total condensation heat flux be given by

$$\dot{Q} = \dot{q}\pi DL = O(2\rho\lambda L\delta U^{c}) = O(\rho\lambda HSW^{c}) \tag{4.57}$$

where W^{c} is the longitudinal velocity scale in a pool of maximum depth H. The properties of a circle require that $S \simeq (HD)^{1/2}$ for a shallow pool, and therefore eqn (4.57) provides one suitable relation between the two unknowns W^{c} and H. The other relation is provided by the equation of motion for the pool. Following arguments similar to those which led to eqn (4.47), it may be shown that

$$W^{c} = O\left(\frac{\Omega^{2}\mathfrak{R}H^{3}\Delta\rho}{\mu L}\right) \tag{4.58}$$

and hence

$$H = O\left(\frac{\mu\pi\dot{q}L^{2}D^{1/2}}{\Omega^{2}\mathfrak{R}\lambda\rho\Delta\rho}\right)^{2/9} \tag{4.59}$$

which may be compared with the film thickness by using eqn (4.54). Thus,

$$\frac{\delta}{H} = \left(\frac{\mu\pi\dot{q}D^{2}}{8L^{4}\Omega^{2}\mathfrak{R}\lambda\rho\Delta\rho}\right)^{1/9} = K_{1}\left(\frac{D}{L}\right)^{4/9}\left(\frac{Ja}{Ar_{D}Pr}\right)^{1/12} \tag{4.60}$$

where $K_{1} = O(1)$. It is thus apparent that the pool depth will normally be much greater than the film thickness, consistent with the tacit assumption that condensation on the pool may be ignored. From eqn (4.54) the film thickness may also be expressed by

$$\frac{\delta}{D} = K_{2}\left(\frac{Ja}{Ar_{D}Pr}\right)^{1/4} \tag{4.61}$$

in which $K_{2} = O(1)$. Also the pool depth, as a fraction of the tube diameter, is given by

$$\frac{H}{D} = \frac{K_{2}}{K_{1}}\left(\frac{L}{D}\right)^{4/9}\left(\frac{Ja}{Ar_{D}Pr}\right)^{1/6}. \tag{4.62}$$

These last three equations affirm the model depicted in Fig. 4.26 under the condition $Ar_{D}Pr/Ja \gg (L/D)^{8/3}$, at least in the condenser, where the detailed calculations by Gi and Maezawa (1990) are shown to be in agreement with experiment. Similar arguments apply to the evaporator but are essentially limited to surface evaporation from the pool itself and thus have more limited scope.

In both the condenser and the evaporator the lateral dimensions of the pool (H and S) decrease in the direction of the temperature gradient, but this is especially true in the evaporator where the pool is fed less, if at all, by the side wall film, and loses mass by evaporation. For small filling charges, the evaporator pool may be little more than a thin rivulet of liquid which is incapable of transmitting large heat fluxes by surface evaporation alone. The work of Negishi and Sawada (1983) on the horizontal thermosyphon suggests that under these conditions the evaporator pool bears only a qualitative resemblance to the laminar stream analysed above. Its surface will not remain smooth as the heat flux increases, and any turbulent mixing will likely be accompanied by nucleate boiling which, if it becomes vigorous enough, may lead to the complete disruption of the pool and the frequent bursting of vapour bubbles. Such a quencing mechanism provides some compensation for the absence of a continuous film by creating a spray of droplets which impinge on the curved superheated surface opposite where they are able to evaporate by a process resembling drop condensation in reverse; it is a form of spray cooling. Without this, the tube would rapidly experience dryout over most of the evaporator surface.

The high-speed data of Gi and Maezawa (1990) throw some light on the problem. Figures 4.27 and 4.28 show their measurements of heat transfer coefficient in the evaporator and condenser, respectively. It is clear from these figures that as the acceleration ratio increases above 5:1, the two sets of data diverge. In the condenser, the increased centrifugal force produces a thinner circumferential film, as eqn (4.54) would suggest, but no such change occurs in the evaporator where the heat transfer coefficient remains unaltered beyond the value obtained $\eta_R = 5$. This suggests an evaporator heat transfer mechanism which is not only different but largely independent of the centrifugal force field, at least in the range covered. The behaviour may be attributable to simple 'forced convection' in the evaporator pool but with such a small charge the quenching process mentioned above may provide an alternative explanation, provided the velocity of the ejected droplets is high enough for them to reach the tube surface opposite. Figure 4.27 suggests a slight decline in evaporator performance with speed, particularly at the highest speeds, and this may be the result of a centrifugal restraint which would have little effect until the droplets began to fall back into the vigorously boiling tongue. However, it is important to recall that an increased speed may also lead to the suppression of nucleate boiling.

Figures 4.27 and 4.28 also reveal a dramatic change in behaviour at low speed, specifically near $\eta_R = 1$; the axis of rotation was horizontal. Since the change occurs in both the condenser and the evaporator it is reasonable to suppose that it has a common cause; and since the existence of an extensive evaporator film is in some doubt, it is likely that the explanation is to be found in pool behaviour. Figure 4.29 suggests the motion and distribution of the

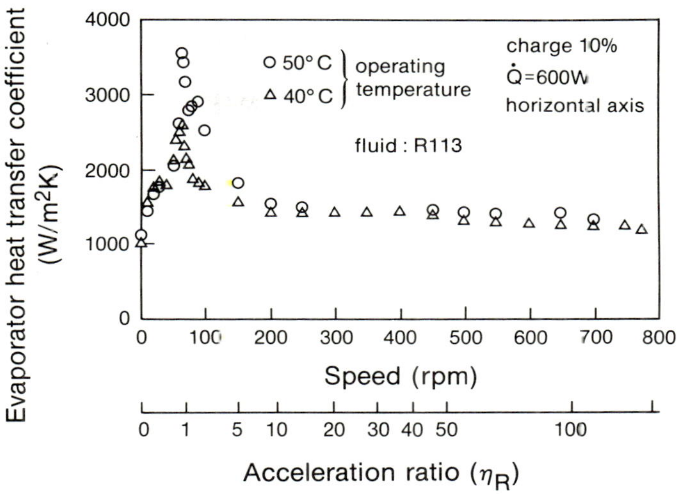

Fig. 4.27. Effect of speed on evaporator heat transfer coefficient in a horizontal, eccentrically rotating thermosyphon (after Gi and Maezawa 1990).

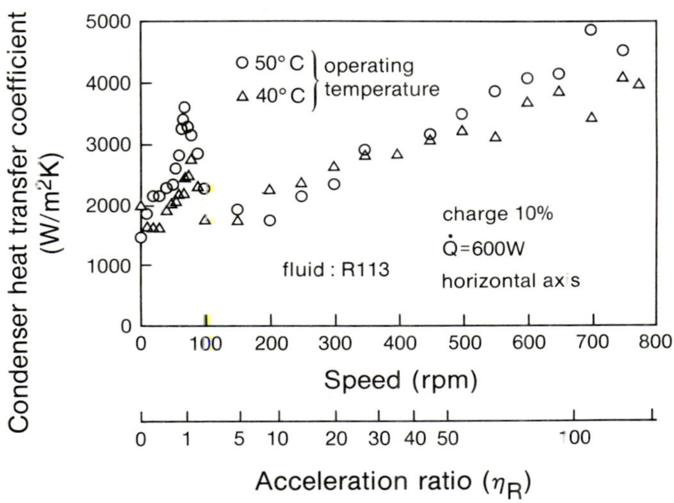

Fig. 4.28. Effect of speed on condenser heat transfer coefficient in a horizontal, eccentrically rotating thermosyphon (after Gi and Maezawa 1990).

pool liquid at various speeds above and below $\Omega = (g/\mathfrak{R})^{1/2}$. At very low speed, the pool simply rotates around the tube circumference in a quasi-steady manner, its direction being opposite to that of the shaft; this is illustrated in Fig. 4.29(c). On the other hand, the pool remains fixed on the radially outermost side of the tube when the speed is very high; this is seen in Fig. 4.29(a). The heat transfer rate in the evaporator at very low speed is essentially the same as that observed at very high speed. It thus appears that these rates are controlled by the 'stationary' pool.

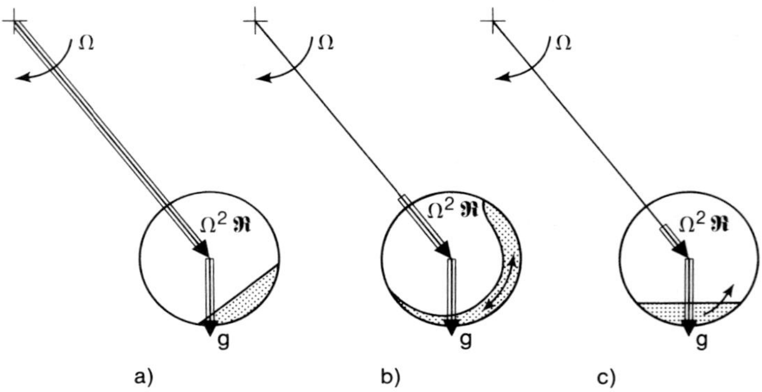

Fig. 4.29. Effect of rotational speed on liquid pool distribution in an eccentric tube: (a) high speed: $\Omega^2\mathfrak{R} \gg g$; (b) low speed: $\Omega^2\mathfrak{R} \gtrsim g$; (c) very low speed: $\Omega^2\mathfrak{R} \ll g$.

As the rotational speed increases from zero, the circumferential motion of the pool creates a washing effect which increases steadily up to $\Omega \simeq (g/\mathfrak{R})^{1/2}$. Likewise, as the speed is decreased from $\Omega \gg (g/\mathfrak{R})^{1/2}$, the amplitude of an oscillatory wash increases until the speed is again given by $\Omega \simeq (g/\mathfrak{R})^{1/2}$. The resulting peak in heat transfer may not be explained in terms of a transient laminar flow, as examination of the governing differential equations shows. Instead, the maximal effect might be described as a type of separation in which the oscillatory wash depicted in Fig. 4.29(b) ultimately reaches half-way around the tube, the liquid there falling away from the radially innermost surface, to be reattached later in the cycle. This behaviour has been observed visually by Gi and Maezawa (1990), but its effect on heat transfer is not fully understood. In part, it may represent the growth of a vigorous recirculation, as often happens in separated flows; that is, turbulent mixing may be induced in a spreading pool. Alternatively, or perhaps additionally, it may reflect the temporary detachment of the pool from the tube wall, with an attendant cascade or waterfall effect.

When the eccentricity is small, the situation changes yet again, as the high-speed work of Schneller *et al.* (1982) has revealed. Figure 4.30 demonstrates that while the interface is essentially flat when $\mathfrak{R} \gg D/2$, it is more generally described by a circle centred on the axis of rotation. If the interfacial radius R_c corresponds to a concentric film of thickness δ_c, then

$$\delta_c = \frac{D}{2} - R_c.$$

Thus, when $\mathfrak{R} \gg \delta_c$, the interface creates a crescent-shaped pool which becomes an eccentric annulus when $\mathfrak{R} < \delta_c$. In the evaporator, which is most likely to exhibit the controlling thermal resistance in the eccentric tube, this model may be applied to surface evaporation using the familiar Nusselt analysis. The result is shown in Fig. 4.31. For a very small eccentricity, which might be produced by manufacturing errors or by bending of a hollow shaft, for example, the heat transfer rate actually improves, but beyond a sharp maximum it declines rapidly to beneath the concentric value. Very small eccentricities accentuate the trend noted earlier for large eccentricities in the presence of a continuous film; namely, that the decrease in heat transfer attributable to a thicker film over the surface furthest from the axis of rotation is more than offset by the gain created by a thinner film nearer the axis. When $\mathfrak{R} = \delta_c$, the net effect is maximized with an incipient dryout condition beyond which the surface area of the film progressively shrinks with increasing eccentricity.

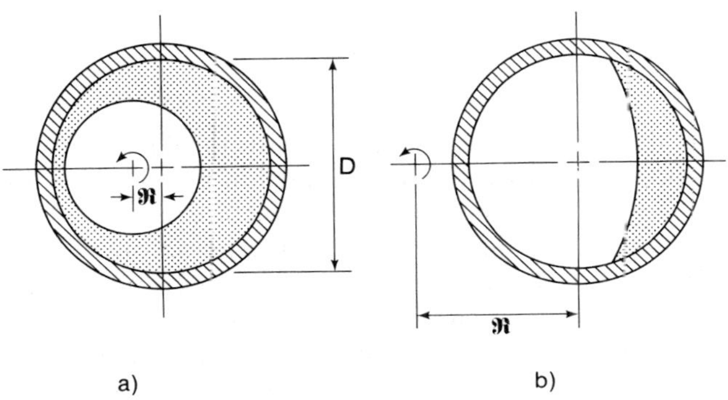

Fig. 4.30. High-speed film shape for small (a) and large (b) eccentricities (after Schneller *et al.* 1984).

Finally, it is worth briefly mentioning the effect of tilt on the axial thermosyphon. This is illustrated in Fig. 4.26(b) where an eccentric tube is shown tilted (in a radial–axial plane) at an angle α to the axis of rotation. With the hot end being further from the axis than the cold end, the longitudinal

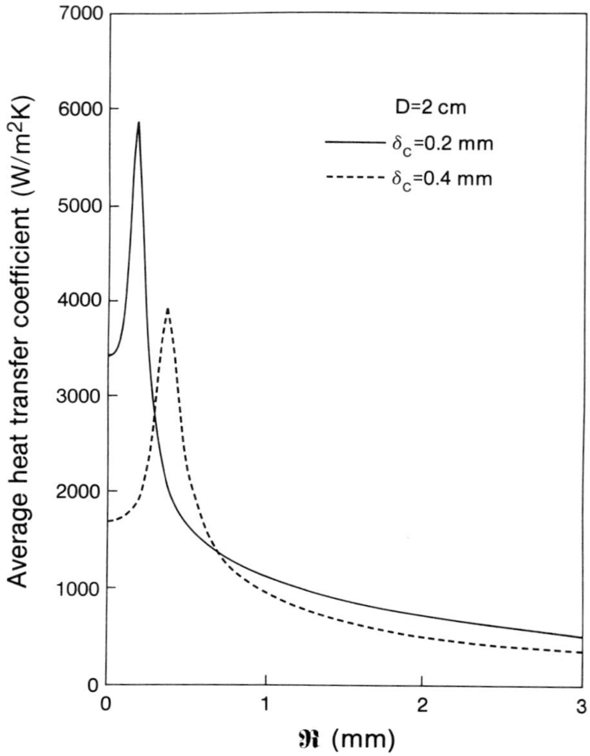

Fig. 4.31. Effect of eccentricity on the average evaporator heat transfer coefficient (after Schneller *et al.* 1984).

component of the centrifugal acceleration assists in bifilamental circulation in much the same way as when the tube is tilted in a gravitational field. Thus, although the Coriolis acceleration alters the picture near the closed ends, the observations made in Section 3.4 may be applied broadly to the tilted, rotating tube provided the speed and eccentricity are both great enough. It is to be expected that behaviour would be more complicated for small eccentricities and low speeds but few, if any, studies of such conditions have been published.

The work of Chen (1987, 1990) demonstrates that tilting does indeed produce the expected improvement at high speed. The heat transfer coefficients measured for a straight tube, with and without tilt, were compared with those for a slightly bent tube having a tilted condenser but an untilted evaporator. Not surprisingly, the straight, tilted tube performed best, followed by the bent tube. Given the higher heat transfer coefficient in the condensing section, the bent tube performance would probably lie even closer to the straight, tilted tube performance if the vaporator, rather than the condenser, had been the tilted section. Most of Chen's data were for $\Omega > (g/\mathfrak{R}_m)^{1/2}$, where $\mathfrak{R}_m$ is the mean eccentricity, and reveal the same trend found in untilted tubes; specifically, an *overall* heat transfer coefficient which decreases monotonically with increasing speed. However, the data presented do not show a maximum in the heat transfer coefficient below $\Omega(=g/\mathfrak{R}_m)^{1/2}$, at least in the speed range displayed. When this curious fact is combined with the observation that the heat transfer rate was found to be negligible below a reported 'starting speed', it suggests that a peak heat transfer rate did occur when $\Omega < (g/\mathfrak{R}_m)^{1/2}$. The precise explanation may lie in the oscillatory effect of the gravitational acceleration component $g \sin \alpha$ at very low speeds but further clarification is required because the washing phenomenon discussed earlier will be felt first in the evaporator (where the eccentricity is greatest) as the speed rises, but decay last in the condenser; the maximum region may thus be flattened and extended.

Even at high speed, an increasing tilt may be expected to modify behaviour in an eccentric thermosyphon substantially. Firstly, the pool at the end of the evaporator will deepen, thus exacerbating the problem of subcooling promoted by the radial pressure gradient; at the same time, however, the wetted surface area in the evaporator will increase, providing some compensation. Secondly, the condensate film will be subject to an increasing longitudinal acceleration driving it towards, and into, the evaporator. Overall, the increased circulation will bring improved thermal performance as axial velocities increase and the condensate film wets more of the tube surface. The observations of Niekawa *et al.* (1982), together with similar observations on the concentric tube, suggest that this improvement could be further enhanced by longitudinal grooving which introduces capillary effects. When the inclination becomes very large (i.e. $\alpha \to \pi/2$), the behaviour again changes dramatically. The device is then more appropriately described as a radial thermosyphon.

4.6 The evaporative, radial thermosyphon

In section 4.2 it was observed that the single-phase, radial thermosyphon has certain similarities with the single-phase, gravitational thermosyphon if the latter is tilted. This comparison was facilitated through the Coriolis tilt α_c which was shown to vary over a wide range. With small values of α_c (i.e. $\alpha_c \ll 1$), the effect of the Coriolis acceleration was to improve the heat transfer rate through the development of a secondary flow. For large tilts, on the other hand, the Coriolis effect produced a reduction in performance because of its influence on the primary flow. It is natural to expect that comparisons between the evaporative, radial thermosyphon and its tilted, gravitational counterpart woud lead to similar conclusions but, as noted in Section 4.4, a centrifugal field is also capable of altering the boiling process. Moreover, a more appropriate definition of the Coriolis tilt is required for evaporative behaviour.

As a starting point, it is worth adapting the model of the vertical, evaporative thermosyphon to radial conditions, ignoring the fact that the centrifugal force will be higher in the evaporator than in the condenser. Figure 4.32(a) provides a representation of the Holzwarth–Schmidt system originally proposed for the cooling of gas turbine blades, that is for high-speed application. As indicated, liquid flows into the blade cavity from inside a rotor or drum, radius $\mathfrak{R}$, where it collects under the influence of centrifugal force. The annulus of liquid thus flowing into the blade will be eccentric, with the leading face exhibiting the smallest thickness; that is, the Coriolis acceleration introduces a secondary motion which tends to create a thicker film along the trailing face. After evaporation, the liquid returns as a vapour core on which the Coriolis acceleration acts in the opposite direction, thus reinforcing the eccentric shape of the annulus. At first glance, this model is reminiscent of the slightly tilted gravitational thermosyphon, keeping in mind that the secondary flow is caused not by lateral buoyancy forces but by lateral inertia forces attributable to the primary flow. Figure 4.32(b) shows the corresponding model of the closed system, the main difference being in the external use of a secondary coolant to remove heat from the inward-projecting condensing section.

It is clear from Chapter 3 that the details of this model, and the associated hydrodynamic and thermal behaviour, will vary enormously. More will be said about these later. At this point, it is only necessary to recall that the most dominant feature of the evaporative thermosyphon is the liquid–vapour interface which, as Fig. 4.32 suggests, principally occupies the surface of an extensive liquid film running over the full length of the condenser and part way into the evaporator. A measure of the effect of the Coriolis acceleration acting on the film beneath the interface is provided by the Coriolis tilt which, under these conditions, may be defined by

$$\alpha_c = \frac{2\Omega U^c \rho}{\Omega^2 \mathfrak{R} \Delta\rho} \tag{4.63}$$

where U^c is the longitudinal (radial) velocity scale in the liquid film. This is analogous to the definition used for the single-phase system in eqn (4.7) which incorporates the thermal buoyancy $\Omega^2\mathfrak{R}\beta\theta^c$ rather than the Archimedean buoyancy $\Omega^2\mathfrak{R}\Delta\rho/\rho$. To determine the Coriolis tilt it is first necessary to find the velocity scale from the equations governing the motion of the film.

Consider the flow of liquid over the flanking (side) walls where the Coriolis acceleration may be expected to have its greatest effect on secondary motion. Let X, Y and Z be the rotating coordinates representing the radial, tangential, and axial directions, respectively. From the radial equation of motion under

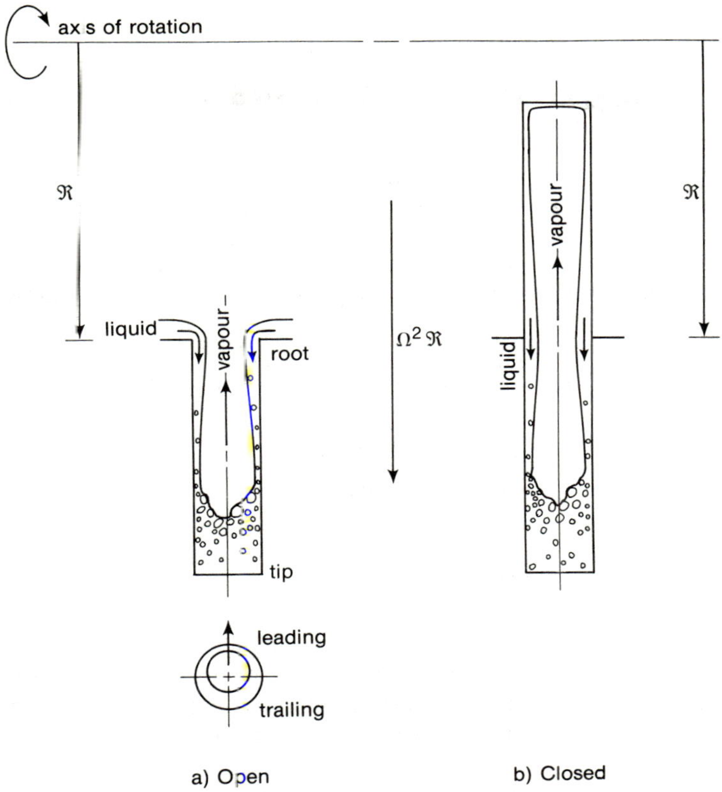

Fig. 4.32. Basic models of open (a) and closed (b) radial, evaporative thermosyphons.

laminar conditions, a balance between the Archimedean buoyancy and principal viscous forces acting on the film requires that

$$U^c = O\left(\frac{\Omega^2 \Re \Delta\rho \delta^2}{\mu}\right), \tag{4.64}$$

where δ is the unknown film thickness. In the tangential equation of motion, on the other hand, the viscous force is balanced by the Coriolis force, and hence the tangential velocity scale is given by

$$V^c = O\left(\frac{2U^c \Omega \delta^2}{\nu}\right). \tag{4.65}$$

The axial equation of motion adds no further information but the interfacial heat balance yields the axial velocity scale as

$$W^{c} = O\left(\frac{\dot{q}}{\rho\lambda}\right). \tag{4.66}$$

Finally, the unknown scales are related through the continuity equation which may be represented by

$$W^{c} = O\left(\frac{\delta U^{c}}{L}\right) + O\left(\frac{2\delta V^{c}}{\pi D}\right) \tag{4.67}$$

in which L and D are the tube length and diameter, respectively. Assuming the radial term on the right-hand side of eqn (4.67) to be larger, it follows that

$$\frac{\delta}{L} = O\left(\frac{\mu k \theta^{c}}{\Omega^{2}\mathfrak{R}L^{3}\lambda\rho\Delta\rho}\right)^{1/4} = O\left(\frac{Ja}{Ar_{L}Pr}\right)^{1/4} \tag{4.68}$$

and

$$\bar{N}u_{L} = \frac{\dot{q}L}{k\theta^{c}} = O\left(\frac{\Omega^{2}\mathfrak{R}L^{3}\lambda\rho\Delta\rho}{\mu k\theta^{c}}\right)^{1/4}$$

or

$$\bar{N}u_{D} = O\left(\frac{D}{L}\cdot\frac{Ar_{D}Pr}{Ja}\right)^{1/4}, \tag{4.69}$$

where $Ar_{D} = \Omega^{2}\mathfrak{R}D^{3}\Delta\rho/\rho\nu^{2}$ and $Ja = c_{p}\theta^{c}/\lambda$. That is, the film thickness and heat transfer rate take the Nusselt forms seen previously in Section 3.1.

Under these conditions, the Coriolis effect is very weak. The tangential velocity ratio may be estimated from eqn (4.65). Thus

$$\frac{V^{c}}{U^{c}} = O\left(\frac{2\Omega\delta^{2}}{\nu}\right) = O\left[\frac{2}{Ek_{L}}\left(\frac{Ja}{Ar_{L}Pr}\right)^{1/2}\right] \ll 1 \tag{4.70}$$

where $Ek_{L} = \nu/\Omega L^{2}$ is an Ekman number. The film moves predominantly in the radial direction, towards the evaporator, with a weak secondary tangential component 'draining' liquid towards the trailing face of the tube. This is reminiscent of behaviour in a thermosyphon tube tilted slightly from the vertical. For the radial thermosyphon, the corresponding Coriolis tilt is found from eqn (4.63) to be

$$\alpha_{c} = \frac{2}{Ek_{L}}\left(\frac{Ja}{Ar_{L}Pr}\right)^{1/2} = 2\left(\frac{L}{\mathfrak{R}}\cdot\frac{Ja}{Pr}\cdot\frac{\rho}{\Delta\rho}\right)^{1/2}. \tag{4.71}$$

The tilt is thus equal to the velocity ratio V^{c}/U^{c}, both being small. This is the same conclusion that was reached earlier for the single-phase, radial thermosyphon under comparable conditions. However, the analogy is not complete because the heat transfer relations are different: the single-phase system then operates in a Coriolis-enhanced régime whereas eqn (4.69) indicates that the Coriolis acceleration has no effect on the primary flow under these particular phase-change conditions.

Borrowing experience from the gravitational system, it appears that a weak Coriolis effect, creating a secondary flow, would increase the performance of the radial thermosyphon only slightly. That is, eqn (4.69) is a conservative estimate of heat transfer for a system exhibiting 'near-vertical' behaviour in which the primary flow is not characterized by excessive pooling along the trailing face. Should the Coriolis tilt become larger, however, the condenser performance would improve because circumferential variations in the local heat transfer rate would then be significant. It will be recalled from the discussion of the eccentric thermosyphon in the previous section that thinning of the film over one face of the tube is usually accompanied by thickening on the opposite face, the net effect being an increase in heat transfer rate. For this possibility to occur in the radial thermosyphon, at least under laminar conditions, the assumption used to simplify eqn (4.67) must be relaxed; that is, the tangential term in the continuity equation may no longer be neglected in relation to the other two. Taking instead that

$$\frac{2V^{c}\delta}{\pi D W^{c}} = O(1)$$

together with eqns (4.64)–(4.66), it is found that

$$\alpha_{c} = \frac{V^{c}}{U^{c}} = (2\pi)^{1/3}\left(\frac{D}{\mathfrak{R}}\cdot\frac{Ja}{Pr}\cdot\frac{\rho}{\Delta\rho}\right)^{1/3} \tag{4.72}$$

which, by comparison with eqn (4.71), indicates that the Coriolis effect still does not greatly alter the flow field. Under these conditions, however, it is also found that

$$\frac{\delta}{D} = \left(\frac{Ek_{D}Ja}{Ar_{D}Pr}\right)^{1/6} \tag{4.73}$$

and hence

$$\bar{N}u_{D} = O\left(\frac{L}{DEk_{D}}\cdot\frac{DAr_{D}Pr}{LJa}\right)^{1/6}. \tag{4.74}$$

This is analogous to eqn (4.19) which describes Coriolis-enhanced conditions in the single-phase system. By comparison with eqn (4.69), the result immediately above suggests that the overall heat transfer rate will be improved if

$$Ek_{D} \ll \frac{L}{D}.$$

Typically, $Ek_D \ll 1$. Equation (4.73) confirms that such an improvement is the result of creating a thinner liquid film, for example, by increasing the speed of rotation.

It may be recalled from Section 4.2 that the same general conclusion was reached for single-phase conditions; namely, that the performance is enhanced with a decreasing Ekman number provided the Coriolis tilt remains small. In the same section, it was also noted that if the Ekman number became too small, the performance would deteriorate because of a rising adverse effect on the longitudinal pressure field. Whether or not such a maximal feature exists in the radial, evaporative thermosyphon has not been explored. However, since the Coriolis tilt in the Coriolis-enhanced régime is an inverse function of $Ek^2 Ra$ in the single-phase thermosyphon, and a similar inverse function of $Ek^2 ArPr/Ja$ in the evaporative thermosyphon, it is clear that any maximal behaviour in the latter will require much lower values of the Ekman number; it may be recalled that the ratio $ArPr/JaRa = \lambda\Delta\rho/\rho\beta c_p\theta^2$ is typically very large.

Strictly, the above analysis for high rotational speeds applies only to a laminar film, but a similar set of conclusions may be drawn for turbulent flow, given that the molecular diffusivities are replaced by macroscopic exchange coefficients. On the other hand, the existence of drop condensation, at one extreme, or interfacial instability at the other, would render the analysis invalid in the condenser. The open literature does not speak directly to these questions, although it is quite likely that the 'near-vertical' behaviour of gravitational systems would again act as a good guide in estimating the limits of performance, at least for non-metalic liquids.

High-speed behaviour in the evaporator poses a different set of problems. Near the radial tip of the tube, at the closed end of the evaporator, any pool of liquid will usually be undergoing some form of boiling, but the existence of an intense centrifugal force field may create a substantial departure from gravitational behaviour. As noted in Section 4.4, speed-induced hydrostatic pressure in the liquid may suppress nucleation; and Coriolis forces acting oppositely on the counterflow of bubbles and liquid tend to separate them, moving the liquid towards the trailing face and the vapour towards the leading face. The first of these effects may be even more significant in the radial thermosyphon than in the axial thermosyphon, but the second is probably small, as suggested by the ratio

$$\frac{2U_V\Omega}{\Omega^2\mathfrak{R}} \simeq \frac{2\dot{q}_a}{\rho_V\Omega\mathfrak{R}\lambda}$$

where $\dot{q}_a$ is the radial heat flux corresponding to the superficial radial vapour velocity U_V. With water, for example, a speed of 10^3 rev/min at a tip of 1 m gives $\rho_V\Omega\mathfrak{R}\lambda \simeq 2\times10^8$ W/m^2; this is greater than a typical axial heat flux, thereby indicating that the Coriolis effect on bubble trajectories is usually negligible, at least in a typical turbine application. Even so, high radial vapour velocities associated with high heat fluxes may give rise to substantial interfacial effects. Acting at the thinnest point of the evaporating liquid film, that is on the leading face near the evaporator pool, vapour inertia may be sufficient to cause a flow reversal and thus create a dry patch. Coriolis forces may thus exert an indirect influence on dryout failure when the filling charge is small. It is equally likely that they will introduce variations in interfacial instability around the tube circumference and hence play a role in flooding behaviour when the charge is not small. However, this speculation, like others advanced above, needs to be tested experimentally.

The scarce information available on the effect of pool depth or filling charge in radial, evaporative thermosyphons appears to be in line with the well-established trends in gravitational thermosyphons discussed in Chapter 3. As might be expected, dryout has been ascribed to very small charges while burnout is usually given as the cause of failure with larger charges (Cohen and Bayley 1955). Surprisingly, almost no reference appears to have been made to flooding failure even though it is reasonable to expect it to have occurred under certain conditions. Most observations, however, were made during early experiments, before the holdup phenomenon had been clearly identified. Likewise, and presumably for similar reasons, there appear to be no reports of pulse boiling or the geyser phenomenon in longer radial tubes operating at low pressures.

More recently, some attention has been focused on boiling in the thin films associated with high evaporator heat fluxes, again motivated by turbine blade cooling applications (Mudawwar *et al.* 1985; Mudawwar and El-Masri 1988). In view of the anticipated weak Coriolis effect, the nucleate boiling process would not be expected to vary much around the tube circumference provided a continuous film exists. With small charges, however, the leading face in the evaporator may not retain a film; liquid flow is then concentrated in a rivulet running over the trailing face, as suggested in Fig. 4.33. At very high heat fluxes, when bubbles burst at the liquid–vapour interface, droplets are ejected into the returning vapour; the Coriolis acceleration tends to dampen this entrainment. In extreme circumstances, the critical heat flux leads to complete disruption of the film with drops being blown off by escaping vapour; the only restraint is then provided by the Coriolis force which therefore sets the stability limit of the film. Coriolis forces are also believed to influence waviness and turbulence in the film (Kirkpatrick 1980). The heat transfer rate at rotating radial surfaces is also dependent on surface characteristics, as in gravitational systems. The macroscopic feature of roughness, for example, not only acts as a tripping mechanism for thin, rotating laminar films but may destroy the laminar sublayer of thicker films, thereby creating more vigorous stirring and a beneficial effect near nucleation sites; in general, the heat transfer rate increases with roughness.

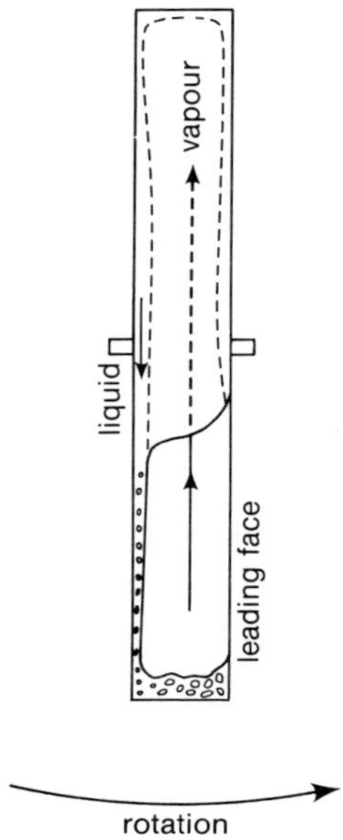

Fig. 4.33. Model of radial, evaporative thermosyphon operating with a small charge and high heat flux.

A typical set of convective heat transfer data for a radial heater are shown in Fig. 4.34. These are for a flat strip set on the trailing face of an open, radial channel in which the flow is driven outwards by the Archimedean buoyancy force; that is, no refluence occurs, but behaviour is still suggestive of that in a closed thermosyphon when neither the end pool nor the returning vapour play a significant role. As indicated, the data reveal the familiar trends extending from single-phase convection up to the critical heat flux; in particular, the arrows indicate that the critical heat flux $\dot{q}_{cr}$ is raised by an increase in pressure. It is interesting to note that, although not given, the Coriolis tilt in these experiments lay in the range $0.03 \lesssim \alpha_c \lesssim 0.1$.

The critical heat flux in thin liquid films, whether driven gravitationally or rotationally, is not usually attributable to conventional film boiling because nucleation frequently occurs in the film interior and the bubbles have limited room for growth and coalescence before they burst. As noted in the previous chapter, the droplets thus ejected tend to promote spray cooling of the opposite surface; they also provide a source of nucleation through re-entrainment. Overall behaviour is not unlike that observed by Katto and Monde (1974) who described it as a 'jump phenomenon'. A schematic representation of events is given in Fig. 4.35 which shows how the critical breakup point B marks the division between two regions: upstream, where

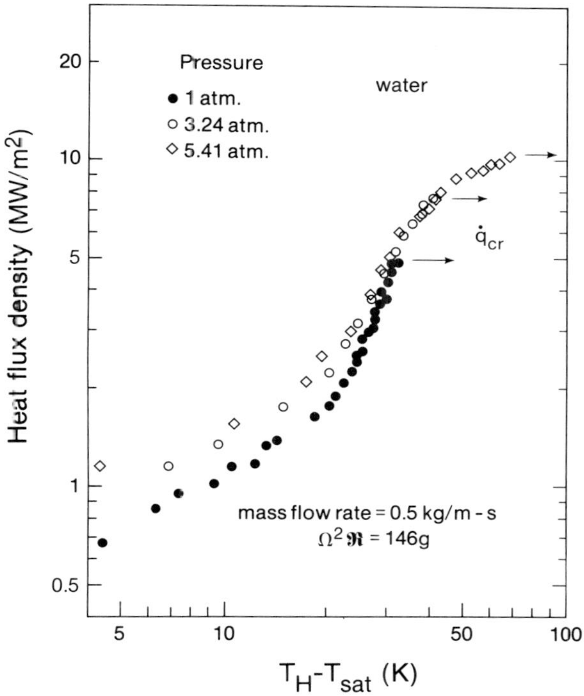

Fig. 4.34. Effect of pressure on boiling heat transfer in a radial film (after Mudawwar *et al*. 1985).

bubble bursting creates a fine mist of droplets but leaves the film otherwise intact; and downstream, where the rate of vapour generation is great enough to shatter the film into small and large drops. It is the behaviour of the latter, and in particular the range of their flight in relation to the remaining length of heated surface, that determines whether the surface downstream will be wetted or dry out. Mudawwar *et al*. (1985) argued that the critical condition would be obtained when the drops were close to mechanical equilibrium, the lateral drag force exerted on them by the leaving vapour then balancing the restraining Coriolis force; that is,

$$\frac{1}{2}\rho_V U_V^2 C_d \cdot \frac{\pi d^2}{4} \simeq 2\Omega U_L \rho_L \cdot \frac{\pi d^3}{6} \tag{4.75}$$

in which C_d is the drag coefficient caused by exiting vapour with superficial velocity U_V flowing around a droplet of diameter d. Estimating the droplet diameter from the critical Weber number We_{cr} at which large droplets shatter,

$$\rho_V \frac{U_V^2 d}{\sigma} = O(We_{cr}),$$

where $\rho_V \lambda U_V = \dot{q}_{cr}$, the critical heat flux, and using the longitudinal equation of motion to determine the liquid velocity scale U_L, eqn (4.75) may be rewritten in the form

$$\dot{q}_{cr} = C\lambda\rho_V^{1/2}(\Omega^2\mathfrak{R})^{1/4} \tag{4.76}$$

in which the coefficient C is a function of fluid properties, surface geometry, and superheat; C_d is taken to be constant.

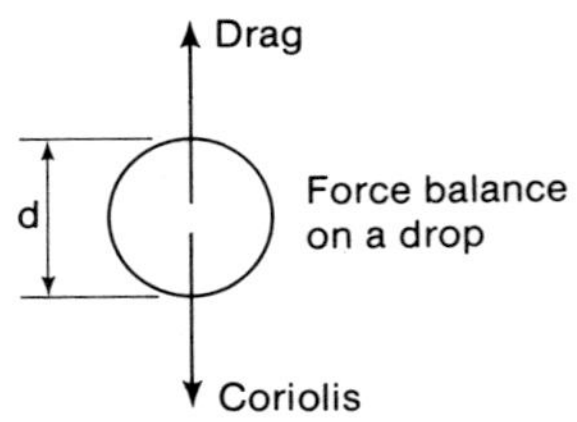

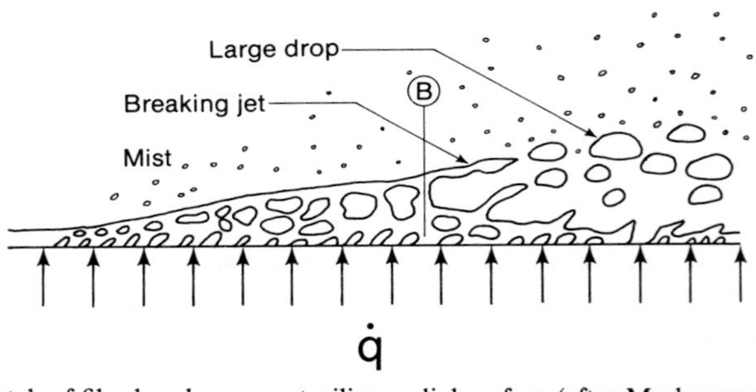

Fig. 4.35. Sketch of film breakup on a trailing radial surface (after Mudawwar *et al.* 1985).

This type of prediction is in accord with the observations of Mudawwar *et al.* illustrated, for example, by Fig. 4.34. However, it cannot be applied to the high-speed radial thermosyphon without qualification. Under refluent conditions, the drops created would likely be entrained by the returning vapour which would also exert an adverse shear force on the liquid film, thereby reinforcing the tendency of the film to separate from the heated surface. In addition, the accompanying reduction in liquid radial velocity reduces the Coriolis acceleration and thus lessens the restraining force. Overall, the effect of refluence will be to reduce the critical heat flux. On the leading surface of the tube, where the Coriolis acceleration tends to destabilize the film, should it exist, the critical heat flux will be considerably smaller. The possibility of film disruption, entrainment, and dryout is much greater along the leading surface. Even with a small Coriolis tilt, failure will likely originate along this surface: by burnout or flooding if the charge is not small or, if it is, by a form of dryout which may vary from total evaporation, at low heat fluxes, to film rupture at high heat fluxes. Against the limiting effects of refluence must be set some compensations. In general, the tendency of the Coriolis effect to stratify high void fraction flows may actually improve circulation near the maximum heat flux, at least for large charges. With small charges, the disruptive process on the trailing face creates a spattering of drops over the leading face. Experience with gravitational thermosyphons suggests that this provides some compensation for the absence of a film.

All of this behaviour undergoes a fundamental change if the thermosyphon is not rotating at high speed. Although no study of the low speed range has appeared in the open literature, a few brief observations may be in order. Considering first the situation where rotation is about a vertical axis, it is clear that slowing from a high speed will produce alterations in the primary flow which are similar to those observed when a vertical tube is tilted towards the horizontal; here the tilt above the 'horizontal' is equal to $\Omega^2\mathfrak{R}/g$. Thus, as the longitudinal (centrifugal) acceleration decreases, the lateral (gravitational) acceleration becomes relatively more important through the development of a secondary flow symmetric about a vertical, radial plane containing the tube axis. The situation is similar to that found in the corresponding single-phase, radial thermosyphon. As noted in Section 4.2, the buoyancy-driven (rising) secondary flow is superimposed on another (subsiding) secondary flow which, being caused by the Coriolis acceleration, is symmetric about the central plane of rotation. The resultant secondary flow thus shifts about the tube axis as the speed reduces. At the same time, the relative importance of the longitudinal pressure gradient also increases until, when the tube is at rest, it provides the main drive for the primary flow; that is, the high-speed annular refluent flow gradually changes into a weaker bifilamental (stratified) refluent flow. Since the lateral accelerations modify only the secondary flow, however complex, and the longitudinal pressure gradient is a much weaker indirect effect, it is to be expected that changes in the primary circulation and the heat transfer rate would not occur rapidly, except in the very slow-speed range when $\Omega \ll (g/\mathfrak{R})^{1/2}$, that is as the tilt above the 'horizontal tends to zero.

Slow-speed rotation about a horizontal axis leads to even more complex behaviour which varies from the steady to the quasi-steady, passing through a transient régime. For very low speeds, the cyclic sequence in the circulation pattern is illustrated schematically in Fig. 4.36. With the thermosyphon directly below the axis of rotation the pattern is obviously that of the vertical thermosyphon; a corresponding observation applies to the horizontal position. Thermally induced circulation in a rising tube, however, is

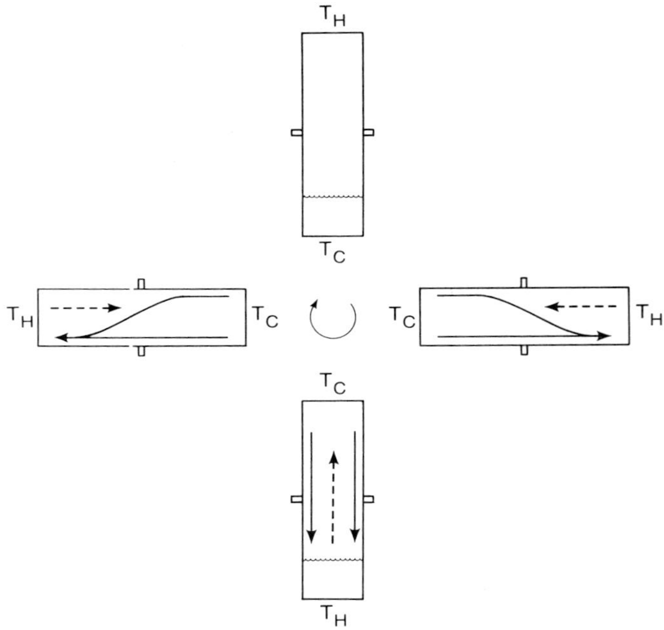

Fig. 4.36. Schematic development of circulation pattern in a radial thermosyphon slowly rotating about a horizontal axis.

principally in the same direction as the shaft rotation whereas it is in the opposite direction in a falling tube. An oscillatory circulation therefore takes place between the two vertical positions, the upper one representing a temporary system shutdown, and change in sign, as indicated by the absence of circulation. Superimposed on this is the washing effect of the evaporator pool which rotates oppositely to the shaft. The pool thus augments the thermally induced circulation during a downsweep but opposes it on an upsweep. At high speeds, the entire oscillatory washing mechanism is merely a perturbation on the basic, centrifugally induced flow field; the greater the length–diameter ratio of the tube, the less significant the oscillation. As the speed falls from a high value, the oscillation process grows until $\Omega = O(g/\mathfrak{R})^{1/2}$ when the system may act more like a cocktail shaker dashing liquid from one end to the other. Given that the thermosyphon owes its performance to a unique convective ability, this dashing process may generate particularly high evaporative heat transfer rates. In this connection, it is interesting to speculate on how the performance would be altered by pulse boiling which was deliberately tuned to match the speed of rotation.

5
RELATED DEVICES

It is evident from the material covered thus far that the tubular thermosyphon may be used in a wide variety of circumstances. Hopefully, the coverage will have led to an understanding of the device which is sufficiently basic and general to enable the reader to envisage other uses and other circumstances. Frequently, these will entail simple adaptation; no new ideas or principles will arise. Occasionally, however, the circumstances may be sufficiently unconventional to require modifications to the operating principles. This chapter is devoted to the main forms of modification, their physical basis, and the performance alterations which they introduce.

A different operating principle, acting alone, implies a different device, and indeed the systems considered below are often regarded in that light. Clearly, the treatment of other devices in any depth is beyond the scope of this book, and the reader seeking a more detailed discussion of these must consult the appropriate references listed in the Bibliography. The following constitutes only an introduction devoted specifically to those devices which are related in application to the thermosyphon. That is, even though they incorporate different operating principles, their ultimate purpose is the same: the creation of large convective fluxes in a tube. Particular attention will therefore be given to situations in which thermally induced buoyancy can be augmented or replaced by other effects. We begin with a description of capillary-driven circulation.

It may be recalled that several of the studies cited in previous chapters were concerned, in part, with the effect of grooves cut inside the tube wall. In general, this type of capillary assistance improves thermal performance, although it also increases manufacturing costs. More commonly, capillary effects are created by using a wick lining to construct the device usually known as a heat pipe.

5.1 The heat pipe

In its simplest form, the heat pipe is just a closed tube lined with a porous wick, as illustrated schematically in Fig. 5.1. The liquid saturating the wick evaporates in the heated section and the resultant vapour flows along the axis of the tube to the cooled section where it condenses. Operation is not unlike that in the evaporative thermosyphon. The distinguishing feature of the heat pipe is that the condensate returns to the heated section by capillary action in the wick, and this is both a simplification and a complication. Fluid circulation is simplified by comparison with that of a gravitational thermosyphon; the pattern is essentially axisymmetric and consists of only

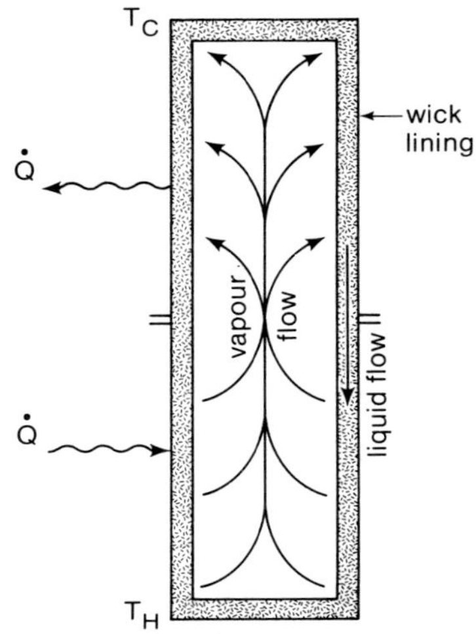

Fig. 5.1. Schematic representation of a heat pipe.

two elements, vapour flowing in an axial core and liquid returning in a porous annulus. It is the latter which complicates the circulation through the requirement that surface tension forces must drive the liquid through a bed of small pores.

In practice, it is found that there are many wick materials (Dunn and Reay 1982) with properties that enable them to transmit liquid fluxes large enough to match the corresponding vapour fluxes in the core. Indeed, the heat pipe is often capable of transmitting higher heat fluxes than a comparable thermosyphon. Moreover, its ability to transmit this heat is largely independent of the gravitational field, or any other body force field, although these fields may provide assistance. The heat pipe thus has the advantage of being less sensitive to the direction of the temperature gradient. It is therefore a natural candidate for applications in space, where the prevailing body force field may be weak or non-existent. On the other hand, this property prevents it from being a natural thermal diode.

To understand the performance characteristics and limits of a heat pipe it is first necessary to understand the behaviour of the wick and assess its ability to provide the pressure head which drives the circulation. This ability may be attributed to the intermolecular forces which operate in the vicinity of phase boundaries and, in particular, within the thin liquid film which creates a meniscus in the wick pores. In such a film, the chemical potential may be written in the form

$$\mu = \mu_\infty - \mu'(Y) \tag{5.1}$$

where μ_∞ is the value in the liquid bulk with which the film connects, and $\mu'(Y)$

represents the departure caused by the surface force field at a distance Y from the surface. For the bulk liquid,

$$d\mu_\infty = v\,dP - s\,dT$$

which if v and s are both weak functions of P and T, may be integrated to give the approximate result

$$\mu_\infty - \mu^o = v(P - P^o) - s(T - T^o) \tag{5.2}$$

where T^o, P^o are reference values such that

$$\mu_\infty(T^o, P^o) = \mu^o.$$

Equation (5.1) may thus be restated as

$$\mu - \mu^o = v(P - P^o) - s(T - T^o) - \mu'(Y). \tag{5.3}$$

Now if P^o, T^o are the saturation pressure and temperature, respectively, it follows that when the bulk liquid is held at these conditions $\mu_\infty = \mu^o$. If the adjacent thin film is also at this temperature the entire liquid will be isothermal. But stable quilibrium requires that $\mu = \mu^o$ throughout, and therefore.

$$\mu'(Y) = v[P(Y) - P^o]. \tag{5.4}$$

It is thus apparent that the chemical potential departure and the *disjoining* pressure $P(Y)$ have essentially the same profile in the neighbourhood of saturation conditions.

The form of this profile is unknown *a priori*, but is evidently rooted in intermolecular forces which vary inversely as some power of the intermolecular distance: for short range Van der Waals forces this power is seven while for electrostatic (ionic) forces it is two. Since the ionic forces are usually the greater, eqn (5.4) may be restated using the empirical form

$$P(Y) - P^o = \frac{a}{Y^n} \tag{5.5}$$

where $n \geq 2$. This provides a useful representation of the liquid disjoining pressure associated with the electrical double layer formed near a solid surface when it is charged oppositely to ions in the liquid. Although the double layer as a whole is electrically neutral it establishes a strong local electrostatic field which has a binding effect on ionic and polar molecules.

The pressure gradient thus formed in a thin liquid film is normal to the surface. In essence, it is a 'hydrostatic' pressure reflecting mechanical equilibrium in the electrostatic force field; the resulting liquid velocity normal to the surface is zero. But if the pressure field is distorted by non-uniformity in the film thickness, a component of the pressure gradient is established parallel to surface. This gradient then drives the liquid over the surface, that is from thick film regions to thin regions. A liquid film with a positive disjoining pressure thus tends to spread over the surface. This is the basis of the wetting[1] phenomenon by which liquid from the bulk is drawn over the surface until mechanical equilibrium establishes a meniscus in which the macroscopic effect of the local surface force field, manifest as surface tension, is balanced by the effect of bulk pressure. The most familiar situation is when a wetting liquid rises in a fine-bore tube standing vertically; for stable equilibrium, the surface tension force, exerted uniformly around a ring surface, is then balanced by the gravitational force acting on the column of liquid below. This illustrated in Fig. 5.2.

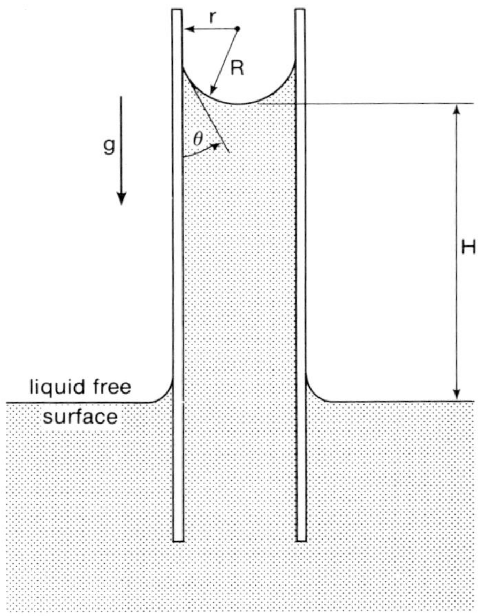

Fig. 5.2. Capillary rise in a fine-bore tube.

In equilibrium, the surface energies σ at liquid–solid, liquid–vapour, and solid–vapour interfaces are related by the force balance

$$\sigma_{sv} = \sigma_{ls} + \sigma_{lv} \cos \theta \tag{5.6}$$

where θ is the contact angle of the meniscus, as defined in Fig. 5.2. Henceforth, only the liquid–vapour energy σ_{lv} will be considered and will be written without the subscripts. As noted above, the wetting is attributable to intermolecular forces capable of moving the liquid over the surface. The

[1] A non-wetting liquid exhibits a negative disjoining pressure and thus behaves oppositely.

velocity with which this occurs depends upon the film thickness and the liquid viscosity, the latter in turn depending upon the molecular structuring caused by the intermolecular force field. Macroscopic manifestations of the intermolecular forces are evident in the heat of wetting and the contact angle, both of which exhibit the hysteresis which accompanies irreversible phenomena. For example, it has been noted that the contact angle is often greater when the meniscus advances than when it retreats, although the general explanation of this has yet to be found; for a porous medium, it has been attributed to the 'bottleneck' hypothesis, alluding to the fact that the radius of a pore is generally greater than the radii of necks connecting it with other pores (Defay *et al.* 1966).

The simplest way to determine the capillary rise in a fine-bore tube of radius r is to appeal to mechanical equilibrium which requires that the rise of liquid H shown in Fig. 5.2 should have a weight, $H\pi r^2 \rho g$, which is equal and opposite to the surface tension force, $2\pi r\sigma \cos\theta$. Thus

$$H\rho g = \frac{2\sigma\cos\theta}{r} = \frac{2\sigma}{R} \tag{5.7}$$

represents the hydrostatic pressure difference between the base and the top of the rise if $R = r/\cos\theta$ is the interface radius. It therefore follows that the liquid pressure just below the meniscus is given by

$$\Delta P_{\text{cap}} = \frac{2\sigma}{R} \tag{5.8}$$

in relation to the ambient (and base) pressure, providing we adopt the sign convention that R is positive when the centre of curvature is in the liquid. In a non-circular pore, this expression must be replaced by Laplace's equation

$$\Delta P_{\text{cap}} = \frac{2\sigma}{R_{\text{m}}}$$

in which R_{m} is found from the principal radii R_1, R_2 of the meniscus; that is

$$\frac{2}{R_{\text{m}}} = \frac{1}{R_1} + \frac{1}{R_2}.$$

Alternatively, the capillary pressure in real pores of unknown dimensions may be represented by

$$\Delta P_{\text{cap}} = \frac{2\sigma\cos\theta}{r_{\text{p}}} = \frac{2\sigma}{R_{\text{p}}} \tag{5.9}$$

where r_{p} is the effective pore radius and R_{p} the effective interfacial radius. This empirical expression assumes that σ and θ are both known and independent of

r_p, and is convenient for media with pores which vary in size and are difficult to measure.

The pressure difference across a curved liquid–vapour surface also alters the saturation vapour pressure, but this change is negligible unless the pores have a very small radius, for example less than 10 μm (Ivanovskii *et al.* 1982). More significant is the effect of evaporation or condensation illustrated in Fig. 5.3.

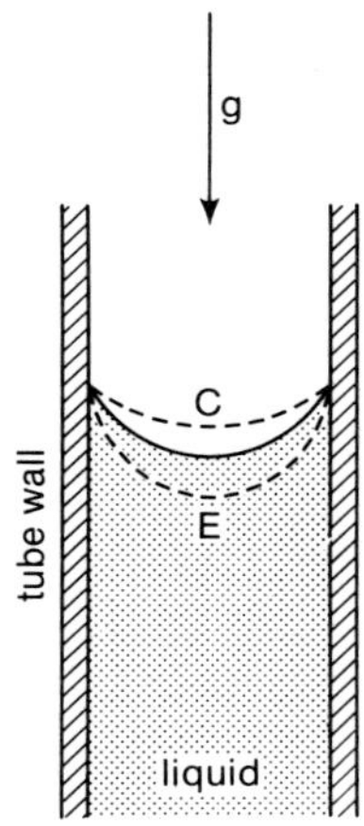

Fig. 5.3. Condensation and evaporation in a fine-bore tube.

In the absence of phase change, but at the saturation temperature, the meniscus has the form indicated by the solid line. When evaporation occurs, it does so both at the edge of the meniscus (at the surface of the thin film) and on the rest of the curved surface. In most situations, the contribution from the film may be neglected. The effect of evaporation is then to increase the curvature of the meniscus, as suggested by the dashed curve E. Conversely, condensation creates the meniscus curve C which has a lower curvature. It is now apparent that evaporation at one end of a wick, while condensation occurs at the other, will create a difference in capillary pressure capable of driving the liquid through the wick towards the evaporator where the meniscus radius (and capillary pressure) is smaller. This is an ideal model of events as they would occur in a homogeneous wick having a uniform pore size. In practice, behaviour is modified by non-uniformity which causes the smaller pores to be active first. Thus condensation tends to shift activity to larger pores once the smaller pores are filled and the interfacial radius in them has become infinite (Defay *et al.* 1966).

The capillary pressure difference is not caused by any body force field, other than the local surface field which causes the disjoining pressure. In the absence of (latent) heat transfer, saturated conditions establish the same interfacial curvature in each of the (uniform) pores along the exposed wick surface. The liquid capillary pressure[2] in the wick interior is then uniform and given by eqn (5.9). Subsequent evaporation at any point on the wick lowers the liquid pressure (increases the suction) thereby drawing liquid from regions where evaporation occurs at a lower rate or not at all. In other words, variations in the rate and sign of the phase change over the wick surface create a liquid flux distribution in the wick interior. The capillary pressure difference in a heat pipe must be capable of driving more than the liquid flow, however. It must also provide the means by which the evaporated liquid is propelled back from the heated section to the cooled section.

Figure 5.4 illustrates the pressure distribution in both liquid and vapour pathways during phase change. In the porous wick, the flow is usually laminar and is governed by Darcy's law

$$\mathbf{V} = -K\nabla H \tag{5.10}$$

in which $\mathbf{V}$ is the local fluid velocity and H is the piezometric head[3]. The quantity K is usually known as the hydraulic conductivity, and may be written

$$K = \frac{cgd^2}{\nu}$$

where d is the mean pore diameter, ν is the kinematic viscosity of the fluid, and c is a non-dimensional coefficient. Alternatively, the intrinsic permeability of the medium may be defined by $k = cd^2$, in which case $K = kg/\nu$.

For a horizontal heat pipe, eqn (5.10) reduces to

$$\mathbf{V} = -\frac{k}{\mu_L}\nabla P_L \tag{5.11}$$

which may be applied to a thin wick lining where radial and tangential pressure variations are usually negligible in comparison with the axial variation displayed in Fig. 5.4. This poses a quasi-one-dimensional problem in which the axial liquid flux $\rho A V$ at any section of the heat pipe is given by

$$\dot{m} = -\pi D t \rho_L \frac{k}{\mu_L}\frac{dP_L}{dX} \tag{5.12}$$

where D is the tube diameter and t is the wick thickness. However, this mass flux varies with position in the heated and cooled sections because of evaporation and condensation, repectively. Thus, although the liquid pressure

[2] It is, of course, a suction.

[3] The corresponding pressure is thus $H\rho g = P + Z\rho g$ where P is the fluid pore pressure and Z is the elevation of the free surface.

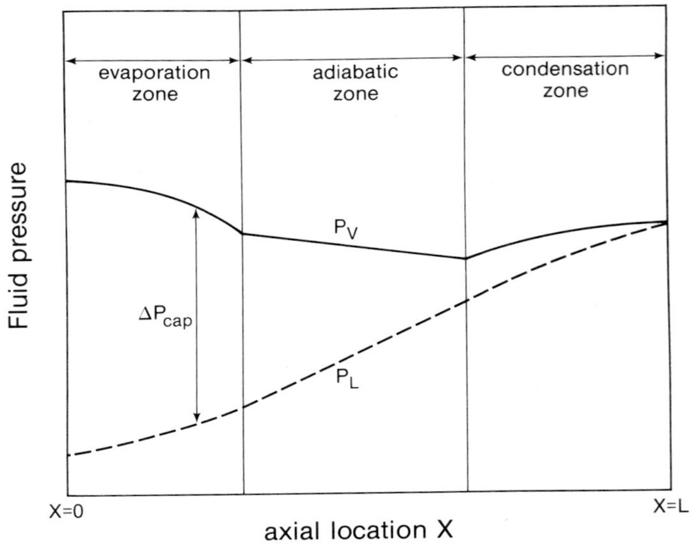

Fig. 5.4. Liquid and vapour pressure distribution in a horizontal heat pipe.

field is one-dimensional its solution depends upon the continuity equation given by

$$\frac{d\dot{m}}{dX} = -\pi(D-2t)\dot{m}_E$$

where $\dot{m}_E$ is the surface evaporation rate. Hence

$$\frac{d^2 P_L}{dX^2} - E(X) = 0 \tag{5.13}$$

where

$$E(X) = \frac{\dot{m}_E \nu_L}{k}\left(\frac{D-2t}{Dt}\right).$$

In the heated section, $E > 0$ and the liquid pressure profile is concave upwards; conversely, $E < 0$ in the cooled section where the curve is concave downwards. In an isothermal adiabatic section, $E = 0$ and the curve is linear. This explains the sigmoidal shape of the liquid pressure distribution indicated in Fig. 5.4. The local capillary pressure, when added to the liquid pressure, establishes the local vapour pressure.

Under steady conditions, the axial mass flow of vapour at any section must be equal and opposite to that of the liquid, and since the vapour is very much less dense than the liquid it is to be expected that vapour velocities would typically be higher than liquid velocities, although the difference is partly offset by the respective cross-sectional areas through which they flow. The pressure gradient driving the vapour must overcome two main forces: a frictional force caused by the no-slip boundary condition at the wick surface; and an inertial force arising from acceleration of the vapour stream whose mass increases through evaporative loss from the wick. The resultant profile is suggested in Fig. 5.4. Typically, the inertial effect dominates the frictional effect and hence the vapour pressure profile is concave downwards in the heated section but remains linear in the adiabatic section. In the cooled section, the inertial effect may lead to a partial pressure recovery, as indicated. The vapour pressure in the cooled section may thus increase towards the closed end, while the capillary pressure difference decreases. Where the vapour pressure equals the liquid pressure, the liquid–vapour interface is flat; this point is often called the *wet* point, in distinction to the closed end of the evaporator which is conventionally designated as the *dry* point.

The solution of the vapour flow problem is more difficult to obtain than that of the liquid flow. It too must account for evaporation and condensation which may be incorporated through boundary conditions of blowing and suction, respectively, but the complexity is not so much in the boundary conditions as in the governing equations themselves. The full Navier–Stokes equations must be solved under conditions which may be laminar, transitional, turbulent, or simultaneous combinations of these (Yuan and Finkelstein 1956; Cotter 1965; Busse 1967). In addition, it has been observed that as the vapour pressure of the system is lowered the effect of compressibility, as indicated by the Mach number M, becomes increasingly important; M may approach and exceed 1.0 (Levy 1968; Levy and Chou 1971; Bystrov and Popov 1976; Levitan and Vasiliev 1976; Busses 1973; Ivanovskii *et al.* 1970). As will be seen later, the pressure profile in the cooled section under supersonic conditions exhibits the familiar jump across a normal shock wave before returning to the subsonic pressure recovery evident in Fig. 5.4.

When the frictional pressure loss in the cooled section is initially greater than the pressure recovery, for example, if the heat pipe is particularly long, the vapour pressure continues to fall, as illustrated in Fig. 5.5, until it is equal to the liquid pressure at $X = L_w$. This condition again defines the wet point, which is no longer at the end of the tube; in fact this point is not restricted to any particular location in the cooled section. Should the vapour pressure fall slightly below the liquid pressure, the liquid–vapour interface becomes slightly convex and exerts little further effect on the fluid circulation rate. In general, circulation is maintained providing the overall capillary pressure difference is not less than the combined pressure losses; that is,

$$(\Delta P_{\mathrm{cap}})_{X=0} \geq \int_0^{L_w} \left(\frac{\partial P_L}{\partial X}\right) \mathrm{d}X + \int_{L_w}^{0} \left(\frac{\partial P_V}{\partial X}\right) \mathrm{d}X. \tag{5.14}$$

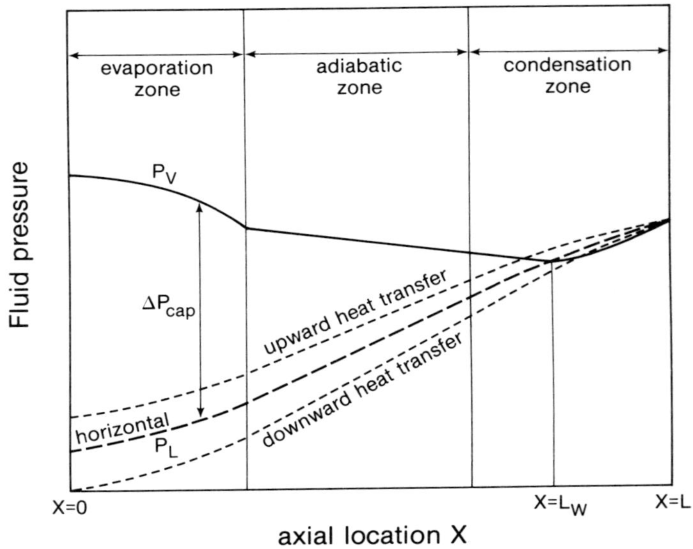

Fig. 5.5. The effect of skin friction and gravity on the pressure distribution in a heat pipe.

This statement ignores two factors. The first of these is phase change which, in transferring fluid between the liquid and vapour streams, creates a momentum exchange at the interface. This may be ignored unless the evaporation and condensation rates become exceptionally high. The second factor is gravity, which affects the liquid pressure distribution through the full Darcy equation (5.10). Thus, unless the heat pipe is horizontal[4], the liquid pressure profile will be displaced by an amount which varies linearly with elevation. For a vertical tube, this additional pressure contribution is given by

$$P_g = \pm g(L-X)(\rho_L - \rho_V), \qquad (5.15)$$

assuming that the wet point is fixed at $X=L$: the plus sign applies when the cooled section is on top. These shifts are illustrated in Fig. 5.5. It is evident that Archimedean buoyancy assists the circulation when the cooled section is above; the required capillary pressure difference at any given elevation is then reduced. Clearly the opposite is true when the pipe transmits heat vertically downwards.

Whatever the equilibrium pressure distribution in a heat pipe, it is established and maintained by the heat transfer processes which accompany evaporation and condensation. It has been tacitly assumed that these take place only at the wick surface in those pores which communicate directly with

[4] Gravity also plays a role in a horizontal heat pipe because the net suction will be greater along the higher region of the wick where dryout will likely occur first.

the central vapour space. This implies that evaporative heat transfer may be represented locally by an equation of the form

$$\dot{q}_s = k_s\left(\frac{\partial T}{\partial R}\right)_s = \lambda \dot{m}_E. \tag{5.16}$$

That is, the heat conducted radially and steadily inwards to the wick surface is balanced by the local evaporation rate. A similar expression applies to condensation. The temperature gradient is almost constant in a thin wick and may be approximated by $(T_w - T_I/t$, where T_I is the temperature at the liquid–vapour interface. The thermal conductivity of the wick is, in general, a complex function of the geometry and interconnectedness of its pores; it will include contributions from diffusion and dispersion. In practice, however, it may be represented by an empirically obtained mean $\bar{k}$. Hence eqn (5.16) may be approximated by

$$\dot{m}_E = \frac{\bar{k}}{\lambda}\frac{(T_w - T_I)}{t} \tag{5.17}$$

in which $\dot{m}_E$ is based on unit surface area of the wick[5].

The permissible range of conditions covered by this heat balance is limited not only by the temperature gradient but by the rate at which evaporation may occur; that is, by the permissible net flux of vapour that results from the simultaneous molecular processes of escape and deposition which continually take place at the interface. For small departures from thermodynamic equilibriums the rate has been determined by Labuntsov and Muratova (1967) to be

$$\dot{m}_E = \frac{P^o - P}{(2\pi R T^o)^{1/2}} \cdot \frac{f}{1 - 0.4f} \tag{5.18}$$

where f is the fraction of incident molecules which are retained by the interface. Subbotin *et al.* (1969) provide a rough estimate of the evaporation rate during large departures from equilibrium. This is given by

$$\dot{m}_E = \frac{8P}{(2\pi R T)^{1/2}}\left\{\frac{1}{f} - \frac{1}{2} - \left[\frac{1}{f^2} - \frac{1}{f} + \frac{P_o}{4P}\left(\frac{T}{T_o}\right)^{1/2}\right]^{1/2}\right\}. \tag{5.19}$$

During evaporation of a pure liquid $f = 1.0$, but in practice it is found that f is less than, and even much less than, 1.0. This has been attributed to the effect of impurities.

High heat fluxes may produce boiling in the wick. This is more likely to

[5] Alternatively, it may be based on the pore surface area if this is known as a fraction of the total.

occur in non-metallic liquids, which have a relatively low value of thermal conductivity and nucleate more readily[6]. The wick boiling process is not yet fully understood, but appears to have features in common with pool boiling (Tien 1975; Cornwell and Nair 1976; Groll *et al.* 1984). Nucleation within individual pores is found to be beneficial in reducing the radial temperature graident in the wick, but only so long as the vapour bubbles can find their way into the vapour stream, either directly by radial migration or indirectly via large pores which are interconnected axially. When the boiling rate inside the wick exceeds the rate at which the bubbles may escape, interconnecting pores become filled with vapour (Davis and Ferrell 1974). Whether this continuous layer of vapour is adjacent to the tube wall or not is a moot point but, if so, it produces an effect similar to film boiling, with evaporation being restricted to the surface which the layer shares with the remaining liquid.

In the cooled section, condensation in the pores of the wick is, broadly speaking, the reverse of evaporation. However, this condensation only takes place up to the wet point, beyond which the wick is flooded, bearing in mind the effect of varying pore size mentioned earlier. It is thus evident that condensation in the region between the wet point and the closed end of the condenser is essentially the same process as that which occurs in a thermosyphon. A condensate film then forms on the surface of the wick, its behaviour being governed by the hydrodynamic and buoyancy considerations discussed in Chapter 3. For liquid metals, condensation may be accompanied by an interfacial thermal resistance (Ivanovskii *et al.* 1976*b*) but this is usually negligible for non-metallic liquids. The presence of incondensable gases in a thermosyphon was briefly discussed in Sections 3.3 and 3.4; their role in a heat pipe is very similar.

The thermal and hydrodynamic features of the heat pipe, taken together, reveal a simple device with a complex behaviour. We conclude this section with a review of its principal peformance limits. Since the precise details of these vary from device to device, and fluid to fluid, they are shown only schematically in Fig. 5.6 using an illustrative plot of maximum heat flux versus operating (saturation) temperature. At very low[7] temperatures and vapour pressures, viscous drag acting on the vapour stream restricts the fluid circulation rate (and heat transfer rate) to very low values (Busse 1973; Vinz and Busse 1973). For higher temperatures and pressures, the flow is dominated by inertial forces. Under these conditions, the vapour exhibits the well-known characteristics of a compressible fluid accelerating (through mass addition) along the heated section (Busse 1973). Ultimately its velocity at the end of the heated section may reach the sonic velocity thereby ‘choking’ the

[6] As indicated by the smaller size of their critical radii for spontaneous (homogeneous) nucleation (Ivanovskii *et al.* 1982).

[7] Low for a given fluid. For long, sodium-filled heat pipes, for example, the temperature range is typically 450°C to 530°C, well above the operating range of many non-metallic fluids.

tube; that is although supersonic velocities may be reached in the cooled section, the vapour *mass* flow rate is limited by the choking velocity where $M = 1$. Figure 5.7 shows axial variations of temperature, pressure, and Mach number obtained in a sodium heat pipe (Ivanovskii *et al.* 1982). The choked condition is evident at entry to the cooled section, while a normal shock wave is found further downstream.

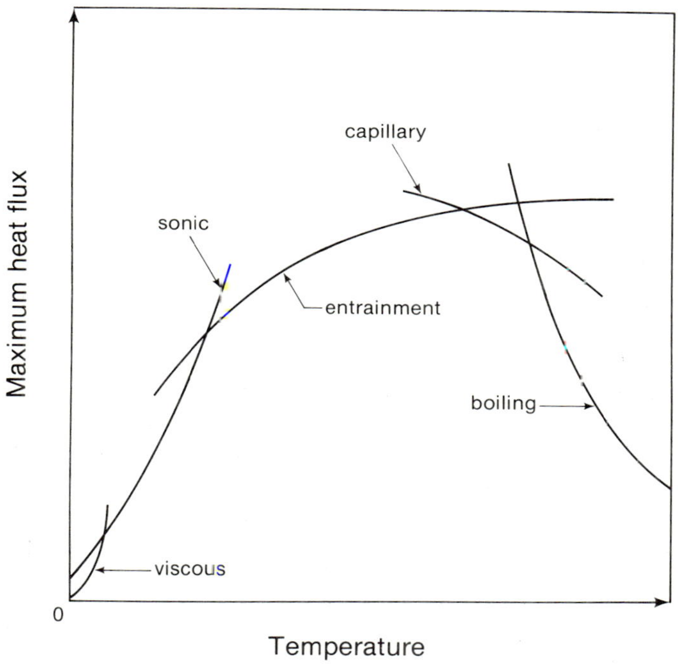

Fig. 5.6. Thermal performance limits of a heat pipe (after Ivanovskii *et al.* 1982).

It may have been noticed that viscous and sonic limits are found in both the heat pipe and the thermosyphon. The reader is therefore referred to Chapter 3 for further details of behaviour under these low-temperature conditions. A similar comment may be applied to higher-temperature conditions when performance is compromised by entrainment. This is again associated with interfacial instabilities driven by vapour velocities high enough to create numerous droplets. Entrainment is particularly noticeable in the cooled section, especially beyond the wet point, but is also exhibited in the heated section when vapour bubbles on the surface of the wick burst, spraying the vapour stream with droplets (Bähr *et al.* 1968). As noted earlier, boiling in a wick is not unlike pool boiling, and hence the boiling limit curve in Fig. 5.6 is

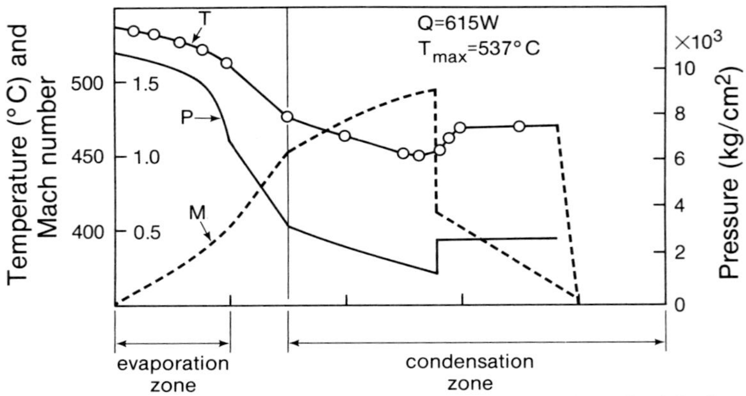

Fig. 5.7. Temperature, pressure, and Mach number profiles in a sodium heat pipe (after Ivanovskii *et al.* 1982).

not unlike the burnout limit in a thermosyphon under similar conditions; that is, at very high pressures when numerous bubbles coalesce. The capillary limit, on the other hand, is unique to the heat pipe. It is met whenever the overall condition (5.14) is not satisfied, and therefore varies with those factors which influence either the capillary rise or the pressure loss; that is, it is a function of pore size, surface tension, saturation pressure, tube diameter, tube length, and orientation. In particular, it depends upon the hydrodynamic characteristics of the wick.

Almost all of the discussion in this section has been based on the simplifying assumption that the wick is a homogeneous lining. For the sole purpose of providing capillary assistance to a thermosyphon, the wick may be unnecessary (Kamotani 1976; Sasin *et al.* 1984). Longitudinal or spiral grooves in the tube wall serve a similar purpose, especially when the filling charge is small or is unevenly distributed around the circumference as a result of tilting, etc. However, such measures are ineffective if the grooves are flooded; a similar limitation holds for a wick. More generally, the wick may take a compound form designed to raise the capillary limit. This is done by separating the requirement of high capillary pressure difference (demanding small pores) from the requirement of low hydraulic resistance (demanding large pores). Figure 5.8 illustrates a number of ways in which this may be accomplished. At one behavioural extreme (high capillary pressure and high resistance) is the layered screen, Fig. 5.8(a), while the opposite behaviour is exhibited by the open channels, Fig. 5.8(b). The remaining schemes represent various attempts to combine the effect of the screen with the effect of the channels.

Heat pipes are employed in many situations where thermosyphons could be used, and vice versa. This is particularly true of terrestrial applications

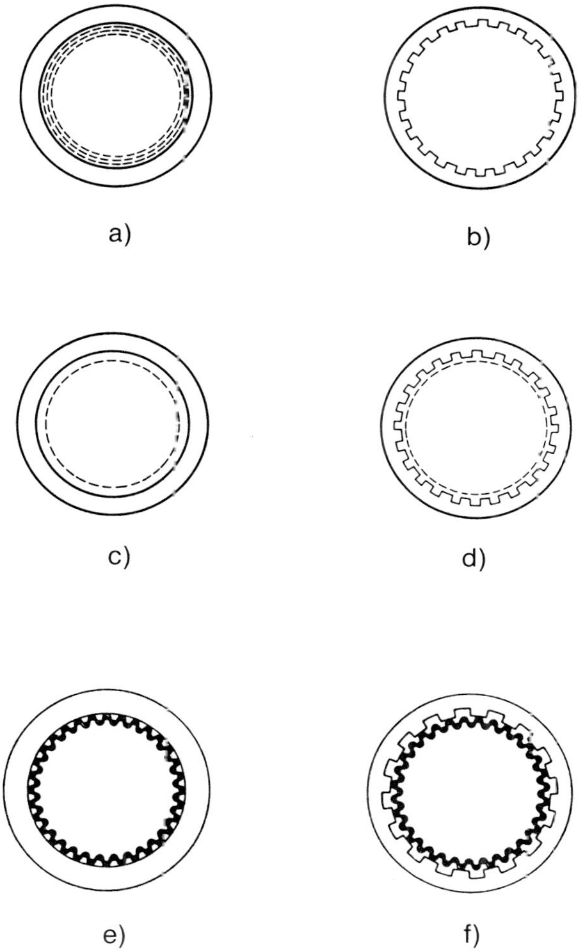

Fig. 5.8. Simple and compound wicks: (a) screen wick; (b) open channels; (c) annulus behind screen; (d) channels covered with screen; (e) corrugated screen; (f) channels and corrugated screen.

when the temperature ranges within 100 degrees or so above and below the ambient. For temperatures very much lower or higher than this, however, the heat pipe is often the superior alternative (Dunn and Reay 1982), especially when the gravitational field is absent. In a terrestrial context, the temperature range below 100 K is evidently accessible using liquefied gases while the range in excess of 2000 K is equally practical if the working fluid is a metal For the latter, heat flux densities in excess of 10 MW/m^2 have been recorded. As with the thermosyphon, the heat pipe may also be used in a variety of thermoregulatory capacities, most notably by employing an incondensable gas to alter the overall thermal resistance in some prescribed manner. And, of course, capillarity and buoyancy can always be called upon to assist each other in a hybrid design.

5.2. The aerosyphon

As a result of a 1972 study on low water levels in the Peace–Athabasca Delta region of northern Alberta, the feasibility of a downstream ice 'dam' was considered. The proposed structure had the attraction of low cost and the merit of being environmentally benign. It was to consist of a row of thermosyphon, or cryosyphon, tubes positioned across the river, as shown earlier in Fig. 1.13. During the winter period, the cryosyphons permit cold air to remove heat from the water flowing beneath the ice cover, and since the water is usually close to its freezing point the heat lost is largely latent heat. As indicated schematically in Fig. 5.9, an annulus of ice begins to form on the submerged section of each of the cryosyphons with the eventual result that these annuli merge and thereby block the local flow of water. The ice veil thus formed helps limit the low water levels upstream without creating a permanent obstruction. Each summer the ice melts, and in the ensuing winter the veil forms anew.

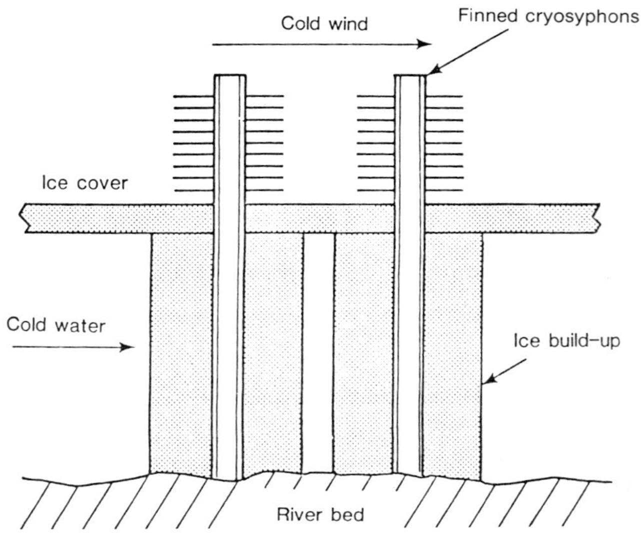

Fig. 5.9. Build-up of ice on submerged cryosyphons.

The absence of experience in such a scheme eventually prompted the initiation of the BIVA[8] project with the specific objective of obtaining data on the operating characteristics of appropriate cryosyphon devices (Lock 1986*a*,*b*). In particular, consideration was given to the concept of an aerated thermosyphon, or *aerosyphon*, in which air is bubbled through a conventional thermosyphon in such a way as to augment the thermally induced buoyancy (Lock and Maezawa 1975). The scheme is illustrated in Fig. 5.10. Since the density difference between air and most liquids is several orders of magnitude greater than the density variations created through temperature gradients, it is to be expected that the Archimedean buoyancy manifest in bubbling will create a significant increase in the heat transfer rate. Aeration obviously requires an external power supply but the amount is small, typically a few per cent of the overall heat transfer rate.

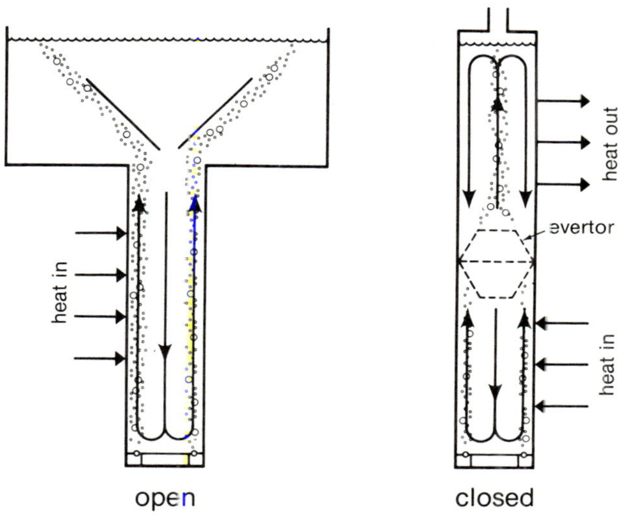

Fig. 5.10. Aerosyphon schematic (after Lock and Maezawa 1975).

Early experiments were conducted on the open aerosyphon apparatus shown in Fig. 5.11. An orifice chamber in the base of the apparatus ensured that air was delivered to a ring of orifices immediately beneath the aerosyphon tube, thus producing the flow pattern and bubble trajectories suggested in Fig. 5.10. The flow pattern was confirmed using streak photography which also revealed some details of the bubble trajectories. Experiments were also conducted on the closed aerosyphon using an apparatus which differs from

[8] Built Ice Veil Assessment.

that shown in Fig. 5.11 in two ways: the reservoir and cooling coil were replaced by an inverted, water-cooled tube; and an eversion device similar to that shown in Fig. 2.30 was inserted at the junction between the heated and cooled sections of the tube.

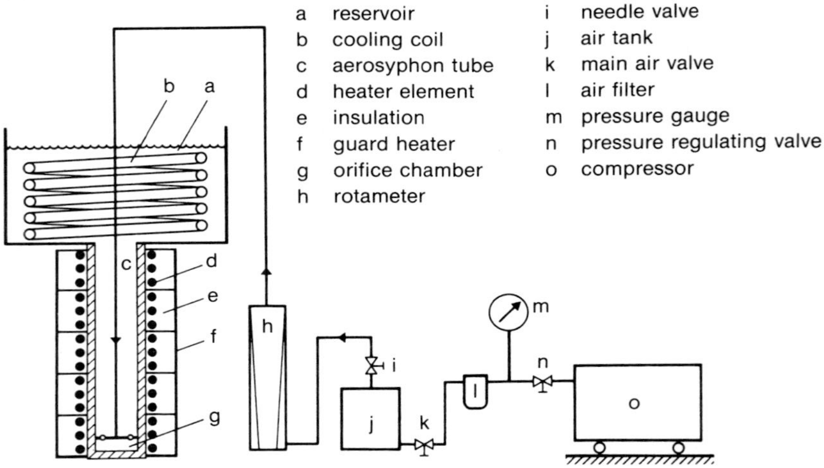

Fig. 5.11. Open aerosyphon apparatus (after Lock and Maezawa 1975).

The evertor has the effect of converting the aerated annulus, in the heated section, into the aerated core in the cooled section above. This is demonstrated with streak photographs obtained above diametrically opposed orifices in Fig. 5.12 which also illustrates the the tendency of the bubbles to spread radially inwards from the lower tube wall. A more careful construction of the bubble mean trajectories is shown in Fig. 5.13 for both the open and closed forms of aerosyphon. More will be said about these flow patterns later; at this point it is sufficient to note the radial spread above and below the evertor. Above the evertor, the outward tendency is very strong and is essentially completed within about one tube diameter; below the evertor, the tendency is not only in the opposite direction but typically requires three or four diameters for completion.

At first thought, it may appear that a ring of orifices beneath the tube would present no particular difficulty in the systematic generation of air bubbles in the liquid above. The preliminary findings of Lock and Abdurahman (1988), however, revealed that low air flow rates produced unstable behaviour: the liquid had a tendency to drip back, or weep, into the air-filled orifice chamber, eventually creating a partial obstruction to the flow of air and producing an erratic 'firing' pattern around the orifice ring. A subsequent study was therefore undertaken to determine the range of conditions over which this effect occurred and to experiment with orifices having various lengths and

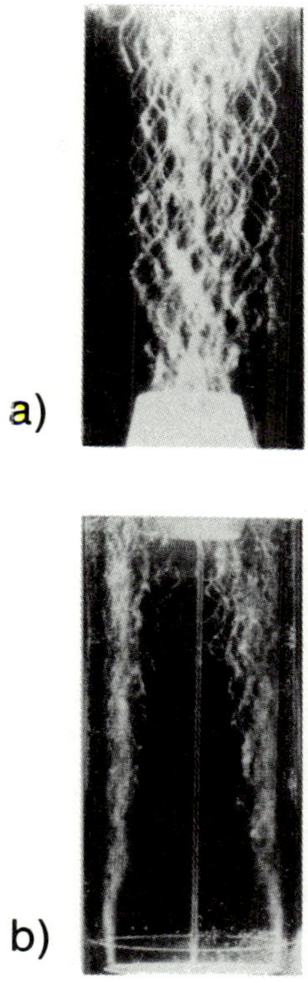

Fig. 5.12. Air bubble trajectories in water-filled aerosyphon tubes (a) upper tube, (b) lower tube (after Lock and Maezawa 1975).

diameters. Unlike the usual remedy, which consists of using a small-diameter orifice to create a relatively large pressure drop, it was found that the optimal arrangement consisted of holes which were countersunk from beneath, as indicated in Fig. 5.14. The displacement cone was used to ensure that air entering the chamber automatically filled a toroidal space beneath the orifice ring. This configuration eliminated the appearance of drip-back for all but the lowest, and least important, air flow rates.

The effect of aerator geometry had been systematically studied by Lock and Abdurahman (1988) using methanol as the working liquid in a closed aerosyphon. Three parameters were considered: the number of orifices in the range $24 < N < 60$; the orifice diameter ratio in the range $0.0048 < 2r/D < 0.0196$; and the orifice ring diameter ratio in the range $0.032 < \delta/D < 0.250$, where δ is the distance from the tube wall to the orifice centre. The aerator geometry would not be expected to influence conditions significantly in the (upper) cooled section of the device, above the evertor, and therefore its

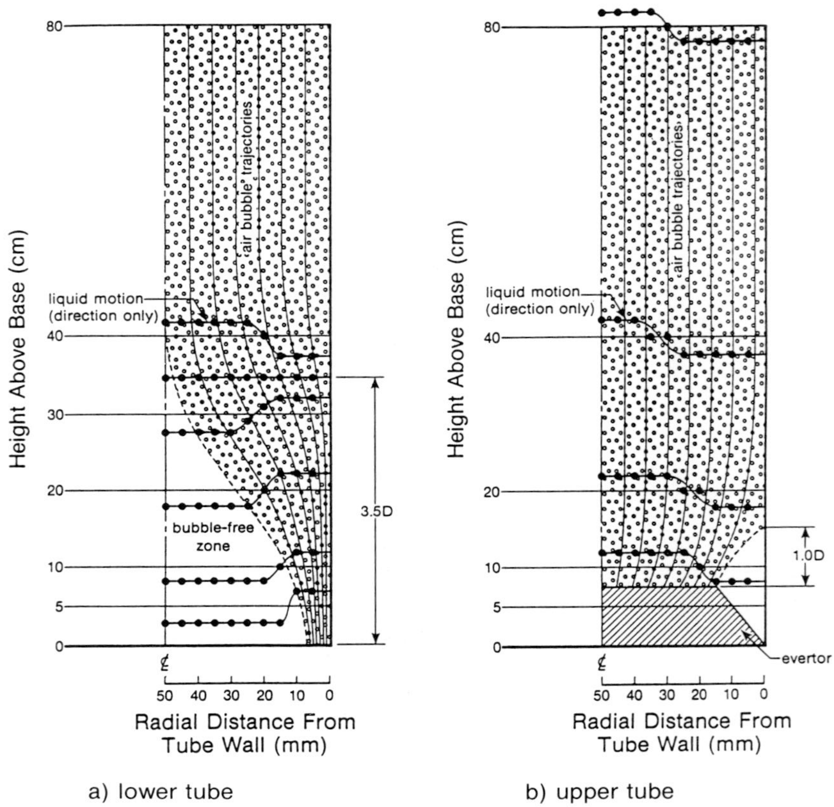

Fig. 5.13. Flow patterns in aerosyphon tubes (after Lock and Kirchner 1989*a*): (a) lower tube; (b) upper tube.

effect will be felt mainly in the heated section. It was found that the effect of orifice ring diameter ratio was negligible in the range considered. The orifice diameter and number of orifices, on the other hand, showed a modest but systematic influence, the heat transfer rate increasing with both. Not surprisingly, the heat transfer rate also increased with the air flow rate. The reasons for this are best understood in terms of the physical behaviour of the bubbles.

To a first approximation, bubbling in the aerosyphon produces a similar effect to boiling in the evaporative thermosyphon, at least in the lower reaches of the nucleate boiling régime. However, there are important differences. Obviously, the effect of latent heat is absent in the aerosyphon, and the air bubbles expand with height principally because of a decreasing hydrostatic head, not as the result of evaporative growth produced by liquid superheat,

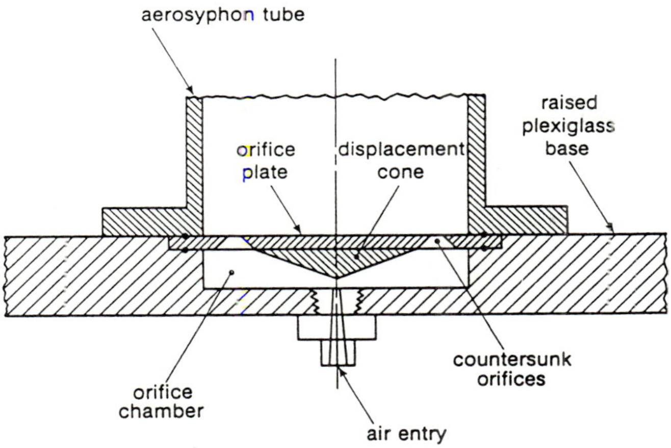

Fig. 5.14. Aerator configuration (after Lock and Abdurahman 1988).

which has no meaning in this context. Another important distinction is the independent control of bubble characteristics exercised through aerator geometry which determines both the size and frequency of the bubbles.

The behaviour of individual bubbles in a quiescent liquid may be described on the régime map displayed in Fig. 5.15. This shows the Reynolds number–Eötvös number domain in which

$$Eo = \frac{4R^2 g \Delta\rho}{\sigma}$$

has the same structure as the Bond number used in previous chapters. It is desirable to obtain a bubble Reynolds number in the range 10^2–10^4 where the non-linear trajectory of the bubble and turbulence in its wake both serve to increase the heat transfer rate. Immediately above the orifice, the bubble diameter is not usually much different from the orifice diameter and the Eötvös number may be quite small (e.g. $O(1)$), but as the bubble rises its pressure and density both decrease while its mean diameter increases; the result is a substantial increase in Eo and a tendency towards entering the spherical cap régime. Figure 5.12 reveals that bubble trajectories under these conditions are typically non-linear, beyond the spherical régime. Stroboscopic photographs of individual bubbles confirmed the wobbling behaviour and spherical-cap shapes.

Wallis (1974) has developed relationships between the velocity and radius of a single bubble, as incorporated in the terminal Reynolds number

$$Re_\infty = \frac{2U_\infty R}{\nu} = 2u_* r_*$$

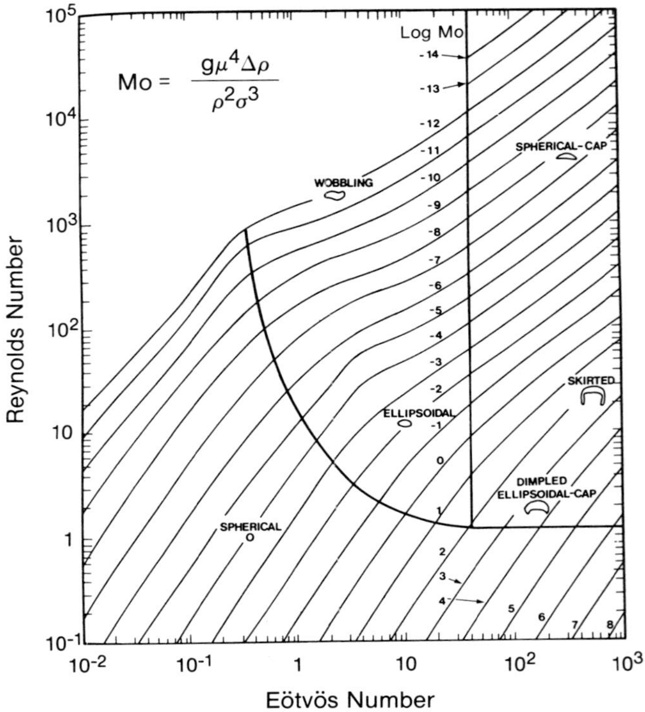

Fig. 5.15. Archimedean bubble régimes (after Clift *et al.* 1978).

where

$$u_* = U_\infty \left(\frac{\rho_L^2}{\mu_L g \Delta\rho} \right)^{1/3}$$

and

$$r_* = R \left(\frac{\rho_L g \Delta\rho}{\mu_L^2} \right)^{1/3}.$$

In the bubbly flow régime, for example,

$$u_* = a r_*^m, \tag{5.20}$$

where a increases from 0.3 to 2.5 and m decreases from 1.21 to 0.5 as the bubble velocity increases. In this range, the wake of the bubble gradually changes; firstly vortices are formed and shed, and then a turbulent wake develops. Beyond this range, the bubble cannot be treated as a sphere, except in calculating the mean radius of its volume, and behaviour changes radically. To begin with, the trajectory of the bubble becomes a spiral and the exponent in eqn (5.20) becomes negative in a régime where the Weber number $We = \rho_L U_\infty^2 R/\sigma = O(1)$. This implies a local maximum in the $u_* \sim r_*$ curve near the point where inertial forces balance surface tension forces, and where

$$u_* = \left(\frac{8}{Mo}\right)^{1/6} / r_*^{1/2} \tag{5.21}$$

in which Mo is the Morton number introduced in Fig. 5.15. Immediately following this régime is another in which the Kutateladze number[9] $Ku = U_\infty^2 \rho_L/(\sigma g \Delta\rho)^{1/2} = 2$, and

$$u_* = \left(\frac{64}{Mo}\right)^{1/12} \tag{5.22}$$

This corresponds to a local minimum beyond which the bubbles have a spherical cap and obey the relation

$$u_* = r_*^{1/2}. \tag{5.23}$$

The behaviour of single bubbles at their terminal velocity provides a useful aid in the interpretation of the heat transfer characteristics of the aerosyphon but this information cannot be applied directly to actual devices. There are several reasons for this. Firstly, the bubbles do not, in general, form and move at their terminal velocity, although the discrepancy may be quite small. Secondly, the decreasing pressure exerted on the rising bubble will cause alterations in the coefficient and exponent in eqn (5.20) along the bubble path, even assuming a quasi-steady development. Thirdly, the presence of impurities in the liquid may greatly alter bubble behaviour (Wallis 1974). In particular, the negative exponent régime represented by eqn (5.21) may be suppressed if the impurities are surfactants and substantially alter the surface tension by congregating near the rear stagnation point. However, the régimes above this, that is those represented by eqns (5.22) and (5.23), are largely unaffected by impurities.

The interaction between bubbles ascending in clouds is also very important. At the very least, their proximity causes overlap in their tubulent wakes; turbulence is thus enhanced along with overall heat transfer effectiveness. The bubbles may also collect in swarms (Lock and Kirchner 1989*a*). It has already been noted that bubble clouds tend to spread and may therefore fill the entire cross section of a tube if sufficient elevation is allowed. This is particularly noticeable when the bubbles enter through the core region but, as Fig. 5.13(b) indicates, it does not eliminate the core–annulus structure of the liquid flow;

[9] Here defined by $Ku = (Fr\,We)^{1/2}$. More commonly it is defined by $Ku = (Fr\,We)^{1/4}$.

the primary circulation still continues, though perhaps with reduced vigour. On the other hand, when the bubbles enter through the annular region, their migration towards the tube axis is slow enough to permit a bubble-free zone at the base of the aerosyphon, as indicated in Fig. 5.13(a); liquid descends in this zone and rises near the wall. Above the bubble-free zone, the liquid circulation pattern reverses, thereby creating a natural eversion zone several diameters above the base. In water, this 'coupling' zone was found to be as unstable as its counterpart in the single-phase thermosyphon, but may contain a convective coupling mechanism in more viscous liquids. The reader may wish to consult Section 2.3 for more information.

Since liquid circulation velocities are determined by Archimedean buoyancy forces, it is natural to expect the (longitudinal) convective flux to increase with the void fraction, bearing in mind that as this fraction approaches unity the liquid circulation would tend to zero. Large void fractions create entirely different conditions, ranging from plug flow through churn flow to annular flow. Experience with the evaporative thermosyphon suggests that aerosyphon performance would improve with the onset of churn flow; improvements may also occur with plug flow, though perhaps intermittently. The same experience also indicates that performance will be limited by entrainment and flooding.

The behaviour of bubbles driving the circulation is crucially important, but control may only be exercised externally through the air flow rate. The liquid injection characteristics recorded in Fig. 5.16 suggest that increases in air flow rate would not simply continue to increase bubble diameter and frequency. Evidently, the bubble frequency may become great enough to permit the chain of bubbles leaving the orifice to form a continuous jet, and if the flow rate is increased even further this jet is likely to break up into a dense cloud of very fine bubbles. It therefore appears that aerosyphon operation is best served by air flows rates below the critical value corresponding to jet formation. In practice, this implies void fractions which are much less than 1.0.

Under such conditions, an increased flow rate gives rise to an increased void fraction and an increased bubble diameter. The first effect implies an increased (mean) liquid velocity while the second, taken together with the $u_* \sim r_*$ equations discussed earlier, implies a higher level of turbulent mixing. It is thus to be expected that increasing the air flow rate would increase the heat transfer rate. This expectation has been confirmed by the work of Lock and Abdurahman (1988) who sought a relation between the tube Nusselt number $Nu_D = hD/k_L$ and the bubble Reynolds number $Re_b = U_\infty d_b/\nu_L$ based on the $u_* \sim r_*$ relations and the observations of Peebles and Garber (1953). Figure 5.17 illustrate their findings for air in liquid methanol, with 24, 48, and 60 orifices. This reinforces the basic physical description of bubbly flow in short aerosyphon tubes and reveals that the bubble Reynolds numbers fell mainly in the range $10^2 < Re_b < 10^3$; that is, the bubbles were spiralling and non-

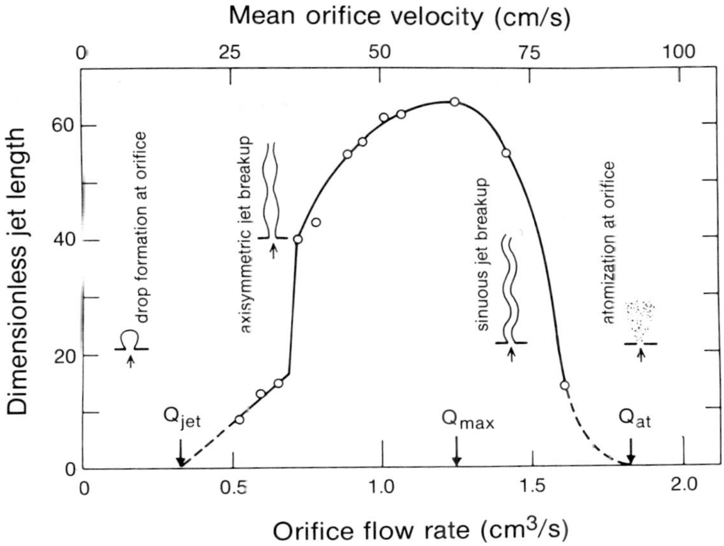

Fig. 5.16. Liquid injection characteristics (after Meister and Scheele 1969).

spherical, according to Fig. 5.15. The correlation could be extrapolated to lower air flow rates where interaction effects would be reduced; however, the data leave the question of extrapolation to higher bubble Reynolds number unanswered.

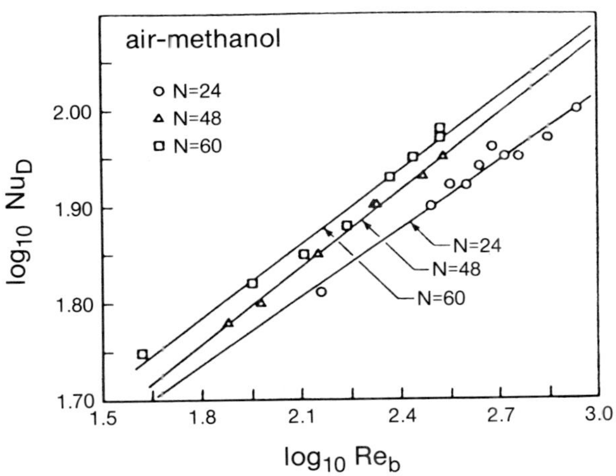

Fig. 5.17. Effect of bubble Reynolds number and number of orifices on heat transfer in an aerosyphon (after Lock and Abdurahman 1983)

In this form, the data are of limited value to the designer who is seldom in a position to know the bubble Reynolds number *a priori*. To obviate this difficulty, the results may be presented with the abscissa as the orifice Reynolds number $Re_o = U_o d_o/v_a$, where d_o is the orifice diameter and U_o is the mean air velocity at the orifice. Assuming that the effects of air flow rate, orifice diameter, and number of orifices may all be represented by power law forms, the data correlate within experimental uncertainty. The result is shown in Fig 5.18 from which it is apparent that the orifice Reynolds number exponent is approximately 0.14.

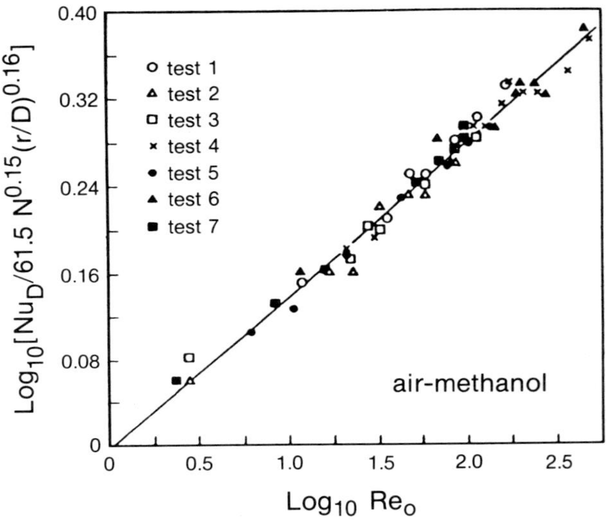

Fig. 5.18. Heat transfer behaviour of an aerosyphon (after Lock and Abdurahman 1988).

Perhaps the most important conclusion to be drawn from the above results is the confirmation that heat transfer coefficients in an aerosyphon are greater than those in the corresponding single-phase thermosyphon; typically, about twice. The discussion of bubble physics leads to the suggestion that the overall effect of aeration is one of intense augmentation manifest through increases in convection and the level of turbulence. To test this hypothesis further, Lock and Kirchner (1989*b*) extended the methanol data to include the effect of tube geometry, as defined by the length–diameter ratio L_H/D and the heated–cooled length ratio L_H/L_C. To begin with, an exploratory experiment was undertaken with $L_H/D = 30$ and $L_H/L_C = 3$ so as to examine the effect of deliberate measures to keep bubbles out of the core in the heated section. This was done using wire mesh deflectors at 1.0 m and 2.0 m above the base of a 3.0 m heated section. The heat transfer data thus obtained were found to be indistinguishable, within experimental error, from those obtained with a bare tube. Visualization studies (Lock and Kirchner 1989*a*) suggest that the explanation may lie in the natural eversion process above each deflector; that is, over most of the distance separating deflectors the liquid may have been everted, the core moving up and the annulus moving down, as in Fig. 5.13(a). If so, the use of a concentric, tubular flow divider separating core and annulus may be necessary if further improvements in heat transfer are sought. This measure was discussed earlier in Chapter 3 and will be taken up later in this section and again in the next section.

Of the two geometrical parameters, the heated–cooled length ratio has the least effect, with the exponent in a power law correlation being 0.34. Not unexpectedly, increasing L_H/L_C causes a decrease in heat transfer rate, a fact which may be attributed to either or both of two effects: an increase in thermal resistance of the (decreased) cooled section, and an increase in hydraulic resistance of the (increased) heated section. An increase in length–diameter ratio also creates a decrease in heat transfer effectiveness, the exponent of 1.04 revealing the strong influence of hydraulic resistance. The two geometrical effects appear to be independent of each other, and it is interesting to note that the L_H/D exponent is approximately three times that of the L_H/L_C exponent. The effect of tube geometry may thus be represented by the single parameter $L_H/L_C\ (L_H/D)^3$, precisely the same non-dimensional group that arises under single-phase conditions (see Section 2.3). The overall correlation is shown in Fig. 5.19 which reveals that this power law form is adequate for orifice Reynolds numbers greater than about $10^{1.2}$. The figure also indicates that the exponent of Re_o is about 0.18, which is slightly higher than the value found in Fig. 5.18. This suggests that, in general, it may not be possible to represent the effect of bubble physics on heat transfer fully by means of single exponent of the Reynolds number.

It is not difficult to see how aerosyphon performance might be improved within and beyond the orifice Reynolds number range discussed above. Most obvious, perhaps, is the fact that the data are limited to a single, annular ring of orifices located in the base. This restriction is rather like limiting boiling in the evaporative thermosyphon to a narrow ring around the edge of the base. At the very least, the distribution of orifices across the base is worthy of further exploration. A central, rather than a peripheral, concentration, for example, may lead to a single, ascending liquid column in both the heated and cooled sections; that is, it may replace the figure-of-eight circulation pattern with the single loop pattern of the evaporative thermosyphon. Alternatively, the distribution of orifice rings along the length of the wall may facilitate loop circulation in the opposite direction. To the extent that thermal buoyancy is dominated by Archimedean buoyancy, the precise direction would appear to be unimportant in either section of the tube. Multiple rings may serve to intensify the annular upflow, and may even reduce the undesirable radial

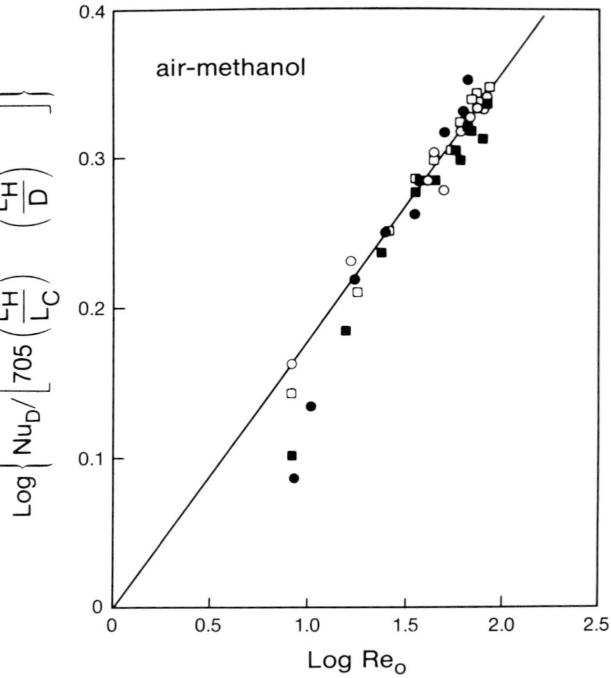

Fig. 5.19. Effect of tube geometry on heat transfer in an aerosyphon (after Lock and Kirchner 1989*b*).

spread of bubbles. As the number of rings was increased, the behaviour would tend towards a limit corresponding to a perforated or porous wall.

While an increased orifice density is likely to boost performance, it will also increase the cost of manufacture and the complexity of operation. The use of flow dividers, on the other hand, may be an attractive alternative. For example, if the flow divider in a vertical aerosyphon consists of a perforated or mesh tube which simply ensures that aerated liquid, moving upwards, is kept separate from pure liquid, moving downwards, then it offers a simple and inexpensive method of circulation control. It may permit the use of higher local void fractions without the adverse circulation-blocking effects that can accompany large bubbles. However, here again is an unexplored area. It is not known if large gas plugs do reduce the heat transfer rate in a bare aerosyphon tube; one can only speculate on their effect in the presence of a flow divider (Bezrodnyi and Antoshko 1990). In any event, the circulation pattern is best suited to the vertical, or near-vertical, orientation for which the device was originally conceived. Unlike the heat pipe, the aerosyphon is not easily adapted to tilted or non-linear configurations; but neither is it restricted to a closed form. Its open form combines simplicity with a thermal effectiveness approaching that of the closed evaporative thermosyphon.

Finally, it is worth noting that, apart from its heat transfer performance, the aerosyphon differs from its single-phase and evaporative counterparts in another fundamental way. When the heated section lies beneath the cooled section, a thermosyphon acts as a one-way heat valve or thermal *diode*. The aerosyphon also exhibits this one-way behaviour, but the control of air flow rate enables the device to operate as a thermal *triode*. This feature may be used in a feedback loop to regulate the thermal resistance of the device. In this connection, it is important to recognize that the response time of an aerosyphon undergoing changes in air flow rate is much smaller than that ordinarily associated with thermal equipment. The effect of a sudden change in air flow rate on overall temperature difference in the aerosyphon studied by Lock and Abdurahman (1988) is shown in Fig. 5.20. The time constant in this and other similar tests was found to be slightly greater than a minute. This suggests a rapid response in temperature or heat flux, and thus lends itself to effective thermal control.

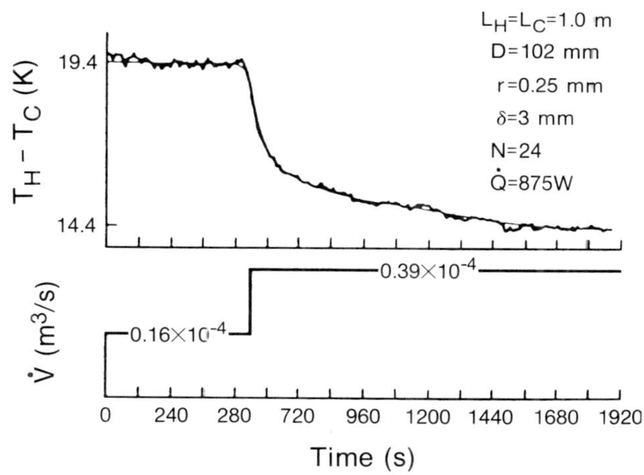

Fig. 5.20. Transient response of an aerosyphon (after Lock and Abdurahman 1988).

5.3 The bayonet tube

In the heat pipe, the existence of a liquid–vapour interface enables surface tension to be exploited in the creation of a capillary head which drives the circulation. The absence of a natural interface in the single-phase thermosyphon is circumvented in the aerosyphon by means of aeration which creates Archimedean buoyancy forces that drive the circulation. Capillary action in the heat pipe is associated with a latent heat transfer but buoyancy action in the aerosyphon is not. A third strategy, which is not restricted to either single-phase or two-phase operation, is to drive the circulation with a pressure differential. Like the aeration method, this allows the device to be open but, in contrast, a pressure-driven flow is not limited to sensible heat transfer.

In essence, this method is simply forced convection superimposed on natural convection or boiling (and condensation), but the context is significant. Characteristically, the device is some type of bayonet tube. Depicted schematically in Fig. 5.21, this finds extensive use in situations where the medium to be heated or cooled is either too large to be treated in its entirety or is readily accessible from one side only. The medium may be solid ground, a gas stream, etc., and is penetrated by the insertion of bayonet tubes, in much the same way that is employed with thermosyphon tubes.

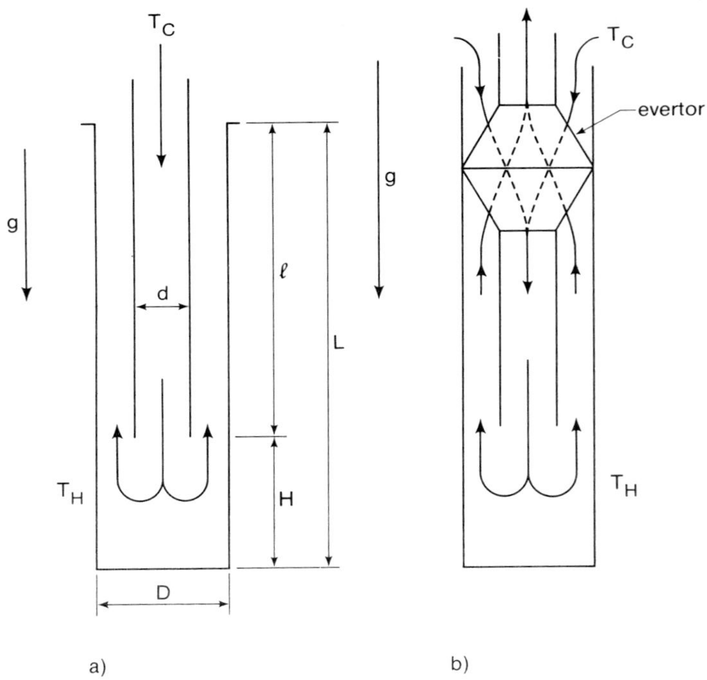

Fig. 5.21. Schematic of bayonet tube: (a) central entry; (b) annular entry.

The earliest industrial uses of the bayonet tube were in heat exchangers (Gel'perin 1939; Hurd 1946) where it simplifies installation and maintenance by permitting independent, external assembly of the tubes. Filling, and later inspection or replacement, may thus be undertaken without having to shut down the entire heat exchanger. Typically, the tubes are subject to lower bending, compressive, and thermal stresses than conventional heat exchanger tubes, and thus simplify design. More recent industrial applications include a potassium condenser–water boiler in a binary cycle power plant (Fraas 1973) and a counterflow heat exchanger in chemical processing plants (Zhang and Zhang 1988).

In a geophysical context, the bayonet tube may be used to extract geothermal heat for use in the generation of steam or hot water above the ground surface (Fukusako *et al.* 1987). Like the cryosyphon, it also finds use in the creation or maintenance of frozen ground through the circulation of refrigerants or cold winter air (Jahns *et al.* 1973; Haynes and Zarling 1982; Lock *et al.* 1989). Similarly, it may be inserted downwards through the winter ice cover on a river to help build a submerged wall of ice, thereby facilitating the control of spring runoff (Lock 1986*b*).

Other uses of the bayonet tube vary as widely as those of the thermosyphon; perhaps two examples will be sufficient here. In outer space, the tubes are evidently suitable for the collection of solar radiation (Beekly and Mather 1978); presumably this use extends to the earth. On a very small scale, the device has also found service as a fine probe at the tip of which precise control of low temperatures has facilitated the development of cryosurgery; in particular, it has made possible the treatment of medical conditions declared otherwise inoperable (Rowbotham *et al.* 1959; Cooper 1967; Onik *et al.* 1987).

In the bayonet tube, as in the heat pipe and the aerosyphon, it is preferable for the added driving mechanism to reinforce the natural thermosyphon circulation which would exist in its absence. This is evident in Fig. 5.21(a) where the wall temperature T_H is taken to be greater than that of the fluid T_C entering the insert tube at the top. The same result is achieved in Fig. 5.21(b) where the fluid is shown entering an annular mouth in which it is everted before flowing down into the tube proper. In these situations, we encounter an example of mixed convection in which the forced flow aids the natural flow. Of course, when the forced flow dominates, the bayonet tube is no longer a flow-enhanced thermosyphon, and the direction of the flow becomes less important. The general situation is one to which the familiar régime map of Metais and Eckert (1964) for circular tubes does not strictly apply. The régime map for refluent flows does not yet exist.

A general zonal model of the primary flow pattern is suggested in Fig. 5.22. This divides the tube into four principal zones, here illustrated with axial entry down an insert tube of length l:

1. Entrance zone. Fluid moving down inside the insert tube creates an entry flow for the 'open' area beneath. The insert tube flow itself may be developing or developed, depending primarily upon the insert–diameter ratio l/d and the insert Reynolds number Re_d.

2. Turning zone. Before returning and entering the annulus, the flow must

double back in an effective length which will depend upon fluid inertia and the tube geometry.

3. Thermosyphon zone. At the bottom of the turning zone, conditions approximate those at entry to a pure, open thermosyphon. The effective length of this end zone is an inverse function of the insert Reynolds number, and its internal details depend mainly upon the tube Rayleigh number Ra_D.

4. Annular zone. This is a mixed convection zone in which the flow details depend upon the Reynolds number, Rayleigh number, and length–width ratio of the gap $(D-d)$. From a heat transfer standpoint, this is usually the most important zone.

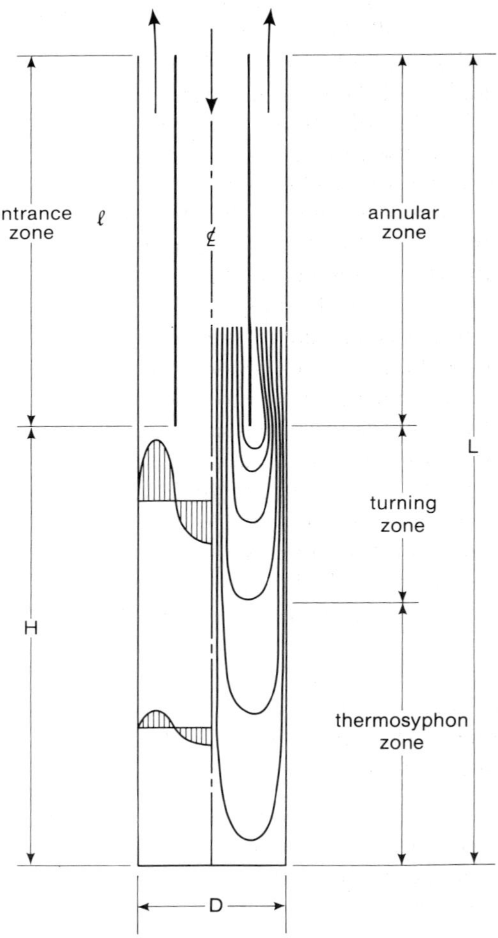

Fig. 5.22. Zonal model of bayonet tube (after Lock and Kirchner 1990).

The zonal model lends itself to the calculation of pressure drop and heat transfer rate by assuming that each zone may be treated separately in a one-dimensional description. The pressure losses, for example, may each be represented by the empirical form

$$\Delta_j P = K_j \dot{V}^2 \tag{5.24}$$

where $\dot{V}$ is the volumetric flow rate and K_j is a loss coefficient defined in terms of the geometry and local conditions to which it refers. The overall pressure drop may thus be written

$$\Delta P = \sum_j \Delta_j P = K\dot{V}^2 \tag{5.25}$$

where $K = \Sigma_j K_j$ is the overall loss coefficient, assuming the fluid is incompressible.

A representative version of this model is given by Lock *et al.* (1989) for an air-filled bayonet tube 6.0 m long and 0.3 m diameter. With an insert tube 5.4 m long and 0.212 m diameter, this corresponds to $d/D = 0.7071$, $l/L = 0.9$, and $L/D = 20$. Figures 5.23 and 5.24 are results from a sensitivity analysis around this 'standard' configuration for which the overall loss coefficient is signified by K_s. Figure 5.23 reveals that the effect of the diameter ratio d/D is to create a minimum pressure drop for a given flow rate; under these conditions the calculated minimum corresponds to the cross-sectional area of the annulus being equal to that of the insert tube, as might be expected. Using this 'optimum' diameter ratio, the influence of length–diameter ratio is explored in Fig. 5.24 from which it is evident that the effect is monotonic and substantial.

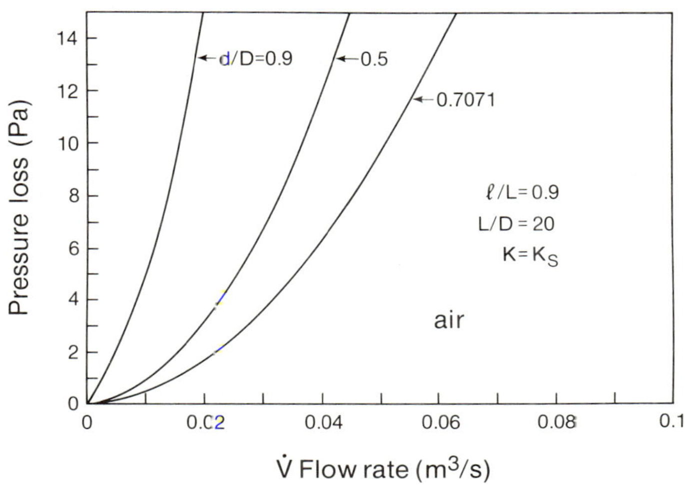

Fig. 5.23. Effect of diameter ratio on pressure drop in a bayonet tube (after Lock *et al.* 1989).

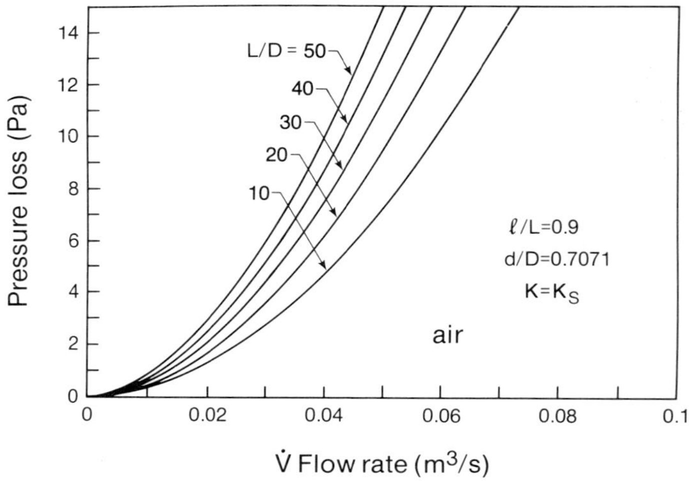

Fig. 5.24. Effect of length–diameter ratio on pressure drop in a bayonet tube (after Lock *et al.* 1989).

The one-dimensional, pressure-loss model is useful for comparative purposes but suffers from two limitations. Firstly, the loss coefficients are often approximations based on the nearest geometry for which empirical data are available; variability in K_j may be as much as $\pm 50\%$. Secondly, the assumed flow behaviour in each zone, and its independence of behaviour in contiguous zones, may be incorrect; for example, separation may or may not occur as expected, and stability may not exist. Recent experimental data of Lock and Wu (1991*a*) for air provide some quantitative comparisons under laminar conditions. Of particular interest here are the effects of diameter ratio and tube length ratio on the Euler number $Eu = \Delta P/\frac{1}{2}\rho\bar{U}^2$. These are shown in Figs 5.25 and 5.26, respectively, with the diameter ratio replaced by the area ratio F_i/F_a while the length ratio is replaced by its complement, the clearance ratio H/D. Since $F_i = \pi d_i^2/4$ and $F_a = \pi(D^2 - d_o^2)/4$, where d_i and d_o are the inner and outer diameters, respectively, of the insert tube, the area ratio F_i/F_a automatically takes account of the thickness of the insert tube wall. The clearance ratio H/D focuses attention on the turning zone for each of two cap geometries: hemispherical and cylindrical, the latter referring to the use of a plane end flange.

It is immediately obvious from Fig. 5.25 that the minimum pressure drop does not occur at $F_t/F_a = 1.0$ but nearer to half of this value, independently of the end cap geometry. This indicates that the inertial contribution to pressure drop, which is theoretically zero when $F_i = F_a$, is not the only one; in particular, the effect of the refluent flow pattern and the cross-sectional areas created in the end region must be accommodated. This leads quite naturally to the

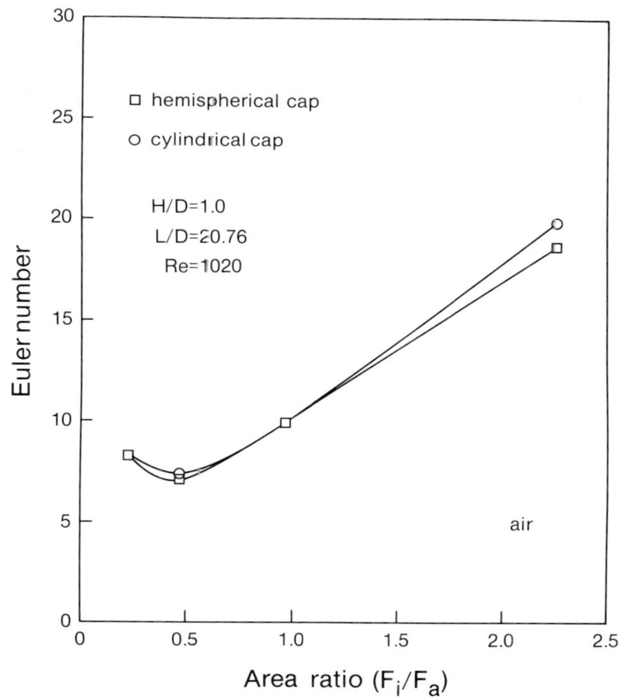

Fig. 5.25. Effect of area ratio on Euler number in a bayonet tube (after Lock and Wu 1991*c*).

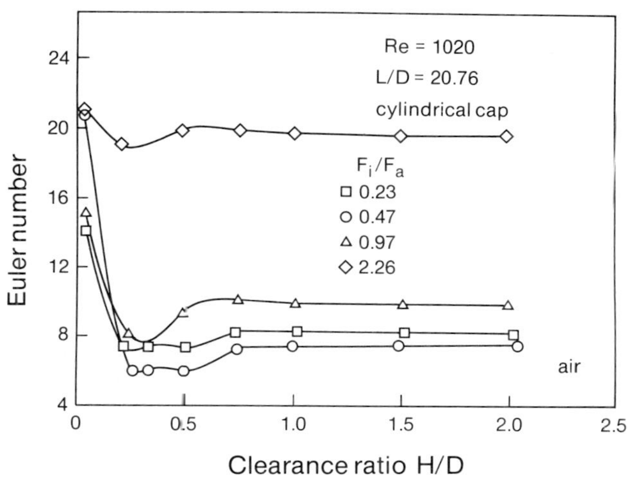

Fig. 5.26. Effect of end clearance ratio on Euler number in a bayonet tube (after Lock and Wu 1991*a*).

discussion of behaviour in the turning zone, and to the effect of clearance ratio on this behaviour. Figure 5.26 reveals it to be very complex.

For very small clearances (e.g. $H/D<0.1$), the effect of any further reductions in clearance ratio is to create large increases in the pressure loss which is then attributable to two causes: separation and reattachment of the flow, as discussed below; and the constriction imposed by a reduced flow area. On the other hand, for large clearances (e.g. $H/D>1.0$), further increases in clearance ratio produce little effect, as might be expected when the closed end of the tube recedes far enough from the mouth of the inner tube.

Within the intermediate range $0.1<H/D<1.0$, the behaviour is not so easily explained, and is best understood using the flow visualization results presented in Figs 5.27 and 5.28. The first of these, corresponding to F_i/F_a less than the minimal value of 0.47, indicates that a stable toroidal separation zone forms on the *outside* surface of the insert tube when H/D is very small; this torus gradually moves downwards with the end of the main tube as H/D is increased. Under similar conditions, this separation ring has been detected numerically (Yang and Hsieh 1988). As Fig. 5.27(a) shows, the separation ring stands off the end of the insert tube when $H/D=0.25$–0.50, creating a minimum in the corresponding curve. Further increases in H/D evidently introduce greater viscous losses associated with an increased surface area and thus raise the Euler number towards an asymptotic value.

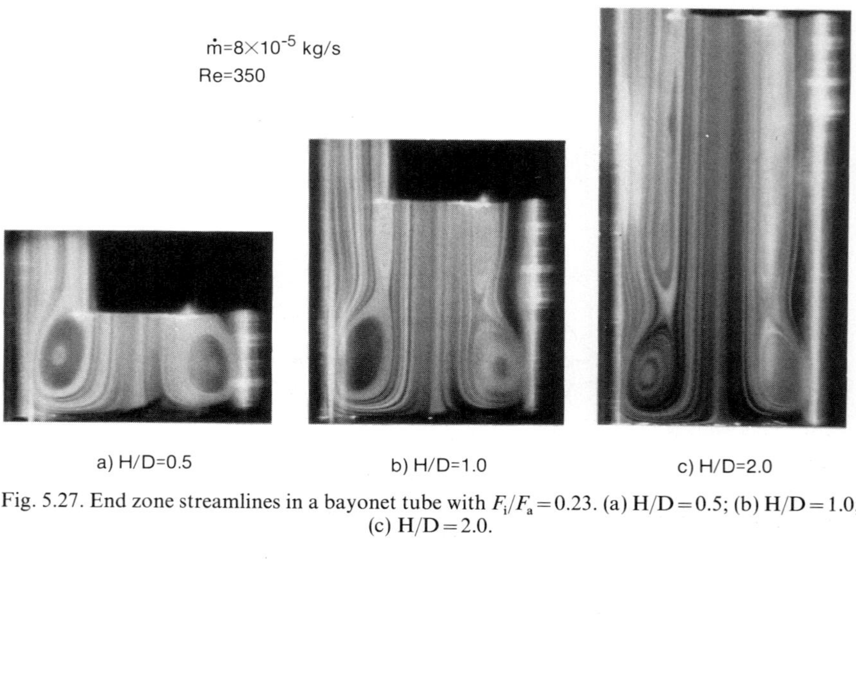

Fig. 5.27. End zone streamlines in a bayonet tube with $F_i/F_a=0.23$. (a) H/D = 0.5; (b) H/D = 1.0; (c) H/D = 2.0.

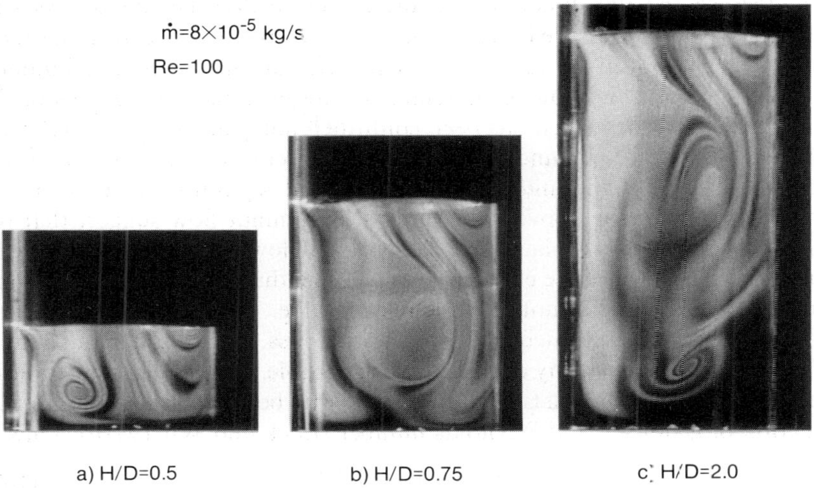

Fig. 5.28. End zone streamlines in a bayonet tube with $F_i/F_a = 2.26$. (a) H/D = 0.5; (b) H/D = 0.75; (c) H/D = 2.0.

The patterns in Fig. 5.28 correspond to a value of F_i/F_a greater than the minimal value, and reveal very different behaviour. Under this condition, the ring of separation appears on the *inside* surface of the insert tube, but this flow is evidently unstable in axisymmetric form, despite the symmetry of the imposed boundary conditions; however, it is stable in the three-dimensional form displayed in Fig. 5.28. The flow might best be described as a necked jet which issues from within the insert tube and develops a sinuous path before tranposing itself into a complementary return flow. Whether or not this flow is a bifurcation is unknown but it is clear that it is a stable form.

In view of the fact that the general shape of the $Eu \sim H/D$ curves shown in Fig. 5.26 is unchanged by F_i/F_a, the arguments advanced above for $F_i/F_a = 0.23$ evidently apply for larger values; in particular, for $F_i/F_a = 0.47$, the minimal value, which marks the division between the two types of flow patterns seen in Figs 5.27 and 5.28. The minimal value of F_i/F_a evidently corresponds to a ring of separation which is associated with neither the inside nor the outside surface of the insert tube. This suggests that a detached vortex spanning the lower edge of the insert tube represents the least possible dissipation of energy in the end region of a bayonet tube.

Difficulties also arise in the use of the one-dimensional, pressure-loss model under turbulent conditions. The pioneer study of Idel'chik and Ginzburg (1968) covers a wide range of the geometrical parameters H/D and F_i/F_a, with the Reynolds number remaining in the vicinity of 10^5. Their work demonstrates not only that H/D exerts a very strong influence on pressure loss but does so through fundamental alterations in flow behaviour. As with laminar flow, the pressure loss rises sharply as $H/D \rightarrow 0$ and tends to level out for $H/D > 2.0$, but in the range $0.2 < H/D < 2.0$ various maxima and minima are observed and the flow is characteristically unstable. The details of this complex behaviour, which has been confirmed independently by Zhang and Zhang (1988) and Lock and Wu (1991*b*), have yet to be clarified. The visual observations of Zhang and Zhang again reveal separation at entry to the annular zone but the above observations on laminar flow suggest that this alone is insufficient to explain three-dimensional flow patterns which evidently vary with the shape of the end cap. It is possible that separation on the inner surface again creates a motion which is unstable. It is certain that the flow alterations create large fluctuations in pressure loss, and thus map out a range of conditions where steady operation is impossible. Despite the uncertainties and instabilities, the mean Euler number may still be expressed as an empirical function of geometry and Reynolds number (Lock and Wu 1991*b*). Thus

$$Eu = 6.57 f_h / Re^{0.18} \tag{5.26}$$

for a hemispherical cap, and

$$Eu = 3.34 f_c / Re^{0.15} \tag{5.27}$$

for a cylindrical cap. In these expressions,

$$f_h = \left(\frac{L}{D}\right)^{0.63} \left[0.01 + \left(\frac{F_i}{F_a} - 0.67\right)^2\right]^{0.27} \left[0.001 + \left(\frac{H}{D} - 0.47\right)^2\right]^{0.048} \left[\frac{D}{H}\left(\frac{F_a}{F_i}\right)^{0.5}\right]^{0.29}$$

and

$$f_c = \left(\frac{L}{D}\right)^{0.68} \left[0.01 + \left(\frac{F_i}{F_a} - 0.71\right)^2\right]^{0.27} \left(\frac{H}{D}\right)^{0.25} \left[\frac{D}{H}\left(\frac{F_a}{F_i}\right)^{0.5}\right]^{0.32}$$

in which the optimum area ratio, $F_i/F_a \simeq 0.7$, is noticeably higher than the value for laminar flow, and an optimum end clearance ratio, $H/D = 0.47$, is evident only in the hemispherical-cap data. Equations (5.26) and (5.27) correlate the experimental data within $\pm 18\%$.

Turning now to heat transfer behaviour, the discussion will follow similar lines beginning with Fig. 5.22. This indicates that three zones participate in the overall heat transfer rate. In the thermosyphon zone, natural convection alone is operative and local heat transfer coefficients are therefore relatively low. In the annular zone, mixed convection ensures higher heat transfer coefficients. Flow uncertainties in the shorter turning zone make a general assessment there more difficult. Once again, a one-dimensional flow model may be used to estimate the individual heat transfer rates, but this approach is rejected in favour of empirical data, limited though they are.

As a general observation, it is worth pointing out that heat transfer rates in a bayonet tube are invariably greater than in an empty tube under otherwise identical conditions. In any given situation, this fact must always be weighed against the corresponding observation that the pressure drop is greater. As noted earlier, the principal flow alterations responsible for these changes are found in the turning zone, and it is therefore the parameters which most affect the flow in this zone that will be responsible for the details of heat transfer there. However, this does not imply that the *overall* heat transfer rate is set by the turning zone.

The *aeropile*, a bayonet tube using the natural speed and temperature of winter wind to maintain soil in frozen condition, provides a useful example (Lock *et al.* 1989). With a turbo-fan arrangement, in which a large wind turbine drives a generator which in turn provides power to drive one or more fans supplying air to the aeropile, the Reynolds number of the tube may be made as large as necessary. In this situation, natural convection will play only a minor role and heat transfer will occur mainly as the result of turbulent forced convection in the annular zone. The turbo-ventilator provides an alternative arrangement in which a propeller-fan, mounted on the top of tube and driven directly by the wind, draws the cold air through the system. Although simpler, and less expensive, this system is limited by the elevation of the propeller-fan above the ground surface and by the relatively low Reynolds numbers, typically in the laminar range.

Experimental work with small-bore aeropile tubes (Lock and Kirchner 1988, 1990) reveals the complexity of behaviour. This was undertaken in a wind tunnel with the entry of the insert tube being bent to face upstream, thus providing a dynamic head. Under these conditions, the annulus Reynolds number $Re=\bar{U}(D-d)/\nu$ varied in the range $0<Re<2550$ while the tube Rayleigh number $Ra_D=\beta g(T_W-T_\infty)D^3/\nu\kappa$ covered the range $10^{3.1}<Ra_D<10^{4.8}$. According to the map of Metais and Eckert (1964), the data extend over several régimes: laminar natural convection, laminar forced convection, mixed laminar convection, and forced turbulent convection. It is therefore unlikely that a single correlation will fit such data accurately, especially in view of the appearance of chasmic thermosyphon behaviour noted in Chapter 2. The equation

$$Nu_D=2.36\,\frac{l}{L}\,(Gz+8\times10^{-7}t_H^{1.3})^{0.77} \tag{5.28}$$

provides an adequate representation for $Nu_D\geq 10$ when $L/D=10:1$. This form suggests that natural convection, incorporated through

$$t_H=Ra_C D/(L-l),$$

is additive to, but much smaller than, forced convection which is represented by the annulus Graetz number

$$Gz = \frac{\bar{U}(D-d)^2}{\kappa l}$$

where $\bar{U}$ is the mean annular velocity. The index of 0.77 in eqn (5.28) suggests a turbulent flow.

For longer tubes, and large end clearances, the heat transfer rates are generally lower, particular for air under laminar conditions when the chasmic thermosyphon behaviour again reappears (Lock and Kirchner 1990). A typical set of data are shown in Fig. 5.29 from which it is evident that heat transfer rates are substantially above those of the corresponding open thermosyphon. The form of eqn (5.28) does not fit data such as these, at least not to the left of the chasms, but the simpler form

$$Nu = A(Ra)Re^n\left(\frac{D-d}{L}\right)^m \tag{5.29}$$

may be adequate with a fixed value of the end clearance ratio H/D. For a typical bayonet tube, $H/D = O(1)$ whereas $L/D \gg 1$, and hence mixed convection in the annulus is by far the greatest contribution to the overall heat transfer rate. The Nusselt, Rayleigh, and Reynolds numbers are then more logically based on the gap.

In the absence of a superimposed forced flow, the bayonet tube reduces to a concentric tube, open thermosyphon. It may be recalled from Chapter 2 that the separation of annular and core flows then has the potential of avoiding adverse mixing, particularly in the vicinity of the tube mouth. Visual observations confirm this potential, and heat transfer measurements reveal a substantial improvement over the simple thermosyphon (Seki *et al.* 1980). Not surprisingly, the improvement is found to be a function of the area ratio F_i/F_a, with the optimum occurring in the vicinity of $F_i/F_a = 0.35$, noticeably lower than the values observed in the bayonet tube proper.

Representative water data are shown in Fig. 5.30 in which both the Nusselt number and the Rayleigh number are based on a modified form of hydraulic mean radius defined by

$$a = \frac{(d^2 - d_o^2)}{2D}.$$

It is interesting to note from this figure that the familiar thermosyphon régimes are clearly evident, and are adequately described by the single variable aRa/L which largely contains the effect of tube geometry, at least in these experiments. This variable automatically covers simple thermosyphon data as

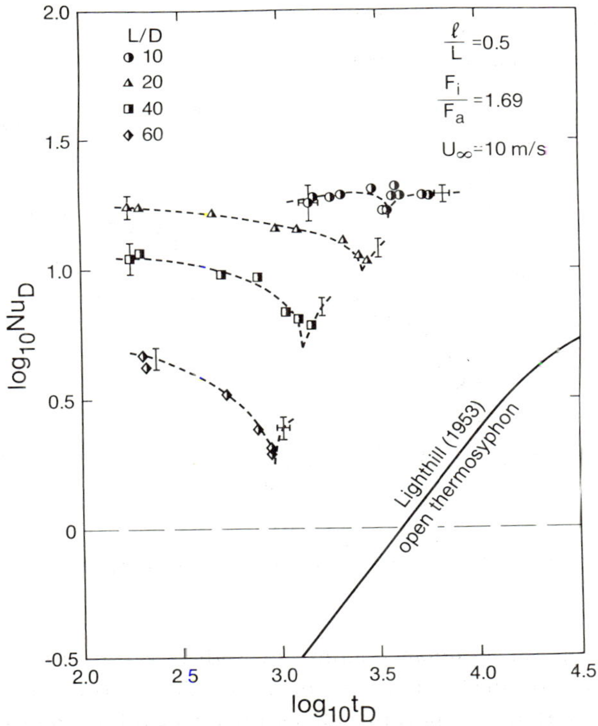

Fig. 5.29. Effect of length–diameter ratio on heat transfer in a wind-driven bayonet tube (after Lock and Kirchner 1900).

$d_o \rightarrow 0$; as indicated, the empty tube data are substantially below the concentric tube data.

The concentric tube, evaporative thermosyphon[10] provides another striking example of a buoyancy-driven bayonet tube. This is shown schematically in Fig. 5.31 which depicts an insert tube extending the full length of the main tube, in this case closed, but allowing communication between the central and annular spaces through perforations near the ends. Natural convection augments heat transfer in the evaporator and reduces it in the condenser but the effect in both is typically unimportant; circulation is maintained in a single, closed loop created by the continuous flow divider. Although not clear from Bezrodnyi *et al.*'s figure, the flow appears to be further divided in the condensing section, perhaps to separate pure liquid, in the inner annulus, from a two-phase mixture next to the cold wall.

[10] Alternatively, this may be described as an evaporative thermosyphon fitted with a flow divider, as described in Section 3.4.

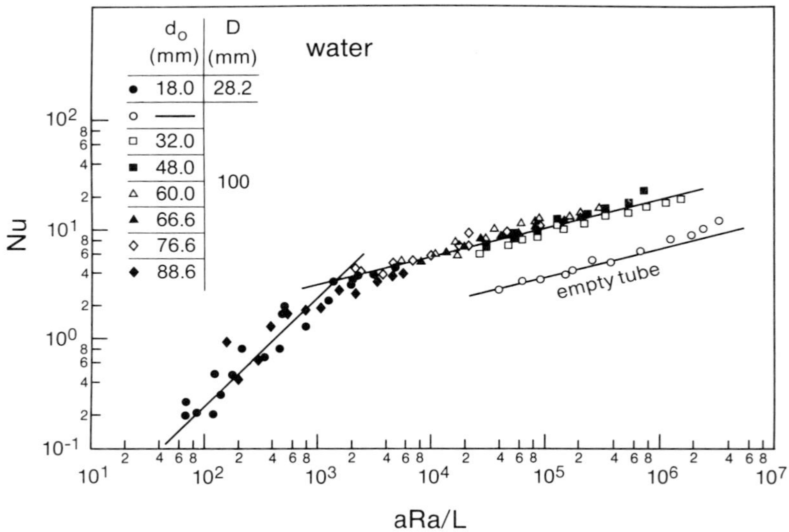

Fig. 5.30. Heat transfer in the single-phase, concentric tube thermosyphon (after Seki *et al.* 1980).

Focusing on the annular regions, Bezrodnyi *et al.* (1980, 1983) have interpreted the limiting behaviour of this system in the terms of two-phase flow characteristics. In the evaporator, increasing the heat flux tends to produce a typical behaviour sequence: bubbly flow followed by plug flow leading to annular flow. In the condenser, on the other hand, the sequence is different: falling film flow followed by flooding leading to annular flow. Obviously, these individual flow characteristics must be combined somehow in the determination of overall behaviour which also depends upon the filling charge. Bezrodnyi *et al.* have identified three principal régimes. With the charge ranging from 35% to 75% of the evaporator volume, bubbly flow is observed in the evaporator until overheating occurs, presumably because of burnout or dryout, though neither were identified. For charges closer to 100% of the evaporator volume, the maximum heat transfer rate is raised with the establishment of a cocurrent liquid film in the annulus; the limit is then determined by entrainment. Surprisingly, when the charge is much greater than the evaporator volume, the performance of the device is improved yet again. Under these conditions, bubbly flow extends over the full length of the annulus, with increased circulation from the additional head of liquid in the insert tube more than compensating for reduced local condensation rates. Presumably, even greater performance would be possible if an evertor were installed in the central region, thereby creating a figure-of-eight loop. The principal effect of geometry was found in the area ratio, as might be expected.

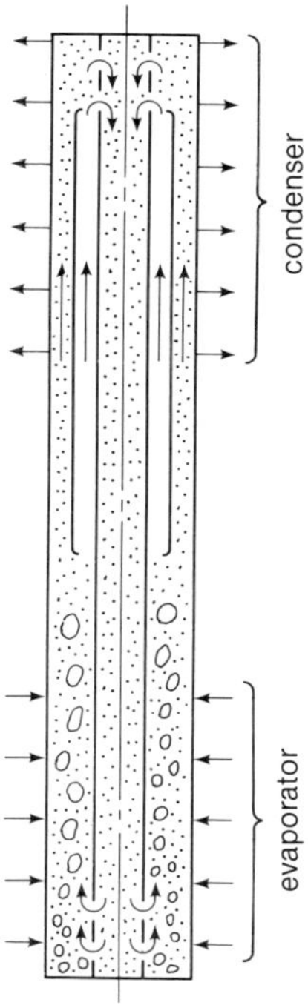

Fig. 5.31. Schematic of evaporative, concentric tube thermosyphon (after Bezrodnyi *et al.* 1983).

The optimum value, which balances the limiting effect of the liquid velocity in the insert tube against the limiting vapour velocity in the annulus, was in the range $0.19 < F_i/F_a < 0.25$, even lower than in the single-phase, concentric thermosyphon.

In closing this section, it is worthwhile returning to the opening idea that circulation and heat transfer rates may be enhanced, and indeed greatly increased, by the application of a pressure differential between the entry and exit of a bayonet tube. Strictly, the improvement results from forced convection superimposed on an annular flow which is physically divided from a replenishing core, but the concept may be extended to other forcing mechanisms. In the simple, concentric tube thermosyphon, for example, the opposing flows are divided but the forcing superimposed is simply thermal buoyancy; in the evaporative, concentric tube thermosyphon, on the other hand, the forcing is created through Archimedean buoyancy forces. Outside of a bayonet tube proper, the actual choice of forcing mechanism is a matter of convenience but is typically a pressure head: static, dynamic, capillary, etc. The role of the flow divider is crucial in all bayonet tubes. Not only does it prevent undesirable mixing, which reduces the net convective flux, it also influences the flow régimes and their interdependence. As we have seen, a judicious combination of insert geometry and forcing mechanism, internal or external, may produce heat transfer rates well above those obtainable in corresponding thermosyphons.

5.4 Anti-gravity devices

It is clear from previous chapters that thermally induced buoyancy forces which create a longitudinal convective flux are intrinsic to the tubular thermosyphon. In a gravitational field, this flux is usually associated with a downward temperature gradient, and is zero when the vector is directed upwards. Such a 'one-way' feature has advantages in certain situations but there are others where it is a disadvantage. Whenever it is necessary to transfer heat downwards there is need of an anti-gravity device. This closing section surveys related devices whose principles of operation make them especially suitable for work against gravity.

Theoretically, it is possible to reverse the circulation in a thermosyphon if the sign of any of the quantities in the buoyancy acceleration is reversed. Thus, if the fluid expansion coefficient β were uniformly negative, a simple tubular thermosyphon would produce a net convective flux vertically downwards when the cooled section lay below. However, this is a rare possibility and may be ignored for most practical purposes. In general, an anti-gravity capability may be created if either, or both, of two other conditions are met: fluid circulation is caused by another appropriate force field, independent of gravity; or circulation is established by external means. Each of the devices described earlier in this chapter provides an example of this capability, as will shortly become more apparent. Under single-phase conditions, the problems of altering the circulation is evenly divided between the heated and cooled sections of the tube; under evaporative conditions, it is essentially reduced to the problem of raising liquid from the condenser to the evaporator above.

The application of a centrifugal field, discussed extensively in Chapter 4, provides a convenient point of departure. Figure 5.32 illustrates two, single-phase, concentric anti-gravity thermosyphons which rotate about their own axes. For both the open and closed arrangements, heat is transferred downwards from T_H to T_C by virtue of the axisymmetric flow pattern created through indirect thermal buoyancy. Depending upon the sign of the local radial temperature gradient, the gravitational field may reinforce or oppose the main, centrifugally driven circulation.

In the open system, fluid enters an annular region at the mouth where the temperature is lowest (T_C). Moving vertically upwards, this annulus of fluid is heated by the wall and hence $\partial T/\partial R > 0$ near the wall; gravity augments the circulation in this region. Near the top of the tube, the fluid turns before

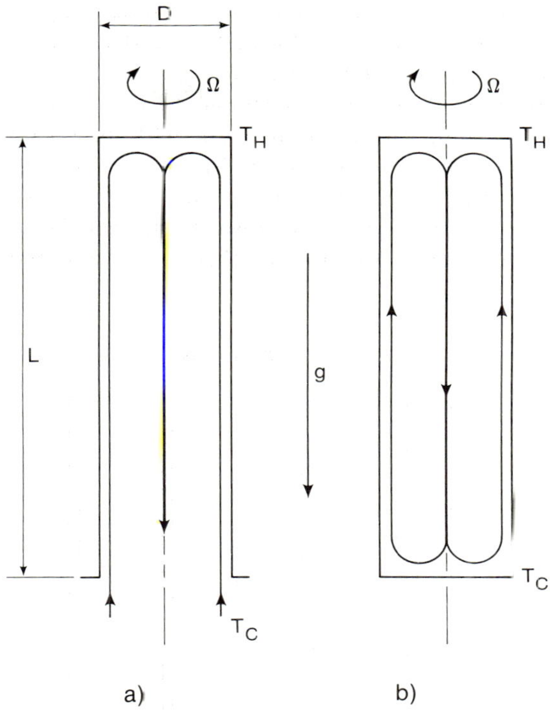

Fig. 5.32. Concentric, anti-gravity thermosyphons: (a) open; (b) closed.

moving downwards towards the mouth, *en route* to which it must pass annular fluid which is cooler. That is, $\partial T/\partial R<0$ in the returning core, where gravity tends to oppose the circulation. Both of these effects may be ignored within the tube when $\Omega^2 D \gg 2g$. At the mouth of the tube, however, gravity may continue to play a role because the warm central discharge naturally tends to spread out radially, thereby denying unrestricted entry to cooler fluid arriving from beneath. Were it not for the influence of Coriolis forces, this situation would be the inverse of that described in Section 2.2; see, for example, Figs 2.10 and 2.18. Presumably, the laminar flow pattern at the mouth in Fig. 5.32(a) would consist of a natural eversion process in which a cool core of fluid approaching from beneath splits into several filaments which later recombine inside the tube to form a rising annulus, while the warm exiting core interdigitates with these filaments before creating a spiralling, radially divergent, horizontal jet. No study of these rotational mouth conditions has yet been reported; nor has the use of eversion devices or tubular inserts been explored, particularly under turbulent conditions, where they would be most effective.

In the closed tube arrangement, illustrated in Fig. 5.32(b), the same general circulation pattern exists except that it is now completely contained within the tube. The possibility of adverse mixing in the mouth is thus eliminated but it is to be expected that the overall heat transfer rate would be lower than that of a comparable open tube because of the absence of a convective flux through the mouth. A similar comparison may be made with the eccentric (or annular) thermosyphon shown schematically in Fig. 5.33, although the circulation pattern is then basically bifilamental. In any of these schemes, the cooler, radially outermost fluid always flows towards the heated end while the return flow occupies the innermost region whether the space is enclosed or not.

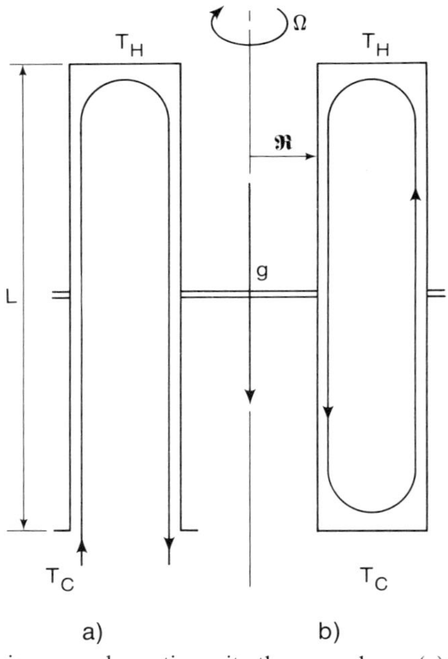

Fig. 5.33. Eccentric or annular anti-gravity thermosyphons: (a) open; (b) closed.

It will be recalled from Section 4.3 that the eccentric thermosyphon exhibits a secondary flow consisting of pairs of thermally induced vortices which are strongly influenced by Coriolis forces near the closed tube ends. When the spaces depicted in Fig. 5.33 are interpreted as an annulus, and not as single tubes, the thermally induced secondary flow ceases to exist, but the Coriolis twist effect remains near the ends; the lower the Ekman number, the greater the effect. The absence of the buoyancy-driven secondary motion presumably implies a lower heat transfer efficiency, much as observed previously in the horizontal thermosyphon when it is infinitely wide. However, for very small

Ekman numbers the two-dimensional nature of the flow may be altered by the appearance of cells, particularly near the end regions. Such a prospect has yet to be explored in both annular and tubular eccentric thermosyphons.

Many variations on the above rotational theme are possible, not least among which are non-linear configurations where the heated section is radial and the cooled section is axial, or vice versa. Of greater interest here, however, are evaporative thermosyphons rotating either concentrically or eccentrically. From Section 4.5, it is evident that indirect Archimedean buoyancy forces are capable of producing substantial convective fluxes along the tube, providing the acceleration ratio is large, that is the speed is high enough. This requirement is easier to meet in eccentric tubes but is often possible with concentric tubes, particularly for high-speed application, for example $\Omega > 150$ rad/s for a 4 cm radius tube.

For either type of evaporative axial thermosyphon, the charge must not be too large or too small. Ideally, the liquid phase should cover the entire inner surface of the tube with a film thick enough to permit high circulation rates but thin enough to promote rapid evaporation and condensation. Because of axisymmetry, this is easier to accomplish in the concentric tube. In the eccentric tube, pooling of liquid on the radially outermost surface makes it difficult to achieve naturally, particularly at high speed. Axial grooves cut along the innermost surface provide a partial remedy through capillarity, as does a slight tilt of the tube axis relative to the axis of rotation, the evaporator being further out.

Surface tension forces, acting in grooves or a wick lining, have a broad role to play in anti-gravity thermosyphons. Whether it is more accurate to ascribe these forces to the near-surface body force field responsible for the disjoining pressure, or simply to treat capillarity as a phenomenon in its own right, is perhaps a moot point, but there is no doubt that the head produced is capable of raising the condensate against gravity. Thus two design options emerge: using capillarity to assist centrifugally driven circulation (Gerasimov *et al.* 1985), and using capillarity alone. It is clear from Section 5.1 that the capillary pressure difference between the heated and cooled sections is independent of gravity and may therefore be regulated to maintain the liquid return when the heated section is above the cooled section (Voronin *et al.* 1976). However, as Fig. 5.5 reveals, this only provides freedom of operation within the capillary limit $(\Delta P_{cap})_{x=0}$ which may be increased by reducing the pore size but only at the cost of increased hydraulic resistance leading to reduced performance. If the height of the tube is too great, the capillary limit may be insufficient to overcome the entire hydrostatic head and the upper reaches of the evaporator will dry out. Likewise, adequate performance over the full length of the evaporator at low heat loads may be compromised at high heat loads when circulation adds a substantial frictional loss in the wick.

Electric fields are also capable of circulating liquids. For example, the imposition of an electrostatic field normal to the tube wall creates in a dielectric liquid a pressure gradient in the same direction. Longitudinal variations in this pressure field produce liquid motion. The electric field thus acts indirectly, moving liquid normal to the imposed field. It does so independently of gravity and is essentially unaffected by buoyancy. Although high radial voltage gradients are not easily maintained, it appears that the mean Nusselt number in a tube may be doubled by this means under suitable conditions; notably for $EfPr \gtrsim 10^5$, where Ef is the ratio of the electric force to the buoyancy force (Kikuchi *et al.* 1990). This suggests that electric fields applied in both the heated and cooled sections could provide an anti-gravity capability in single-phase systems, and would assist capillary action applied to the raising of condensate in evaporative systems.

Alternatively, the electrostatic field may be used to create the capillary pressure directly (Jones 1973; Jones and Perry 1974). Figure 5.34 illustrates how this may be done by applying a potential difference across the gap between the tube wall and an axial ribbon electrode. Upon application of the field, dielectric liquid confined beneath the electrode as shown exhibits a pair of menisci running the length of the ribbon. The capillary pressure thus generated is the same at each axial location and therefore produces no net flow. As in a conventional wick lining, evaporation at one of the ribbon

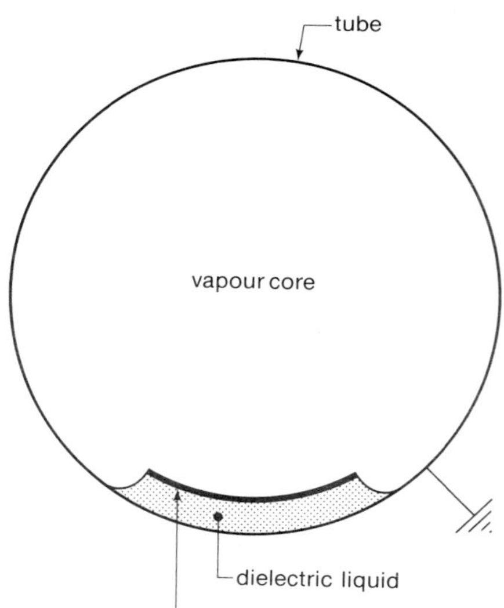

Fig. 5.34. Electro-capillarity.

and condensation at the other create changes in the interfacial curvature thus producing a pressure differential driving the liquid towards the heated end, independently of gravity. This scheme resembles the heat pipe, and therefore suffers from similar disadvantages, in addition to which are the complications caused by high voltages and the progressive degradation of insulating dielectric liquids. As an anti-gravity device, however, it does offer certain advantages (Loehrke and Debs 1975; Loehrke and Sebits 1976; Jones 1976; Kikuchi *et al.* 1984). These include: the reduction of head loss in the liquid return; enhanced boiling attributable to agitation in a non-uniform electric field (Choi 1962); and enhanced condensation attributable to drop-like condensation in a non-uniform field (Choi 1968).

An entirely different method of raising condensate is provided by the osmotic effect. In this, a semi-permeable membrane separates a solution from its solvent, and permits only the solvent to flow across the membrane in the natural tendency to restore chemical equilibrium. The osmotic pressure difference thus created between the solution and the solvent may be used to raise condensate to the heated section where, under steady conditions, the rate of *solvent* evaporated at the higher *solution* saturation temperature equals the rate of solvent condensed at the lower temperature below. Figure 5.35 provides a schematic of two alternatives. In Fig. 5.35(a), the osmotic effect acts alone to create an operating head which is less than the initial osmotic head because of frictional loss incurred in the membrane and the arterial return. The fluid circulation rate, which directly controls the overall heat transfer rate, is given by an expression of the form

$$\dot{m} = \rho k_{\mathrm{m}} A_{\mathrm{m}} (\Pi_{\mathrm{o}} - \Pi_{\mathrm{e}}) \tag{5.30}$$

where k_{m} is the permeability of the membrane and A_{m} is its surface area through which the flow occurs. The subscripts o and e refer to operating and initial (stable equilibrium) conditions, respectively. The osmotic pressure Π is given by

$$\Pi \simeq \frac{\Re T n_{\mathrm{s}}}{V_{\mathrm{L}}} = \frac{\Re T n_{\mathrm{s}}}{H A_{\mathrm{B}}}$$

where n_{s} is the number of moles of solute, H is the osmotic head, and A_{B} is the cross-sectional area of the artery. It is thus evident that the heat transfer rate (circulation rate) is directly proportional to the surface area of the membrane, its permeability, and the concentration increase created by evporation. It is also clear than since $\Pi_{\mathrm{o}} > \Pi_{\mathrm{e}}$, H_{o} must be less than H_{e}. In Fig. 5.35(b), the condensate is lifted in two stages. Firstly, by the osmotic pressure difference which brings it to the bottom of a wick; secondly, through the interior of the wick which is heated over its outer surface, as in a conventional heat pipe. Tandem arrangements such as this help extend the tube length over which anti-gravity behaviour may be produced.

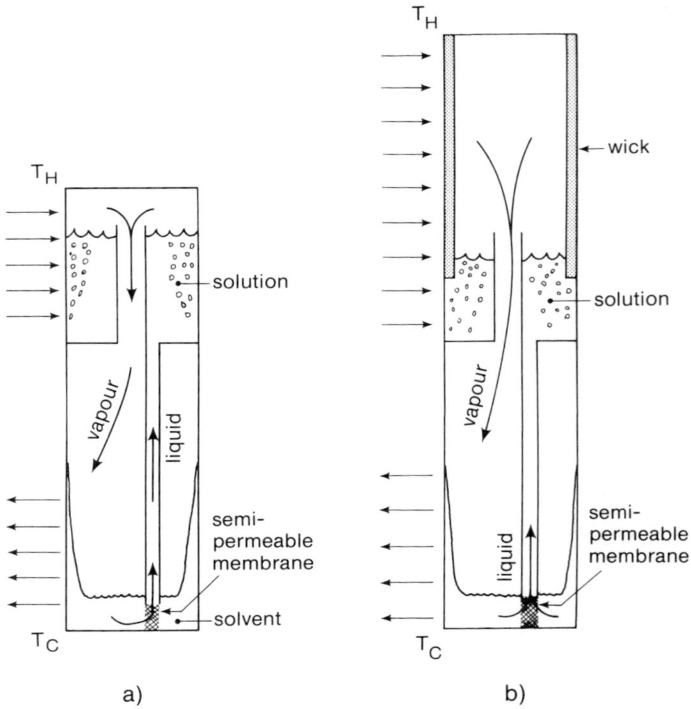

Fig. 5.35. Schematic representation of osmotic, anti-gravity thermosyphons.

It is evidently possible to extend this length even further by applying a longitudinal electric field and thereby inducing the electro-osmotic effect (Dresner 1963; Burgeen and Nakache 1964; Abu-Romia 1971). This is manifest in the near-surface region of a liquid where the disjoining pressure, discussed in Section 5.1, is caused by the electric double-layer field. Ions held in the solution may be moved parallel to the surface by a potential difference applied along it, but the effect is lost in the bulk solution where the electric field ceases to exist. Electro-osmosis is only significant in a bed of fine capillaries or a porous wick which possesses a large surface–volume ratio. The electric field then extends into most, if not all, of the ionic solution contained in the pores, and the resultant charge separation renders the solution responsive to an applied voltage. Abu-Romia (1971) demonstrated that a potential difference of only 20 V created a significant anti-gravity capability.

Finally, we consider the natural Archimedean buoyancy of bubbles as a means of creating a downward convective flux. In a simple thermosyphon, the bubbles may be introduced as an insoluble gas, much as described in Section 5.2. With the heated section above the cooled section, this constitutes an anti-gravity aerosyphon. Figure 5.36 provides two examples. Under conditions where the contribution of thermal buoyancy may be neglected, an annular throughflow establishes the single loop shown in Fig. 5.36(a); the flow divider simply constrains the lateral movement of the bubbles. On the other

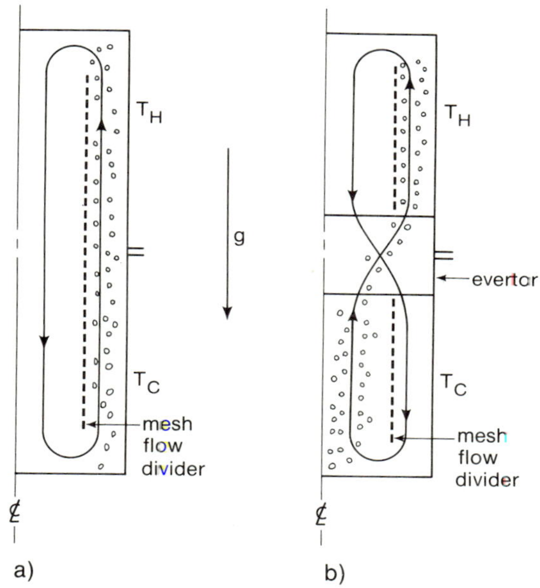

Fig. 5.36. Anti-gravity aerosyphons: (a) through flow; (b) everted flow.

hand, if thermal buoyancy provides a significant augmentation, the flow must be everted between the heated and cooled sections to ensure flow reinforcement in both sections. Figure 5.36(b) illustrates the scheme and the resulting figure-of-eight loop. Regardless of the relative importance of thermal buoyancy, it is always preferable to have the highly turbulent, aerated liquid flowing adjacent to the heated tube wall. As noted in Section 5.2, the necessary supply of external power is usually small.

Power must also be supplied to the evaporative thermosyphon when it is operating against gravity. Here the double problem of maintaining both vapour and liquid circulation may again be reduced to one by creating conditions which ensure continued condensation below; that is, by maintaining the saturated vapour pressure in the condenser low enough to accommodate frictional and hydrostatic heads in the vapour. The condensate, on the other hand, must be pumped upwards. Perhaps the most convenient way of doing this is by using the system's own vapour bubbles to create a lift or bubble pump.

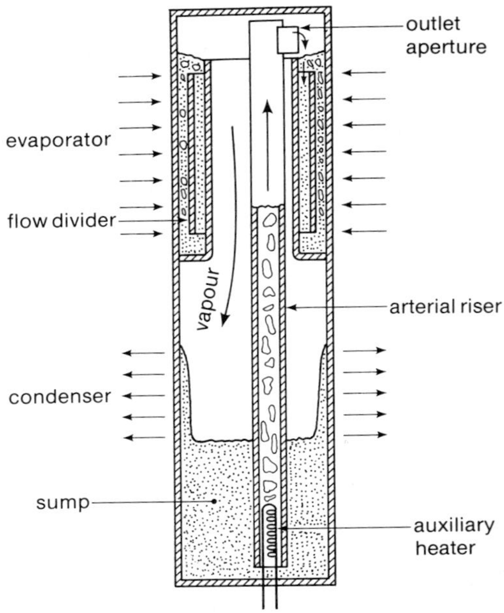

Fig. 5.37. Inverted evaporative thermosyphon (courtesy National Engineering Laboratory, UK).

Figure 5.37 is an illustration of an anti-gravity device sometimes called an inverted thermosyphon (Chisholm 1974). As indicated, condensate in the sump is lifted up to the evaporator by the action of vapour bubbles generated with an auxiliary heat source. The liquid thus elevated is discharged into an annular space on the inner side of a flow divider which separates the downward inflow of cold liquid from the boiling, upward outflow caused by heat supplied through the evaporator wall. The vapour generated flows into the central space whence it flows downwards into the condenser. The arterial riser thus creates a form of bubble pump to overcome the hydrostatic head between the evaporator and the condenser.

The inconvenience accompanying the use of a separate external power supply may be eliminated by tapping the evaporator heat source if suitable alterations are made to the internal plumbing (Grant 1976; Chisholm 1978; Chisholm *et al.* 1978). In the two-way evaporative thermosyphon, illustrated in Fig. 5.38, there are two evaporators, the lower of which has a split function. Above the baffle located near the bottom, any heat supplied to this evaporator has the same function as in a conventional thermosyphon; below the baffle, vapour bubbles are released into the central arterial riser, thereby pumping condensate into the upper evaporator. To secure liquid replenishment beneath

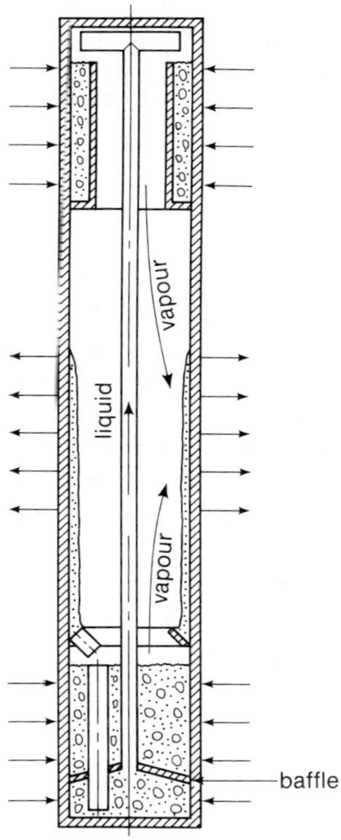

Fig. 5.38. Two-way evaporative thermosyphon (courtesy National Engineering Laboratory, UK).

the baffle, an annular collar at the base of the condensing section collects some of the condensate and pours it into an open supply pipe passing through the baffle. The effect of any vapour bubbles caught in the supply pipe is thus minimized. It is evident that this double arrangement permits some of the heat supplied to the lower evaporator to support the operation of the upper evaporator.

Ingenious variations on this inversion theme are shown in Fig. 5.39. In both of these, the anti-gravity capability is achieved without the use of an additional evaporator situated below. They also eliminate a large sump full of condensate and introduce a head of liquid in the central artery which supplies the evaporator. Liquid covering the base of the condenser is drawn upwards by a vapour bubble pump to a height above the top of the thermosyphon proper before finding its way back down into the evaporator. In Fig. 5.39(a), the vapour bubbles are generated by tapping the evaporator heat supply before and after the condensate flows down through the central artery. Figure 5.39(b)

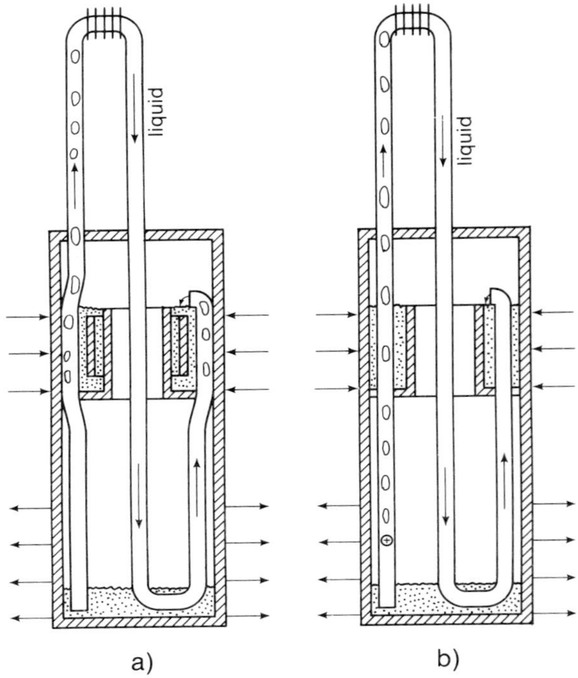

Fig. 5.39. Variations of the inverted evaporative thermosyphon (courtesy National Engineering Laboratory, UK).

shows an alternative arrangement in which the artery is unheated but is left open to the vapour space in the lower reaches of the condenser. Vapour pockets created by surrounding vapour pressure at this riser entry point then become the source of the liquid pumping power. In both arrangements, the central artery is made to run full by a doubling back near the bottom of the condenser.

In closing, it is worth stressing that individual methods for securing circulation under anti-gravity conditions are seldom limited to working alone. The use of osmosis and capillarity in series is one obvious example applicable to evaporative conditions while the simultaneous use of electric and centrifugal fields provides a strategy appropriate to both single-phase and two-phase conditions. In principle, the number of possible enhancement schemes combining series and parallel operation is great. Currently, they belong to a largely uncharted domain, and hence the exploration and evaluation of these schemes provides considerable scope for the future.

SELECTED BIBLIOGRAPHY

ABHAT, A. (1984). High performance heat pipes for a latent heat thermal energy storage application. *Proc. 5th Int. Heat Pipe Conf., Tsukuba*, pp. 321–7.

ABHAT, A. and SEBAN, R. (1974). Boiling and evaporation from heat pipe wicks with water and acetone. *J. Heat Transfer*, **96C**(3), 331–7.

ABIB, A.H. and JALURIA, Y. (1988). Numerical simulation of the buoyancy-induced flow in a partially open enclosure. *Numer. Heat Transfer*, **14**, 235–54.

ABU-ROMIA, M.M. (1971). Possible application of electro-osmotic flow pumping in heat pipes. AIAA Paper #71-423.

ALCOCK, J.F. (1951). General discussion on heat transfer. *Proc. Inst. Mech. Eng.* (*London*), **IV**, 378.

ANDROS, F.E. and FLORSCHUETZ, L.W. (1978). The two-phase closed thermosyphon: an experimental study with flow visualization. In *Two-phase transport and reactor safety*, Vol. 4, pp. 1231–67, (eds T.N. Veziroglu and S. Kakac) Hemisphere, Washingon, DC.

ANDVIG, T.A. (1956). Rotating rig experiments on the effectiveness of the closed thermosyphon system for the cooling of turbine rotor blades. PAMETRADA Contract Rep. #C108.

ANON. (1882). Memoirs of Angier March Perkins. *Proc. Inst. Civil Eng., London*, **67**(1), 417–19.

ANON. (1867). The Paris Exhibition—Perkins' Portable Oven. *The Engineer*, July 7, p. 519.

ANON. (1981). First ship with practical de-icing system. *Zosen*, **26**(2), 26.

BABB, A.L., GODAL, A., WAKEFIELD, A.W., MCKEE, R.E., and STRAND, M.J. (1978). A natural circulation self-refrigerated pile for the direct support of buildings in permafrost regions. *AIChE Symp. Ser.*, pp. 223–34.

BÄHR, A., BURCK, E., and HUFSCHMIDT, W.A. (1968). Liquid-vapour interaction and evaporation in heat pipes. *Proc. 2nd Int. Conf. on Thermionic Electric Power Generation, Stresa*, Paper D-6.

BAKER, O. (1954). Simultaneous flow of oil and gas. *Oil Gas J*.**53**(12), 185–90, 192, 195.

BALLBACK, L.J. (1969). The operation of a rotating wickless heat pipe. M.S. thesis. Naval Postgraduate School, Monterey, CA.

BARANTSEVICH, V.L. and OPRYSHKO, S.I. (1979). Effects of the amount of working fluid on the characteristics of heat pipes with longitudinal slots. *J. Eng. Phys*, **37**, 371.

BARRATT, R.O. and HENDERSON, G.O. (1979). The heat pipe air preheater. Foster Wheeler Publication #SP79-9, Foster Wheeler Corporation, New Jersey.

BATCHELOR, G.K. (1954). Heat transfer by free convection across a closed cavity between vertical boundaries at different temperatures. *Q. Appl. Math.*, **12**, 209–33.

BAUM, A.J. (1978). Optimum design of a bayonet tube heat exchanger. *Proc. 6th Int. Heat Transf. Conf., Toronto*, **4**, 285–9.

BAYLEY, F.J. (1955*a*). Heat transfer characteristics of the closed thermosyphon. Ph.D. thesis. University of Durham.

BAYLEY, F.J. (1955*b*). An analysis of turbulent free-convection heat transfer. *Proc. Inst. Mech. Eng.* (*London*), **169**(20), 361–70.

BAYLEY, F.J. and BELL, N.R. (1957). Improving gas turbine performance—cooling turbine blades. *Engineering, London* **183**, 300–2.

BAYLEY, F.J. and CZEKANSKI, J. (1963). Experimental investigation of free convection in a liquid metal in long tubes. *J. Mech. Eng. Sci.*, **5**(4), 295–302.

BAYLEY, F.J. and LOCK, G.S.H. (1965). Heat transfer characteristics of the closed thermosyphon. *J. Heat Transfer*, **87**, 30–40.

BAYLEY, F.J. and MARTIN, B.W. (1971). A review of liquid cooling of high-temperature gas turbine rotor blades. *Proc. Inst. Mech. Eng.* (*London*), **185**, 219–27.

BAYLEY, F.J., MILNE, P.A., and STODDART, D.E. (1961). Heat transfer by free convection in a liquid metal. *Proc. R. Soc.*, **A265**, 97–108.

BECKMANN, W.A. and MERTE, H. (1965). A photographic study of boiling in an accelerating system. *J. Heat Transfer*, **87**(3), 374.

BEEKLY, D.C. and MATHER, G.R. (1978). Analysis and experimental test of a high performance, evacuated tubular collector. NASA, CR-150874.

BEHNIA, M., MORRISON, G.L., and PARAMASIVAN, S. (1987). Heat transfer and flow in inclined open thermosyphons. *Joint ASME/JSME Thermal Engineering Conf., Honolulu*, **4**, 7–14.

BEJAN, A. (1984). *Convective heat transfer*. Wiley, New York.

BEJAN, A. and TIEN, C.L. (1978). Fully developed counterflow in a long horizontal pipe with different end temperatures. *Int. J. Heat Mass Transfer*, **21**, 701–8.

BEJAN, A., AL-HAMOUD, A.A., and IMBERGER, J. (1981). Experimental study of high-Rayleigh number convection in a horizontal cavity with different end temperatures. *J. Fluid Mech.*, **109**, 283–99.

BELLMAN, R. and PENNINGTON, R.H. (1954). Effects of surface tension and viscosity on Taylor instability. *Q. Appl. Math.*, **12**, 151–61.

BÉNARD, H. (1900). Les Tourbillons cellulaires dans une nappe liquide. *Rev. Gén. Seci.*, **XII**, 1261–309.

BENJAMIN, T.B. and PATHAK, S.K. (1987). Cellular flows of a viscous liquid that partly fills a horizontal rotating cylinder. *J. Fluid Mech.*, **183**, 399–420.

BEREZNAY, I. and ERDEI, J. (1985). Heat pipe waste heat utilizer for boilers and industrial furnaces (in Hungarian). *Veiki Közlemények*, pp. 13–17.

BEZRODNYI, M.K. (1978). On the upper boundary of the maxiumum thermal conductivity of evaporation thermosyphons. *Teploenergetika*, **25**, 63–6.

BEZRODNYI, M.K. and ALEKSEYENKO, D.V. (1977*a*). Boiling heat transfer in closed two-phase thermosyphons. *Heat Transfer—Sov. Res.*, **8**, 52–6.

BEZRODNYI, M.K. and ALEKSEYENKO, D.V. (1977*b*). Boiling heat transfer in closed two-phase thermosyphons. *Heat Transfer—Sov. Res.*, **9**, 14–20.

BEZRODNYI, M.K. and ANTOSHKO, Y.V. (1990). Ascending flow conditions of liquid film and vapour current in vertical annular passages of thermosyphons. *Proc. 7th Int. Heat Pipe Conf., Minsk*, Paper B4.

BEZRODNYI, M.K. and VOLKOV, S.S. (1982*a*). Study of hydrodynamic characteristics of two-phase flow in closed thermosyphons. In *Advances in heat pipe technology*, pp. 115–240. Pergamon, Oxford.

BEZRODNYI, M.K. and VOLKOV, S.S. (1982*b*). The study of hydrodynamic characteristics of a two-phase counterflow in the closed thermosyphon. *Hydrodynamics and heat transfer in condensed media*, pp. 121–7. Thermophys. Inst. Press, Novosibirsk.

BEZRODYNI, M.K., FAYNZIL'BERG, S.N., and KONDRUSIK, Y.E. (1980). Investigation of the maximum heat transfer in annular two-phase thermosyphons. *Heat Transfer—Sov. Res.* **12**(1), 118–23.

BEZRODNYI, M.K., VOLKOV, S.S., and ALEKSEYENKO, D.V. (1983). Maximum heat transfer in thermosyphons with separated uptake and downtake flows. *Heat Transfer—Sov. Res.* **15**(2), 108–14.

BEZRODNYI, M.K., VOLKOV, S.S., IVANOV, B.V., and PETROV, V.N. (1985). New energotechnical system of heat utilization in non-ferrous pyrometallurgical plants. *J. Heat Recovery Syst.*, **5**, 127–31.

BIELIK, J., CAGALA, M., HLAVAČKA, V., KISELEV, V.G., POLÁŠEK, F., ŠTULC, P., VASILIEV, L.L., and ZEMAŇEK, J. (1987). Experience from heat pipe heat exchanger operations in connection with fouling, freezing and other operation conditions. *Proc. 6th Int. Heat Pipe Cont., Grenoble.*

BLANCHARD, D.C. and WOODCOCK, A.H. (1957). Bubble formation and modification in the sea and its meteorological significance, *Tellus*, **9**, 145–58.

BLOEM, H., DEGRŸS, J.C., and DEVAAN, R.L.C. (1982). An evacuated tubular solar collector incorporating a heat pipe. *Philips Tech. Rev.*, **40**, 181–91.

BLYTHE, P.A., DANIELS, P.G., and SIMPKINS, P.G. (1983). Thermal convection in a cavity: the core structure near the horizontal boundaries. *Proc. R. Soc.*, **A387**, 367–88.

BONTEMPS, A., GOUBIER, C., MARQUET, C., and SOLECKI, J.C. (1984). Theoretical analysis of a revolving heat pipe. *Proc. 5th Int. Heat Pipe Conf., Tsukuba*, **1**, 274–9.

BONTEMPS, A., GOUBIER, C., MARQUET, C., SOLECKI, J.C., and NARDI, C. (1987). Performance limits of toluene loaded, closed two-phase thermosyphon. *Proc. 6th Int. Heat Pipe Conf., Grenoble.*

BONTOUX, P., SMUTEK, C., RANDRIAMAMPIANINA, A., ROUX, B., EXTRÉMET, G.P., HURFORD, A.C., ROSENBERGER, F., and DEVAHL DAVIS, G. (1986*a*). Numerical solutions and experimental results for three dimensional buoyancy driven flows in tilted cylinders. *Adv. Space Res.*, **6**(5), 155–60.

BONTOUX, P., SMUTEK, C., ROUX, B., and LACROIX, J.M. (1986*b*). Three dimensional buoyancy-driven flows in cylindrical cavities with differentially heated endwalls. Part I. Horizontal cylinders. *J. Fluid Mech.*, **169**, 211–27.

BOUSSINESQ, J. (1903). *Théorie analytique de la chaleur*, Vol. 2, p. 172. Gauthier-Villars, Paris.

BRODOWICZ, K. and CZAPLICKI, A. (1982). Effect of droplet coalescence in dropwise condensation. *Proc. 7th Int. Heat Transfer Conf., Munich* **5**, 3–7.

BROOKS, I. and OSTRACH, S. An experimental investigation of natural convection in a horizontal cylinder. *J. Fluid Mech.*, **44**, 545–61.

BROST, O., UNK, J., and CANDERS, W.R. (1984). Heat pipes for electric motors. *Proc. 5th Int. Heat Pipe Conf., Tsukuba*, **1**, 151–3.

BROWN, T.W.F. (1950). Some factors in the use of high temperatures in gas turbines. *Proc. Inst. Mech. Eng. (London)*, **162**, 167–85.

BROWN, T.W.F. (1957). *CIMAC Congr., Zurich*, pp. 443–84.

BURGEEN, D. and NAKACHE, F.R. (1964). Electrokinetic flow in ultrafine capillary slits. *J. Phys. Chem.*, **68**, 1084–91.

BUSSE, C.A. (1967). Pressure drop in the vapour phase of long heat pipes. *Proc. IEEE Int. Thermionic Conversion Specialist Conf., Palo Alto*, p. 391.

BUSSE, C.A. (1973). Theory of the ultimate heat transfer limit of cylindrical heat pipes. *Int. J. Heat Mass Transfer*, **16**, 169–86.

BUSSE, C.A. and PRENGER, F.C. (1984). Numerical analysis of the vapour flow in cylindrical heat pipes. *Proc. 5th Int. Heat Pipe Conf., Tsukuba*, pp. 214–19.

Bystrov, P.I. and Popov, A.N. (1976). Characteristics of heat pipes with liquid metal heat carriers under low temperature conditions. *High Temp. (USSR)*, **14**, 556–63.

Casarosa, C. and Dobran, F. (1988). Experimental investigation and analytical modeling of a closed two-phase thermosyphon with imposed convective boundary condition. *Int. J. Heat Mass Transfer*, **31**(9), 1815–33.

Casarosa, C., Latrofa, E., and Shelginski, A. (1983). The geyser effect in a two-phase thermosyphon. *Int. J. Heat Mass Transfer*, **26**, 933–41.

Catton, I. (1978). Natural convection in enclosures. *Proc. 6th Int. Heat Transfer Conf., Toronto*, **6**, 13–30. Hemisphere, Washington, DC.

Chan, Y.L. and Tien, C.L. (1985). A numerical study of two-dimensional laminar natural convection in shallow open cavities. *Int. J. Heat Mass Transfer*, **28**(3), 603–12.

Chan, Y.L. and Tien, C.L. (1986). Laminar natural convection in shallow open cavities. *J. Heat Transfer*, **108**, 305–9.

Chao, L. and Li, T. (1988). Applications of heat pipe for machine tools. *Proc. 2nd National Heat Pipe Conf., Dayong.*

Chao, X., Yin, Z., and Sun, Q. (1988) The development and application of a large heat pipe heat exchanger. *Proc. 2nd National Heat Pipe Conf., Dayong.*

Chato, J.C. (1965). Condensation in a variable acceleration field and the condensing thermosyphon. *J. Eng. Power*, **87**, 355–60.

Cheesewright, R. and Dastbaz, A. (1983). The large scale structure of the flow in the transition region of a natural convection boundary layer. *Proc. 8th Australasian Fluid Mechanics Conf., Newcastle.*

Chen, J. and Lou, D.Y.S. (1990). Investigation of the evaporation heat transfer in the rotating heat pipe. *Proc. 7th Int. Heat Pipe Conf., Minsk.*

Chen, J. and Tu, C. (1987). Condenser heat transfer in inclined rotating heat pipe. *Proc. 6th Int. Heat Pipe Conf., Grenoble.*

Chen, J. and Tu, C. (1988). Investigation of optimum type of heat pipe in rotating pipe heat exchanger. *Advances in phase change heat transfer.* International Academic Publishers, Chongqing.

Chen, J.C. (1966). Correlation for boiling heat transfer to saturated liquids in convective flow. *Ind. Eng. Chem. Proc. Des. Dev.* **5**, 322–9.

Chen, K.S., Humphrey, J.A.C., and Miller, L. (1983). Note on the pulsating nature of thermally-driven open cavity flow. *Int. J. Heat Mass Transfer*, **26**(7), 1090–3.

Chen, S.J. (1986). Turbulent film condensation on a vertical plate. *Proc. 8th Int. Heat Transfer Conf., San Francisco*, **4**, 1601–6. Hemisphere, Washington, DC.

Chen, S.J., Reed, J.C., and Tien, C.L. (1984). Reflux condensation in a two-phase closed thermosyphon. *Int. J. Heat Mass Transfer*, **27**(9), 1587–94.

Chen, Y., Xin, M., and Shen, J. (1988). The industrial test of separate type heat pipe exchanger and waste heat recovery from large blast furnaces. *Proc. 2nd National Heat Pipe Conf., Dayong.*

Cheng, K.C. and Zarling, J.P. (1990). Applications of heat pipes and thermosyphons in cold regions. *Proc. 7th Int. Heat Pipe Conf., Minsk.*

Chi, S.W. (1976*a*). *Heat pipe theory and practice.* MCGraw-Hill, New York.

Chi, S.W. (1976*b*). *Heat pipe theory and practice, A source book.* Hemisphere, Washington, D.C.

Chisholm, D. (1971). *The heat pipe.* Mills and Boon, London.

Chisholm, D. (1974). The anti-gravity thermosyphon. *Symp. on Multiphase Flow Systems, I. Mech E/I Chem E Symp. Ser.38*, Vol. 2. Inst. Chem. Eng., London.

CHISHOLM, D. (1978). Heat pipes with headers and single and two-phase artery systems. NEL (East Kilbride) Report NEL 656 (July).

CHISHOLM, D., GRANT, I.D.R., and DUNCAN, P. (1978). Condenser operated inverse thermosyphons. NEL (East Kilbride) Report NEL 659 (Sept.).

CHOI, H.Y. (1962). Electrohydrodynamic boiling heat transfer. Ph.D. thesis. Massachusetts Institute of Technology.

CHOI, H.Y. (1968) Electrohydrodynamic condensation heat transfer. *J. Heat Transfer*, **90C**, 98–102.

CHU, P.T. and HAMMITT, F.G. (1964). Natural convection inside circular cavity. ASME (WAM) Paper #64, WA-HT-51.

CLEMENTS, B. and LEE, Y. (1981). Additional parameters in two-phase closed thermosyphons: effects of tube diameter and wall thickness. *Int. J. Heat Mass Transfer*, **24**(9), 1554–5.

CLIFT, R., GRACE, J.R., and WEBER, M.E. (1978). *Bubbles, drops and particles.* Academic Press, New York.

COHEN, H. and BAYLEY, F.J. (1955). Heat transfer problems of liquid-cooled gas turbine blades. *Proc. Inst. Mech. Eng. (London)*, **169**(20), 1063–80.

COOPER, I.S. (1967). Cryogenic surgery. In *Engineering in the practice of medicine*, Ch. 12, pp. 122–41. (ed. B.L. Segal and D.G. Kilpatrick). Williams and Wilkins, Baltimore.

COOPER, M.G. (1984). Saturation nucleation boiling—a simple correlation. *1st UK National Heat Transfer Conf.* **2**, 785–93.

CORMACK, D.E., LEAL, L.G., and IMBERGER, J. (1974*a*). Natural convection in a shallow cavity with differentially heated endwalls, Part I. Asymptotic theory. *J. Fluid Mech.*, **65**, 209–29.

CORMACK, D.E., LEAL, L.G., and SEINFIELD, J.N. (1974*b*). Natural convection in a shallow cavity with differentially heated endwalls, Part 2. Numerical solutions. *J. Fluid Mech.*, **65**, 231–46.

CORNWELL, K. and NAIR, B.G. (1976). Boiling in wicks. *Proc. Heat Pipe Forum Meeting*, NEL (East Kilbride) Report 607.

COTTER, T.P. (1965). Theory of heat pipes. Los Alamos Sci. Lab., LA-3246-MS.

COTTER, T.P. (1984). Principles and prospects for micro heat pipes. *Proc. 5th Int. Heat Pipe Conf., Tsukuba*, pp. 328–35.

CRESPO, E., BONTOUX, P., SMUTEK, C., ROUX, B., HARDIN, G., SANI, R., and ROSENBERGER, F. (1987). Three dimensional simulations of convection régimes in cylindrical ampoules. Comparisons with theroretical analyses and experiments. *Proc. 6th Eur. Symp. on Material Sciences under Microgravity Conditions, Bordeaux.*

CUNDY, V.A., NYDAHL, J.E., and PELL, K.M. (1979). Geothermal heating of bridge decks. Snow Removal and Ice Control Research, National Academy of Sciences, Washington, Special Report 185.

CURTILA, R. and CHATAING, T. (1984). Experimental study of a revolving heat pipe. *Proc. 5th Int. Heat Pipe Conf., Tsukuba* **1**, 268–73.

DAKIN, J.T. (1978). Vaporization of water films in rotating radial pipes. *Int. J. Heat Mass Transfer*, **21**, 1325–32.

DANIELS, T.C. and AL-JUMAILY, F.K. (1975). Investigations of the factors affecting the performance of a rotating heat pipe. *Int. J. Heat Mass Transfer*, **18**, 961–73.

DANIELS, T.C. and WILLIAMS, R.J. (1978). Experimental temperature distribution and heat load characteristics of rotating heat pipes. *Int. J. Heat Mass Transfer*, **21**, 193–201.

Davis, E.J. and Anderson, G.H. (1966). The incipience of nucleate boiling in forced convection flow. *AIChEJ.*, **12**, 774–80.

Davis, W.R. and Ferrell, J.K. (1974). Evaporative heat transfer of liquid potassium in porous media. *AIAA–ASME Thermophysics and Heat Transfer Conf., Boston.*

de Vahl Davis, G. (1986). Finite difference methods for natural and mixed convection in enclosures. *Proc. 8th Int. Heat Transfer Conf., San Francisco*, **1**, 190–9. Hemisphere, Washington, DC.

Defay, R., Prigogine, I., and Bellemans, A. (1966). *Surface tension and adsorption.* Wiley, New York.

De Groot, S.R. and Mazur, P. (1962). *Non-equilibrium thermodynamics.* North-Holland, Amsterdam.

Děmčenková, I. (1986). Coolers increase life time of accumulators (In Czech). *Věda a technika v SSSR*, **14**, 28.

Deverall, J.E., Kemme, J.E., and Florschuetz, L.W. (1970). Some limitations and startup problems of heat pipes. LA-4518 (Nov.), Los Alamos Scientific Lab., Los Alamos NM.

Dhir, V. and Lienhard, J. (1971). Laminar film condensation on plane and axi-symmetric bodies in non-uniform gravity. *J. Heat Transfer*, **93**, 97–100.

Dobran, F. (1984). Stability thresholds of a closed two-phase thermosyphon. *Proc. 5th Int. Heat Pipe Conf., Tsukuba*, pp. 240–47.

Dobran, F. (1985). Steady state characteristics and stability thresholds of a closed two-phase thermosyphon. *Int. J. Heat Mass Transfer*, **28**, 949–57.

Dobran, F. (1989). Heat pipe research and development in the Americas. *Heat Recovery Syst. CHP*, **9**(1), 67–100.

Dresner, L. (1963). Electrokinetic phenomena in charged microcapillaries. *J. Phys. Chem.*, **67**, 1635–41.

Dunn, P.D. and Reay, D.A. (1982). *Heat pipes*, (3rd edn). Pergamon, Oxford.

Eckert, E.R.G. and Carlson, W.O. (1961). Natural convection in an air layer enclosed between two vertical plates at different temperatures, *Int. J. Heat Mass Transfer*, **2**, 106–29.

Eckert, E.R.G. and Jackson, T.W. (1951). Analysis of turbulent free-convection boundary layer on flat plate. *Nat. Adv. Comm. Aeronaut.* TC 2207.

Edwards, D.K. and Marcus, B.D. (1972). Heat and mass transfer in the vicinity of the vapour–gas front in a gas loaded heat pipe. *J. Heat Transfer*, **94**, 155–62.

Elder, J.W. (1965*a*). Laminar free convection in a vertical slot. *J. Fluid Mech.*, **23**(1), 77–98.

Elder, J.W. (1965*b*). Turbulent free convection in a vertical slot. *J. Fluid Mech.*, **23**(1), 99–111.

El-Masri, M.A. and Louis, J.F. (1978). On the design of high-temperature gas turbine blade water-cooling channels. *J. Eng. Power*, **100**, 586–91.

Erdei, J. and Bereznay, I. (1985). Waste heat recovery to preheating of air in power stations. *Energiewirtschaft*, **24**, 559–61.

Eschweiler, J.C., Benton, A.M., and Preckshot, G.W. (1967). Boiling and convective heat transfer at high acceleration. *Chem. Eng., Prog. Symp. Ser.*, **63**(79), 66–72.

Espig, H. and Hoyle, R. (1968). The transfer of heat from condensing steam to a cold rotating disk. *Proc. Inst. Mech. Eng.* (*London*), **182**(3H), 406–12.

Evans, A.L. and Reid, R.L. (1982). Heat transfer in an air thermosyphon permafrost protection device. *J. Energy Resour. Technol.*, **104**, 205–10.

FAGHRI, A., CHEN, M-M., and MORGAN, M. (1982). Heat transfer characteristics in two-phase closed conventional and concentric annular thermosyphons. *J. Heat Transfer*, **111**, 611–18.

FARBER, E.A. and SCORAH, E.L. (1948). Heat transfer to water boiling under pressure. *Trans. ASME*, **70**, 369–84.

FELDMAN, K.T and SRINIVASAN, R. (1984). Investigation of heat transfer limits in two-phase closed thermosyphons. *Proc. 5th Int. Heat Pipe Conf., Tsukuba*, pp. 24–9.

FERARA, A. and BRINKMAN, P. (1976). Applying heat pipes to avoid the preferential freezing of highway bridge decks. ASME Paper 76-ENAS-25.

FERRELL, J.K., DAVIS, R., and WINSTON, H. (1973). Vaporization heat transfer in heat pipe wick materials. *Proc. 1st Int. Heat Pipe Conf., Stuttgart*, Paper 6-4.

FETCU, D., HOFFMAN, V., and UNGUREANU, V. (1986). Theoretical and experimental study of wickless heat pipes (in Romanian). *Proc. Lucrările Conferinti de Transmiterea Cáldurii si Optimizarea Utilizárii Energiii Termice*, **1**, 55–62.

FINLAY, I.C. (1975). Heat pipes and two-phase thermosyphons. *Chart. Mech. Eng.*, **22**, 59–63.

FOSTER, A.R., KOWALSKI, G.J., and LOPEZ, G.W. (1985). Solar heat pipe water heater. *Alternative energy sources*, **VI**, 337–51. Hemisphere, New York.

FOSTER, C.V. (1953). Heat transfer by free convection to fluids contained in vertical tubes. Ph.D. thesis. University of Delaware, Newark.

FOYLE, R.T. (1951). General discussion on heat transfer. *Proc. Inst. Mech. Eng. (London)*, **IV**, 382.

FRAAS, A.P. (1973). A potassium–steam binary vapor cycle for better fuel economy and reduced thermal pollution. *J. Eng. Power*, **95**, 53–63.

FRANCIS, J.R.D. (1973). Experiments on the motion of solitary grains along the bed of a water stream. *Proc. R. Soc.*, **A332**, 443–71.

FRECHE, J.C. and DIAGUILA, A.J. (1950). Heat transfer and operating characteristics of aluminum forced-convection and stainless steel natural-convection water-cooled single stage turbines. *Nat. Adv. Comm. Aeronaut* RM E50d0 3a.

FRIEDRICH, R. (1962). Brennstoff, Eine Gasturbine mit gekuehlten Schaufeln fuer Gastemperaturen ueber 1000°C. *Warme-Kraft*, **14**(8), 368–73.

FUKANO, T., KADOGUCHI, K., and TIEN, C.L. (1986). Oscillation phenomena and operating limits of the closed two-phase thermosyphon. *Proc. 8th Int. Heat Transf. Conf., San Francisco*, **5**, 2325–30. Hemisphere, Washington, DC.

FUKUDA, K., YAMAMOTO, A., KONDO, T., and HASEGAWA, S. (1984). Characteristics of closed thermosyphon with internal downcomer tube. *Proc. 5th Int. Heat Pipe Conf., Tsukuba*, pp. 66–73.

FUKUDA, M., TSUCHIYA, F., RYOKAI, K., MOCHIZUKI, M., and MASHIKO, K. (1990). Development of an artificial permafrost storage using heat pipes. *Proc. 7th Int. Heat Pipe Conf., Minsk.*

FUKUSAKO, S., SEKI, N., and YAMADA, M. (1987). Heat removal characteristics of a combined system of concentric-tube thermosyphon and heat pump. *JSME Int. J.*, **30**(264), 936–44.

FUKUSAKO, S., YAMADA, M., and YAMAMOTO, H. (1989). Heat removal characteristics of a combined system of concentric-tube thermosyphon and heat pump in cold climates. *Proc. 2nd Int. Symp. on Cold Regions Heat Transfer*, pp. 233–8.

FULFORD, G.D. (1964). The flow of liquids in thin films. In *Adv. Chem. Eng.* **5**, 151–236. Academic Press, New York.

GALATE, J.W. (1976). Passive refrigeration for arctic pile supports. *J. Eng. Ind. ASME*, **98**, 695–700.

GAY, F.W. (1929). US Patent #1725906 (August 27).

GEL'PERIN, N.I. (1939). Khim. *Mashinostroenie*, **4**, 1 (in Russian).

GERÁK, A., HORVÁTH, L., JELINEK, F., ŠTULC, P., and ZBOŘIL, L. (1987). Examples of heat pipe application in chemical, electrical and other industries. *Proc. 6th Int. Heat Pipe Conf., Grenoble*.

GERASIMOV, Y.F., ANATOV, V.Y., DOLGIREV, Y.Y., and NEPOMNYASHCHIY, A.S. (1976). Certain results of investigations of low-temperature heat pipes operating against gravity. *Inzh-fiz. zhys.* **30**(4), 581–6.

GERASIMOV, Y.F., ANATOV, V.Y., DOLGIREV, Y.Y., and NEPOMNYASHCHIY, A.S. (1985). Performance of a heat pipe with separate vapour and liquid ducts rotating in a gravity field. *Heat Transfer—Sov. Res.*, **17**(4), 127–30.

GI, K. and MAEZAWA, S. (1990). Heat transfer characteristics of parallel rotating heat pipe. *Proc. 7th Int. Heat Pipe Conf., Minsk*.

GIEL, P.W. and SCHMIDT, F.W. (1990). A comparison of turbulence modeling predictions to experimental measurements for high Rayleigh number natural convection in enclosures. *Proc. 9th Int. Heat Transfer Conf., Israel*, **2**, 175–80.

GLAVACKA, V. (1987). *Heat pipe heat exchanger for waste heat recovery*, Preprint N23, Heat and Mass Transfer Institute, Minsk.

GORBIS, Z.R. and SAVCHENKOV, G.A. (1976). Low temperature two-phase closed thermosyphon investigation. *Proc. 2nd Int. Heat Pipe Conf., Bologna*, pp. 37–45.

GOSMAN, A.D., LOCKWOOD, F.C., and TATCHELL, D.G. (1971). A numerical study of the heat transfer performance of the open thermosyphon. *Int. J. Heat Mass Transfer*, **14**(10), 1717–30.

GRAHAM, C. and GRIFFITH, P. (1973). Dropsize distribution and heat transfer in dropwise condensation. *Int. J. Heat Mass Transfer*, **16**, 337–46.

GRANT, I.D.R. (1976). Inverse thermosyphons. *Proc. Heat Pipe Forum*, NEL (East Kilbride) Report NEL 607.

GRAVRILOV, A.F. (1966). The design of an air preheater with an intermediate heat carrier. *Teploenergetika*, No. 8.

GRAY, V.H. (1969). The rotating heat pipe. A wickless hollow shaft for transferring high heat fluxes. ASME Paper #69-HT-19.

GREENSPAN, H.P. (1969). *The theory of rotating fluids*. Cambridge University Press, Cambridge.

GRIGULL, U. (1952). Wärmeübergang bei Filmkondensation. *Forsch. Ing.-Wes.*, **18**, 10–12.

GROLL, M. (1973). Heat pipes as constructional elements in heat and refrigeration technology. *Brennstoff-Wärmekraft*, **25**, 11–14.

GROLL, M. (1989). Heat pipe research and development in Western Europe. *Heat Recovery Syst. CHP*, **9**(1), 19–66.

GROLL, M. and SPENDEL, TH. (1984). Thermal behaviour of high performance two-phase thermosyphons. *Proc. 5th Int. Heat Pipe Conf., Tsukuba*, pp. 30–5.

GROLL, M., BROST, O., MACK, H., and SHEWCHUK, E.N. (1984). Evaporation heat transfer of sodium from capillary structures. *Proc. 5th Int. Heat Pipe Conf., Tsukuba* pp. 164–9.

GROSS, U. and HAHNE, E. (1985). Heat transfer in a two-phase thermosyphon operating with a fluid in the near critical state. *Int. J. Heat Mass Transfer*, **28**(3), 589–601.

GROSS, U. and HAHNE, E. (1986). Reflux condensation inside a two-phase thermosyphon at pressures up to the critical. *Proc. 8th Int. Heat Transfer Conf., San Francisco*, **4**. 1613–19. Hemisphere, Washington, DC.

HAHNE, E. (1965). Wärmetransport durch Natürliche Konvektion in medien nahe Ihrem. *Int. J. Heat Mass Transfer*, **8**, 481–97.

HAHNE, E. and GROSS, U. (1982). The influence of the inclination angle on the performance of a closed two-phase thermosyphon. In *Advances in heat pipe technology*, pp. 125–35. Pergamon, Oxford.

HAHNEMANN, H. (1951). Stationary rig experiments on the heat extracting power of closed thermosyphon cooling holes. NGTE Memo. m128.

HARADA, K., INOUE, S., FUJITA, J., SUEMATSU, H., and WAKIYAMA, Y. (1980). Heat transfer characteristics of large heat pipe (in Japanese). *Hitachi Zosen Tech. Rev.*, **41**, 167–74.

HART, J.E. (1971). Stability of the flow in a differentially-heated inclined box. *J. Fluid Mech.*, **47**, 547–76.

HART, J.E. (1985). Observations of complex oscillations in a closed thermosyphon. *J. Heat Transfer*, **107**, 833–9.

HARTNETT, J.P. and WELSH, W.E. (1957). Experimental studies of free convection heat transfer in a vertical tube with uniform heat flux *Trans. ASME*, **79**(7), 1551–7.

HASEGAWA, S., NISHIKAWA, K., and YAMAGATA, K. (1963). Heat transfer in an open thermosyphon. *Bull. JSME*, **6**(22), 230–50.

HAYNES, F.D. and ZARLING, J.P. (1982). A comparative study of thermosyphons used for freezing soil. ASME Paper #82-WA/HT-40.

HAYNES, F.D. and ZARLING, J.P. (1988*a*). Heat transfer performance of commercial thermosyphons with inclined evaporator sections. *Proc. 7th Int. Conf. on Offshore Mechanics and Arctic Engineering, Houston*, **IV**, 275–80.

HAYNES, F.D. and ZARLING, J.P. (1988*b*). Thermosyphons and foundation design in cold regions. *Cold Regions Sci. Technol.*, **15**, 251–9.

HAYNES, F.D., ZARLING, J.P., QUINN, W.F., and GOOCH, G.W. (1991). Laboratory tests with a hybrid thermosyphons. *Proc. 10th Int. Conf. on Offshore Mechanics and Arctic Engineering*, **IV**, 93–100.

HEUER, C.E. (1979). The application of heat pipes on the Trans-Alaska pipeline. US Army Cold Region Research and Engineering Laboratory, Special Report 79-26.

HEWITT, G.F. (1978). Critical heat flux in flow boiling. *Proc. 6th Int. Heat Transfer Conf., Toronto*, **6**, 143–71.

HEWITT, G.F. and HALL-TAYLOR, N.S. (1970). *Annular two-phase flow*. Pergamon, Oxford.

HIJIKATA, K., CHEN, S.J., and TIEN, C.L. (1984). Noncondensable gas effect in a two-phase closed thermosyphon. *Int. J. Heat Mass Transfer*, **27**(8), 1319–25.

HO, W.K. and TIEN, C.L. (1982). Reflux condensation characteristics of a two-phase closed thermosyphon. In *Advances in Heat Pipe Technology*, pp. 451–8. Pergamon, Oxford.

HOLZWARTH, H. (1938). Die Entwicklung der Holzwarth-Gasturbine, Holzwarth-Gasturbinen. Gmbh Muellheim–Ruhr.

HOOGENDOORN, C.J. (1986). Natural convection in enclosures *Proc. 8th Int. Heat Transfer Conf., San Francisco*, **1**, 111–20. Hemisphere, Washington, DC.

HORVATH, L. and POLÁŠEK, F. (1985). Survey of choice heat pipe applications (in Czech). *Proc. 3rd Heat Pipe Seminar, Prague*, pp. 7–15.

HURD, N.L. (1946). Mean temperature difference in the field or bayonet tube. *Ind. Eng. Chem.*, **38**(12), 1266–71.

IDEL'CHIK, I.E. and GINZBURG, Y.L. (1968). The hydraulic resistance of 180° annular bends. *Teploenergetika*, **4**, 109–14.

IMBERGER, J. (1974). Natural convection in a shallow cavity with differentially heated endwalls, Part 3. Experimental results. *J. Fluid Mech.*, **65**, 2247–60.

IMURA, H. (1980). Optimum fill quantity of liquid in two-phase closed thermosyphons (in Japanese). *Tech. Rep. Kumamoto Univ.*, **29**, 9–18.

IMURA, H. and YOSHIDA, M. (1990). The extraction of geothermal energy by a long double tube thermosyphon. *Proc. 7th Int. Heat Pipe Conf., Minsk*, Paper D23.

IMURA, H., SASAGUCHI, K., KOZAI, H., and NUMATA, S. (1983). Critical heat flux in a closed two-phase thermosyphon. *Int. J. Heat Mass Transfer*, **26**, 1181–8.

ISMAIL, K.A.R. and LIU, C.Y. (1985). Steam generation via solar heat pipes. *Alternative energy sources*, **vi**, 353–7. Hemisphere, New York.

IVANOV, V.L. (1967). Ispol'zovanie vody kak promteplonvsitelya v avtonomnoi sisteme okhalzhdeniya rabochei lopatki gazovoi turbiny. *Izvestia Vysshikh Uchebnykh Zavednii Mashinostroenie*, #10, 70–2.

IVANOVSKII, M.N., ILYN, Y.A., KORNEEV, V.D., PRIVEZENTSEV, V.V., SERBIN, V.I., SERGEEV, Y.Y., SIDORENKO, E.M., and SOROKIN, V.P. (1976*a*). Investigation of non-steady-state processes at cryogenic heat pipe operation. *Proc. 2nd Int. Heat Pipe Conf., Bologna*, **2**, 752.

IVANOVSKII, M.N., SOROKIN, V.P., and SUBBOTIN, V.I. (1976*b*). *Evaporation and condensation of metals*. Atomizdat, Moscow.

IVANOVSKII, M.N., SOROKIN, V.P., SUBBOTIN, V.I., and SHUSTOV, M.V. (1970). Investigation of heat and mass transfer in a heat pipe with sodium coolant. *High Temp. USSR*, **8**, 298–304.

IVANOVSKII, M.N., SOROKIN, V.P., and YAGARDIN, I.V. (1982). *The physical principles of heat pipes*. Oxford University Press, Oxford.

IVEY, H.J. and MORRIS, D.J. (1962). On the relevance of the vapour–liquid exchange mechanism for subcooled boiling heat transfer at high pressure. AEEW-R137.

JAHNS, H.O., MILLER, T.W., POWER, L.D., RICKEY, W.P., TAYLOR, T.P., and WHEELER, J.A. (1973). Permafrost protection for pipelines on permafrost. *2nd Int. Conf. on Permafrost*, pp. 673–84. National Academy Press, Washington.

JAKEŠ, J. OŠLEJŠEK, O., and POLÁŠEK, F. (1984). Intensification of cooling of electrical rotating machines by heat pipes. *Proc. 5th Int. Heat Pipe Conf., Tsukuba*, pp. 280–92.

JALLOUK, P.A. (1969). Experimental investigation of heat transfer in a closed thermosyphon. M.S. thesis. Purdue University, Lafayette.

JALURIA, Y. (1980). *Natural convection heat and mass transfer*. Pergamon, Oxford.

JAPIKSE, D. (1969). Heat transfer in open and closed thermosyphons. Ph.D. thesis. Purdue University, Lafayette.

JAPIKSE, D. (1973). Advances in thermosyphon technology. In *Advances in heat transfer*, **9**, 1–111. Academic Press, New York.

JAPIKSE, D. and WINTER, E.R.F. (1971). Single-phase transport processes in the open thermosyphon. *Int. J. Heat Mass Transfer*, **14**, 427–41.

JAPIKSE, D., JALLOUK, P.A., and WINTER, E.R.F. (1971). Single-phase transport processes in the closed thermosyphon. *Int. J. Heat Mass Transfer*, **14**(7), 869–87.

JOHNSON, R.E. (1988). Steady-state coating flows inside a rotating horizontal cylinder. *J. Fluid Mech.*, **190**, 321–42.

JONES, T.B. (1973). Electrohydrodynamic heat pipes. *Int. J. Heat Mass Transfer*, **16**, 1045–8.

JONES, T.B. (1976). Electrohydrodynamic effects on minimum film boiling. Rep. PB-252 320, Colorado State University.

JONES, T.B. and PERRY, M.P. (1974). Electrohydrodynamic heat pipe experiments. *J. Appl. Phys.*, **45**(5), 2129–32.

JONES, T.B., PERRY, M.P., and MELCHER, J.R. (1971). Dielectric siphons. *Science*, **174**, 1232–3.

KAMINAGA, F., OKAMOTO, Y., ITO, N., TSUYUZAKI, N., SAITO, T., MOCHIZUKI, M., and SUGIHARA, S. (1990). Transient behaviour of a thermosyphon heat pipe under stepwise temperature and heat load changes. *Proc. 7th Int. Heat Pipe Conf., Minsk*.

KAMMINGA, W. (1980). An analytic solution of the film thickness of laminar film condensation on inclined pipes. *Int. J. Heat Mass Transfer*, **23**, 1291–3.

KAMOTANI, Y. (1976). Thermal analysis of axially grooved heat pipes. *Proc 2nd Int. Heat Pipe Conf., Bologna*.

KATSUTA, M., KIGAMI, H., NAGATA, K., SOTANI, J., and KOIZUMI, T. (1984). A study on performance and characteristics of the rotating heat pipe. *Proc. 5th Int. Heat Pipe Conf., Tsukuba*, **1**, 99–103.

KATTO, Y. and MONDE, M. (1974). Study of mechanism of burn-out in a high heat-flux boiling system with an impinging jet. *Proc. 5th Int. Heat Transfer Conf., Tokyo*, **4**, 245–9.

KAWAHARA, Y., GOTO, I., and HAYASHIDA, T. (1984). Heat transfer to a heat pipe evaporator from groundwater flow. *Proc. 5th Int. Heat Pipe Conf., Tsukuba*, pp. 310–15.

KELLER, D. and THIEME, A. (1985). Heat recovery by heat pipe heat exchangers during drying (in German). *Proc. 10th Wärmetechnische Tagung, Trocknung*, p. 17.

KEMME, J.E. (1967). High performance heat pipes. *IEEE Thermionic Specialist Conf., Palo Alto (Oct.)*.

KIKUCHI, K. (1977). Study of the EHD heat pipe. *Technocrat*, **10**(11), 38–41.

KIKUCHI, K., TAKETANI, T., SHIRAISHI, M., and YAMANISHI, T. (1982). Large scale EHD heat pipe experiments. In *Advances in heat pipe technology*, pp. 643–50. Pergamon, Oxford.

KIKUCHI, K., TAKETANI, T., and YABE, A. (1984). Large scale EHD heat pipe experiments. *Proc. 5th Int. Heat Pipe Conf., Tsukuba*, pp. 48–53.

KIKUCHI, K. TAKETANI, T., and YABE, A. (1990). Electroconvection in heat pipe applications. *Proc. 7th Int. Heat Pipe Conf., Minsk*, Paper B15.

KIM, K.H. and LEE, Y. (1984). A study on the use of two-phase closed thermosyphons for latent heat storage system. *Proc. 5th Int. Heat Pipe Conf., Tsukuba*, pp. 316–20.

KING, C.R. (1931). Perkins' hermetic tube boilers. *The Engineer*, **152**, 405–6 (Oct.).

KIRICHENKO, Y.A. (1976). Heat transfer and first critical heat flux in boiling in a centrifugal force field. Low Temperature Institute preprint, Ukrainian Academy of Sciences.

KIRICHENKO, Y.A. (1982). Heat transfer to cryogenic liquids in a centrifugal force field. *Heat Transfer—Sov. Res.*, **14**(4), 117–30.

KIRICHENKO, Y.A., LEVCHENKO, N.M., and RUSANOV, K.V. (1980). Investigation of the effect of subcooling and centrifugal acceleration on nucleate boiling heat transfer to cryogenic fluids. *Heat Transfer—Sov. Res.*, **12**(2), 57–63.

KIRKBRIDE, C.G. (1934). Heat transfer by condensing vapours. *Trans. AIChE*, **30**, 170–86.

KIRKPATRICK, A.T. (1980). Wave mechanics of inclined and rotating liquid films. Ph.D. thesis. Massachusetts Institute of Technology.

KOBAYASHI, Y. (1984). Heat pipe engine and its thermodynamic cycle. *Proc. 5th Int. Heat Pipe Conf., Tsukuba*, pp. 59–65.

KOHTAHA, J. (1987). Development, design and operation of large scale separate type heat pipes. *Proc. 6th Int. Heat Pipe Conf., Grenoble.*

KÖRNER, W. (1970). Influence of higher acceleration on heat flow in boiling. *Chem. Ing. Tech.*, **42**(6), 409–14. (in German).

KUSUDA, H. and IMURA, H. (1973). Boiling heat transfer in an open thermosyphon. *Bull. JSME*, **16**, 1734–40.

KUSUDA, H. and IMURA, H. (1974). Stability of a liquid film on a counter-current annular two-phase flow. *Bull. JSME*, **17**, 1613–18.

KUTATELADZE, S.S. (1969). *Problems of heat transfer and hydraulics of two-phase media.* Pergamon, Oxford.

KUTATELADZE, S.S. and STYRIKOVICH, M.A. (1960). *Hydraulics of gas-liquid systems.* Gosenergoizdat, Moscow, 1958; trans. v.s. Dept. Commerce OTS, Washington, DC.

KUTSUNA, H., AKAGAWA, K., MIZUNO, S., and MATSUI, N. (1984). Thermosyphon heat pipes for long term underground storage system of solar energy. *Proc. 5th Int. Heat Pipe Conf., Tsukuba*, pp. 293–300.

KUZNETSOV, N.V. and GRAVRILOV, A.F. (1964). Air preheater with an intermediate heat carrier. *Teploenergetika*, No. 10.

LABUNTSOV, D.A. and MURATOVA, T.M. (1967). Kinetic analysis of vaporization and condensation processes. *High Temp. USSR*, **7**, 888–96.

LAMB, H. (1932). *Hydrodynamics*, (6th edn). Reprinted by Dover, New York.

LANGMUIR, I. (1912). Convection and conduction of heat in gases. *Phys. Rev.*, **34**(6), 401–21.

LARKIN, B.S. (1971). An experimental study of the two-phase thermosyphon tube. *Trans. Can. Soc. Mech. Eng.*, **14**, 208–15.

LARKIN, B.S. (1982*a*). An experimental study of the temperature profiles and heat transfer coefficient in a heat pipe for a heat exchanger. in *Advances in heat pipe technology*, pp. 177–92. Pergamon, Oxford.

LARKIN, B.S. (1982*b*). A heat pipe for control of heat sink temperatures. *Proc. 7th Int. Heat Transfer Conf., Munich.*

LARKIN, B.S. (1983*a*). A thermosyphon heat exchanger for use in animal shelters. Paper #74–210, Canadian Society of Agricultural Engineers.

LARKIN, B.S. (1983*b*). An experimental investigation of low heat flux wickless heat pipe. *Trans. Can. Soc. Mech. Eng.*, **7**(2), 97–9.

LARKIN, B.S. (1984). Temperature control of electronic systems using inexpensive heat pipes. *Proc. 5th Int. Heat Pipe Conf., Tsukuba*, pp. 346–51.

LARKIN, B.S. (1987). Enhancement of boiling at low temperatures in wickless copper/water heat pipes. *Proc. 6th Int. Heat Pipe Conf., Grenoble.*

LARKIN, B.S. and DUBUC, S. (1976). Self de-icing navigational buoys using heat pipes. *Proc. 2nd Int. Heat Pipe Conf., Bologna*, pp. 529–35.

LARKIN, B.S. and JOHNSTON, G.H. (1974). An experimental field study of the use of two-phase thermosyphons for the preservation of permafrost. *Eng. J.*, pp. 33–7.

LARSEN, F.W. and HARNETT, J.P. (1961). Effect of aspect ratio and tube orientation on free convection heat transfer to water and mercury. *J. Heat Transfer*, **83**, 87–93.

LEE, Y. and BEDROSSIAN, A. (1978). The characteristics of heat exchangers using heat pipes or thermosyphons. *Int. J. Heat Mass Transfer*, **21**(2), 221–9.

LEE, Y. and CLEMENTS, B.R. (1984). An analysis of heat exchangers using finned, two-phase closed thermosyphons. *Proc. 5th Int. Heat Pipe Conf., Tsukuba*, pp. 82–9.

LEE, Y. and MITAL, U. (1971). A two-phase closed thermosyphon. *Int. J. Heat Mass Transfer*, **15**, 1695–707.

LEFEVRE, E.J. and ROSE, J. (1966). A theory of dropwise condensation of steam. *Proc. 3rd Int. Heat Transfer Conf., Chicago*, **2**, 362–75.

LEGRIVÈS, E. and GRÉNOT, J. (1970). Refroidissement des Aubes de Turbines par Métaux Liquides. *Off. Nat. Etud. Rech. Aerosp.* T.P. Na 872.

LESLIE, F.M. (1959). Free convection in the tilted open thermosyphon. *J. Fluid Mech.*, **7**, 115–27.

LESLIE, F.M. and MARTIN, B.W. (1959). Laminar flow in the open thermosyphon, with special reference to small Prandtl numbers. *J. Mech. Eng. Sci.*, **1**, 184–93.

LEVITAN, M.M. and VASILIEV, L.L. (1976). Heat pipes. Theory and prediction. *Proc. 2nd Int. Heat Pipe Conf., Bologna*, pp. 123–30.

LEVY, E.K. (1968). Theoretical investigation of heat pipes operating at low vapour pressures. *J. Eng. Ind.*, **90**, 547–52.

LEVY, E.K. and CHOY, S.F. (1971). The sonic limits in sodium heat pipes. ASME Paper #71-WA/HT-11.

LIGHTHILL, M.J. (1953). Theoretical considerations on free convection in tubes. *Q. J. Mech. Appl. Math.*, **VI**(4), 398–439.

LIN, L., XU, T., ZHANG, Y., and SHI, W. (1988). Investigation of flow patterns and heat transfer mechanism in rotating heat pipe with stepped wall. *Advances in phase change heat transfer*. International Academic Publishers, Chongqing.

LINTHORST, S.J.M., SCHINKEL, W.M.M., and HOOGENDOORN, C.J. (1981*a*). Flow structure with natural convection in inclined, air-filled enclosures. *Proc. ASME-AIChE Conf.*, pp. 39–45.

LINTHORST, S.J.M., SCHINKEL, W.M.M., and HOOGENDOORN, C.J. (1981*b*). Flow structure with natural convection in inclined air-filled enclosures. *J. Heat Transfer*, **103**, 535–9.

LITTWIN, D.A. and MCCURLEY, J. (1982). Heat pipe waste heat recovery boilers. In *Advances in heat pipe technolnogy*, pp. 213–24. Pergamon, Oxford.

LITTWIN, D.A. and WILLIS, D.B. (1984). The use of heat pipes to conserve energy in petroleum refineries. *Energy Prog.*, **5**, 198–202.

LIU JIFU and ANG XUEYIE (1988). The study of the geyser boiling phenomenon in a thermosyphon. In *Advances in phase change heat transfer*. pp. 590–8. Academic Publishers, Chongqing.

LOCK, G.S.H. (1972). Some aspects of ice formation with special reference to the marine environment. *Trans. North East Coast Inst. Eng. Shipbuild.*, **88**, 175–84.

LOCK, G.S.H. (1986*a*). The BIVA Project. *Proc. IAHR Symp. on Ice, Iowa City*, **II**, 269–80.

LOCK, G.S.H. (1986*b*). Control of spring run-off in northern rivers: the ice veil concept. *Polar Rec.* **23**(145), 451–7.

LOCK, G.S.H. and ABDURAHMAN, R.D. (1988). Heat transfer characteristics of the closed tube aerosyphon. *Int. J. Heat Mass Transfer*, **31**(4), 143–52.

LOCK, G.S.H. and FU, JIALIN (1992*a*). The effect of orientation on the cranked tubular thermosyphon. *Proc. 11th Int. Conf. on Offshore Mechanics and Arctic Engineering, Calgary*.

Lock, G.S.H. and Fu, Jialin (1992*b*). Performance of a right-angled, evaporative thermosyphon. *Proc. 11th Int. Conf. on Offshore Mechanics and Arctic Engineering, Calgary.*

Lock, G.S.H. and Fu, Jialin. (1982*c*). Performance of an offset evaporative thermosyphon. *Proc. 11th Int. Conf. on Offshore Mechanics and Arctic Engineering, Calgary.*

Lock, G.S.H. and Han, J.C. (1989*a*). Buoyant laminar flow of air in a long, square-section cavity aligned with the ambient temperature gradient. *J. Fluid Mech.*, **207**, 489–504.

Lock, G.S.H. and Han, J.C. (1989*b*). Flow and heat transfer in a long horizontal, rectangular cavity with differentially-heated ends. In *Transport phenomena in thermal control*, (ed. G.-J. Hwang), pp. 687–700. Hemisphere, New York.

Lock, G.S.H. and Han, J.C. (1990). The effect of tilt, skew and roll on natural convection in a slender laterally-heated cavity. *Math. Comput. Modelling*, **13**(2), 23–32.

Lock, G.S.H. and Kirchner, J.D. (1988). Performance of an air-filled, open thermosyphon tube with particular reference to wind augmentation. *Int. J. Heat Mass Transfer*, **31**(11), 2357–64.

Lock, G.S.H. and Kirchner, J.D. (1989*a*). A visual study of the tubular aerosyphon. *Trans. CSME*, **13**(1/2), 41–4.

Lock, G.S.H. and Kirchner, J.D. (1989*b*). Performance of a closed-tube aerosyphon with large length–diameter ratios. *JOMAE*, **111**, 337–43.

Lock, G.S.H. and Kirchner, J.D. (1990). Wind augmented heat transfer in an open thermosyphon tube with large length–diameter ratios. *J. Heat Transfer*, **112**(1), 71–7.

Lock, G.S.H. and Kirchner, J.D. (1992). Some characteristics of the inclined, closed tube thermosyphon under low Rayleigh number conditions. *Int. J. Heat Mass Transfer*, **35**(1), 165–73.

Lock, G.S.H. and Ladoon, D. (1991*a*). Heat transfer in a right-angled thermosyphon. *Proc. 10th Int. Conf. on Offshore Mechanics and Arctic Engineering, ASME*, **IV**, 261–6.

Lock, G.S.H. and Ladoon, D. (1991*b*). Heat transfer characteristics of a cranked thermosyphon. *Proc. 3rd Int. Symp. on Cold Regions Heat Transfer, Fairbanks*, pp. 417–28.

Lock, G.S.H. and Ladoon, D. (1991*c*). The effect of tilt on the right-angled thermosyphon. *Proc. 4th Int. Symp. on Transport Phenomena in Heat and Mass Transfer, Sydney*, **1**, 311–20.

Lock, G.S.H. and Liu, Y. (1989). The effect of geometry on the performance of the closed tube thermosyphon at low Rayleigh numbers. *Int. J. Heat Mass Transfer*, **32**(6), 1175–82.

Lock, G.S.H. and Maezawa, S. (1975). The aerosyphon: an exploratory study. *Int. J. Heat Mass Transfer*, **18,** 219–26.

Lock, G.S.H. and Park, S. (1991). A numerical study of the right-angled thermosyphon. *Proc. 10th Int. Conf. on Offshore Mechanics and Arctic Engineering, ASME*, **IV**, 113–18.

Lock, G.S.H. and Simpson, G.A. (1989). Performance of the closed-tube thermosyphon with large length–diameter ratios. *JOMAE*, **111**, 22–8.

Lock, G.S.H. and Trotter, F.J. deB. (1968). Observations on the structure of a turbulent free convection boundary layer. *Int. J. Heat Mass Transfer*, **11**, 1225–32.

LOCK, G.S.H. and WU, M. (1991*a*). Laminar frictional behaviour of a bayonet tube. *Proc. 3rd Int. Sym. on Cold Regions Heat Transfer, Fairbanks*, pp. 429–40.

LOCK, G.S.H. and WU, M. (1991*b*). Turbulent frictional behaviour of a bayonet tube. *Proc. 3rd Int. Symp. on Cold Regions Heat Transfer, Fairbanks*, pp. 405–16.

LOCK, G.S.H. and WU, M. (1992). Laminar heat transfer in an air-filled bayonet tube. *Proc. 11th Int. Conf. on Offshore Mechanics, and Arctic Engineering, Calgary.*

LOCK, G.S.H. and ZHAO, LITONG. (1990*a*). The laminar flow field in a near-vertical, closed tube thermosyphon. *Proc. 7th Int. Conf. on Mathematical and Computer Modelling, Chicago*, **14**, 822–5.

LOCK, G.S.H. and ZHAO, LITONG (1990*b*). The effect of inclination on an air-filled, open thermosyphon at low Rayleigh numbers. *Proc. 1990 Joint ASME/AIChE/AIAA Conf., Seattle.*

LOCK, G.S.H. and ZHAO, LITONG (1991). Impeded laminar flow in a short, radial thermosyphon. *Proc. 4th Int. Symp. on Transport Phenomena in Heat and Mass Transfer, Sydney*, **5**, 1811–22.

LOCK, G.S.H. and ZHAO, LITONG (1992). Natural convection in honeycomb wall spaces. *Int. J. Heat Mass Transfer*, **35**(1), 155–64.

LOCK, G.S.H., CHONG, K., DYCKERHOFF, A., HUYER, J., and JONES, B. (1989). On the design of wind-augmented thermosyphons. *Cold Regions Sci. Technol.*, **16**, 11–23.

LOCKWOOD, F.C. and MARTIN, B.W. (1964). Free convection in open thermosyphon tubes of non-circular section. *J. Mech. Eng. Sci.*, **6**(4), 379–93.

LOEHRKE, R.I. and DEBS, R.J. (1975). Measurements of the performance of an electrohydrodynamic heat pipe. AIAA Paper #75–659, *10th Thermophys. Conf.*

LOEHRKE, R.I. and SEBITS, D.R. (1976). Flat plate electrohydrodynamic heat pipe experiments. *Proc. 2nd Int. Heat Pipe Conf., Bologna*, ESA Report SP 112.

LONG, E.L. (1963). The Long thermopile. *Proc. Int. Conf. on Permafrost, National Academy of Sciences, Washington*, pp. 487–91.

LONG, E.L. (1973). Designing friction piles for increased stability at lower installed cost in permafrost. *Proc. 2nd Int. Conf. on Permafrost, Yakutsk*, pp. 693–9.

LORENZ, L. (1881). Über das Wärmeleitvormögen der Metalle für Warme and Elektrizitat. *Wied. Ann. Phys.*, pp. 582–606.

MA, T. and HOU, Z. (1989). Heat pipe research and developments in China. *Heat Recovery Syst. CHP*, **9**(6), 499–512.

MA, T. and JIANG, Z. (1984). Heat pipe research and development in China. *Proc., 5th Int. Heat Pipe Conf., Tsukuba.*

MA, T., ZHANG, Z., and LI, H. (1984). Vaporization heat transfer in capillary wick structures formed by sintered screens. *Proc. 5th Int. Heat Pipe Conf., Tsukuba*, pp. 156–63.

MCADAMS, W.H. (1954). *Heat transmission*, (3rd edn). McGraw-Hill, New York.

MACALLAN, J.V. and BASSANI, C. (1984). Oscillations in gas-controlled heat pipes at low pressures. *Proc. 5th Int. Heat Pipe Conf., Tsukuba*, pp. 136–42.

MCCORMICK, J.L. and WESTWATER, J.W. (1965). Nucleation sites for dropwise condensation. *Chem. Eng. Sci.*, **20**, 1021–36.

MADEJSKI, J. and MIKIELEWICZ, J. (1971). Liquid fin—a new device for heat transfer equipment. *Int. J. Heat Mass Transfer*, **14**, 357–63.

MAEZAWA, S. and ISHIZAKI, K. (1988). High performance two-phase closed thermosyphon with a novel flow separator. *Proc. 3rd Int. Heat Pipe Conf., Tsukuba*, pp. 108–12.

MAJCEN, M and ŠARUNAC, N. (1984). Heat pipe cooled twin airfoil blade as an element for higher efficiency of long life gas turbine, In *Heat and mass transfer in rotating machinery*, (eds D.E. Metzger and N.H. Afgan), p. 689–700. Hemisphere, Washington, DC.

MALLINSON, G.D., GRAHAM, A.D., and DE VAHL DAVIS, G. (1981). Three-dimensional flow in a closed thermosyphon. *J. Fluid Mech.*, **109**, 259–75.

MARCUS, B.D. (1972). Theory and design of variable conductance heat pipes. NASA Report CR-2018, National Aeronautics and Space Administration, Washington.

MARTIN, B.W. (1955). Free convection in an open thermosyphon, with special reference to tubulent flow. *Proc. R. Soc.*, **A230,** 502–30.

MARTIN, B.W. (1959). Free convection heat transfer in the inclined open thermosyphon. *Proc. Inst. Mech. Eng. (London)*, **173**(32), 761–75.

MARTIN, B.W. (1965). Free convection limits in the open thermosyphon. *Int. J. Heat Mass Transfer*, **8**, 19–25.

MARTIN, B.W. and COHEN, H. (1954). Heat transfer by free convection in an open thermosyphon tube. *Br. J. Appl. Phys.*, **5**, 91–5.

MARTIN, B.W. and CRESSWELL, D.J. (1957). Influence of Coriolis forces on heat transfer in the open thermosyphon. *The Engineer*, **204,** 926–30.

MARTIN, B.W. and LOCKWOOD, F.C. (1963). Entry effects in the open thermosyphon. *J. Fluid Mech*, **19**(2), 246–56.

MARTINELLI, R.C. and BOELTER, L.M.K. (1942). The analytical prediction of superimposed free and forced viscous convection in a vertical pipe, *University Calif. (Berkeley) Publ. Eng.*, **5**(2), 23–58.

MARTINI, W.R. and CHURCHILL, S.W. (1960). Natural convection inside a horizontal cylinder. *AIChEJ.*, **6**, 251–7.

MARTO, P.J. (1973). Laminar film condensation on the inside of slender, rotating truncated cones. *J. Heat Transfer*, **95**, 270–2.

MARTO, P.J. (1976). Performance characteristics of rotating, wickless heat pipes. *Proc. 2nd Int. Heat Pipe Conf., Bologna*, pp. 281–91.

MARTO, P.J. (1984). Rotating heat pipes. In *Heat and mass transfer in rotating machinery*, (ed. D.E. Metzger and N.H. Afgan), pp. 609–32. Hemisphere, Washington, DC.

MARTO, P.J. and GRAY, V.H. (1971). Effects of high accelerations and heat fluxes on nucleate boiling of water in an axisymmetric rotating boiler. NASA TN D-6307.

MARTO, P.J. and WAGENSEIL, L.L. (1973). Augmenting the condenser heat transfer performance of rotating heat pipes. *AIAA J.* **17**(6), 647–52.

MARTO, P.J. and WANNIARACHCHI, A.S. (1987). Influence of internal axial fins on condensation heat transfer in co-axial rotating heat pipes. In *Heat transfer and fluid flow in rotating machinery*, (ed. W. Yang), pp. 235–44. Hemisphere, Washington, DC.

MARTO, P.J. and WEIGEL, H. (1982) The development of economical rotating heat pipes. In *Advances in heat pipe technology, Proc. 4th Int. Heat Pipe Conf., London*, pp. 709–23.

MASON, B.J. (1971). *The physics of clouds*. Oxford University Press, Oxford.

MATHUR, A. and SAXENA, S.C. (1983). Heat transfer to a bayonet heat exchanger immersed in a gas-fluidized bed. *Int. Commun. Heat Mass Transfer*, **10**, 241–52.

MEISTER, B.J. and SCHEELE, G.F. (1969). Prediction of jet length in immiscible liquid systems. *AIChE J.*, **15**, 689–99.

METAIS, B. and ECKERT, E.R.G. (1964). Forced, mixed and free convection régimes. *J. Heat Transfer*, **86**, 295–6.

METZGER, D.E. and AFGAN, N.H. (eds) (1984). *Heat and mass transfer in rotating machinery*. Hemisphere, Washington, DC.

MIKHELYOV, A.A. (1949). *Nagrevatelnyye truby dlya khlebopekranykh pechey* (*Heating pipes for baking ovens*). Kiev.

MOORHOUSE, W.E. (1960). Investigation into the reasons causing explosion in steam tubes of bakers' ovens. *Proc Inst. Mech. Eng.* (*London*), **174**(16), 561–74.

MORI, Y. and HIJIKATA, K. (1973). Free convective condensation heat transfer with noncondensable gas on a vertical surface. *Int. J. Heat Mass Transfer*, **16**, 2229–40.

MU, K. (1988). The performance research of a separate type heat pipe for cooling the main spindle system of machine tool. *Proc. 2nd Nat. Heat Pipe Conf., Dayong.*

MUDAWWAR, I.A. and EL-MASRI, M.A. (1988). Boiling incipience in plane rotating water films. *J. Heat Transfer*, **110**, 532–535.

MUDAWWAR, I.A., EL-MASRI, M.A., WU, C.S., and HUSMAN-MUDAWWAR, J.R. (1985). Boiling heat transfer and critical heat flux in high-speed rotating liquid films. *Int. J. Heat Mass Transfer*, **28**(4), 795–806.

MÜLLER, G., NEUMANN, G., and WEBER, W. (1984). Natural convection in vertical Bridgman configurations. *J. Cryst. Growth*, **70**, 78–93.

MÜNZEL, W.D. (1987). Heat pipes for heat recovery from exhaust gas of a diesel engine in a passenger car. *Proc. 6th Int. Heat Pipe Conf., Grenoble.*

NAKAYAMA, W., OHTSUKA, Y., ITOH, H., and YOSHIKAWA, T. (1984*a*). Optimum charge of working fluids in horizontal rotating heat pipes. In *Heat and mass transfer in rotating machinery*, (ed. D.E. Metzger and N.H. Afgan), pp. 633–44. Hemisphere, Washington, DC.

NAKAYAMA, W., OHTZUKA, Y., and YOSHIKAWA, T. (1984*b*). The effects of fine surface structures on the performance of horizontal rotating heat pipes, *Proc. 5th Int. Heat Pipe Conf., Tsukuba*, **2**, 121–5.

NAYFEH, A. (1973). *Perturbation methods*, Ch. 6. Wiley, New York.

NEGISHI, K. (1984). Thermo-fluid dynamics of two-phase thermosyphons. *Proc. 5th Int. Heat Pipe Conf., Tsukuba*, pp. 193–200.

NEGISHI, K. and SAWADA, T. (1983). Heat transfer performance of an inclined two-phase closed thermosyphon. *Int. J. Heat Mass Transfer*, **26**(2), 1207–13.

NEGISHI, K., KANEKO, K., and KUSUMOTO, F. (1987). Analysis of pulsation in two-phase thermosyphons. *Proc. 6th Int. Heat Pipe Conf., Grenoble*, pp. 436–40.

NGUYEN-CHI, H. and GROLL, M. (1982). Entrainment or flooding limit in a closed, two-phase thermosyphon. In *Advances in heat pipe technology*, pp. 147–62. Pergamon, Oxford.

NIEKAWA, J., MATSUMOTO, K., KOIZUMI, T., HASEGAWA, K., KANEKO, H., and MIZOGUCHI, Y. (1982) Performance of revolving heat pipes and application to a rotary heat exchanger. In *Advances in heat pipe technology*, pp. 225–34. Pergamon, Oxford.

NIMMO, B. and LEPERT, G. (1970). Laminar film condensation on a finite horizontal surface. *Heat transfer 1970*. Elsevier, Amsterdam.

NUSSELT, W. (1916). Die Oberflächenkondensation des Wasserdampfes. *Z. Ver. Dtsch. Ing.*, **60**, 541–6 and 569–75.

OGALE, V.A. (1968). On the application of the semi-closed thermosyphon system to gas turbine blade cooling. Dr thesis. Technological University of Delft, Delft.

OGATA, H. and NAKAYAMA, W. (1977). Heat transfer to subcritical and supercritical helium in centrifugal acceleration fields. 1. Free convection régime and boiling régime. *Cryogenics*, pp. 461–70.

ONIK, G., COBB, C., DIAMOND, D., STEELE, G., CADY, B., KANE, R., RUBINSKY, B., and PORTERFIELD, B. (1987). Cryosurgery for unresectable hepatic tumours. *24th Meeting, Society for Cryobiology, Edmonton.*

ORR, F.M. and SCRIVEN, L.E. (1978). Rimming flow: numerical simulation of steady, viscous, free-surface flow with surface tension. *J. Fluid Mech.*, **84**, 145–65.

OSTRACH, S. (1988). Natural convection in enclosures. *J. Heat Transfer*, **110**, 1175–89.

OSTRACH, S. and THORNTON, P.R. (1957). On the stagnation of natural-convection flows in closed end tubes. ASME Paper #57-SA-2.

OSTRACH, S., LOKA, R.R., and KUMAR, A. (1980). Natural convection in low aspect ratio rectangular enclosures. In *Natural convection in enclosures*, HTD-8 ASME, pp. 1–10.

OWEN, P.R. (1964). Saltation of uniforms grains in air. *J. Fluid Mech.*, **20**, 225–42.

PALANIKUMAR, P. (1960). The effect of tube inclination on heat transfer in the closed evaporative thermosyphon. M.Sc. thesis. University of Durham, Newcastle.

PASTUKOV, V.G., MAIDANIK, Y.F., and FERSHTATER, Y.G. (1987). Investigation of the working characteristics of a water 'antigravity' long heat pipe. *Izvestiya Akademii Nauk SSSR. Energetika i Transport*, **25**(4), 142–6.

PEEBLES, F.N. and GARBER, H.J. (1953). Studies on the motion of gas bubbles in liquids. *Chem. Eng. Prog.*, **49,** 88–97.

PENOT, F. (1982). Numerical calculation of two-dimensional natural convection in isothermal open cavities. *Numer. Heat Transfer*, **5**, 421–37.

PERKINS, L. P. and BUCK, W.E. (1892). Improvements in devices for the diffusion or transference of heat. UK Patent #22272, London.

PIORO, I.L. (1983). Maximum heat transfer capability of two-phase thermosyphons. *Heat Transfer—Sov. Res.*, **15**(1), 19–31.

PIORO, I.L. (1986). Influence of the inclination angle of a two-phase thermosyphon on the limiting heat flux. *Heat Transfer—Sov. Res.*, **18**(2), 35–8.

POKORNY, B., POLÁSEK, F., SCHNELLER, J., and STULC, P. (1984). Heat transfer in co-axial and parallel rotating heat pipes. *Proc. 5th Int. Heat Pipe Conf., Tsukuba*, **4**, 259–67.

POLÁSEK, F. (1973). Cooling of a.c. motor by heat pipes. *Proc. 1st Int. Heat Pipe Conf., Stuttgart.*

POLÁŠEK, F. (1984). Heat pipe research and developement in East European countries. *Proc. 5th Int. Heat Pipe Conf., Tsukuba*, **2**, 15–51.

POLÁŠEK, F. (1989). Heat pipe research and development in East European countries. *Heat Recovery Syst. CHP*, **9**(1), 3–17.

PRANDTL, L. (1904). Über Flüssigkeitsbewegung bei sehr kleiner Reibung. *Verhandlungen 3rd Int. Math. Kongr., Heidelberg*, pp. 484–91. Teubner, Leipzig.

PRENGER, F.C. (1984). Performance limits of gravity-assisted heat pipes. *Proc. 5th Int. Heat Pipe Conf., Tsukuba*, pp. 1–5.

PRENGER, F.C. and KEMME, J.E. (1982). Performance limits of gravity-assist heat pipes with simple wick structures. In *Advances in heat pipe technology*, pp. 137–46. Pergamon, Oxford.

PREZIOSI, L. and JOSEPH, D.D. (1988). The run-off condition for coating and rimming flows. *J. Fluid Mech.*, **187**, 99–113.

RAITHBY, G.D. and HOLLANDS, K.G.T. (1984). Natural convection. in *Handbook of heat transfer* (ed. W.M. Rohsenow *et al.*). McGraw-Hill, New York.

RAYLEIGH, LORD. (1892). On the instability of cylindrical fluid surfaces. *Philos. Mag.*, 5, xxxiv, 177.

RAYLEIGH, LORD, (1916). On convection currents in a horizontal layer of fluid when the higher temperature is on the under side. *Philos. Mag.*, Series 6, **32**, 529–46.

REAY, D.A. (1977). *Industrial energy conservation*. Pergamon, Oxford.

REAY, D.A. (1982). The Perkins tube—a noteworthy contribution to heat exchanger technology. *Heat Recovery Syst.*, **2,** 173–87.

REDDY, R.S., VENKATESWARLU, P., and SASTRI, V.M.K. (1990). Effect of inclination on the performance of rotating heat pipe. *Proc. 7th Int. Heat Pipe Conf., Minsk*, Paper B-2P.

REED, J.G. and TIEN, C.L. (1987). Modeling of the two-phase closed thermosyphon. *J. Heat Transfer*, **109,** 722–30.

REICHARDT, H. (1951). Vollstandige Darstellung der turbulenten Geschwindigkeitsverteilung in glatten Leitungen. *ZAMM*, **31,** 208–19.

REID, R.L., TENNANT, J.S., and CHILDS, K.W. (1975). The modeling of a thermosyphon type permafrost protection device. *J. Heat Transfer*, **97**(pt 2), 382–6.

REID, R.L., HUDGINS, E.H., and ONUFER, J.S. (1982). Frost and ice formation in the air convection pile permafrost protection device. *J. Energy Resour. Technol.*, **104,** 199–204.

RIBANDO, R.J., PALMER, J.L., and SCOTT, J.E. (1989). Flow in a partially filled rotating tapered cylinder. *J. Fluid Mech.*, **203,** 541–55.

RIFERT, V.G., BARABASH, P.A., and GOLIYAD, N.N. (1984). Condensation of steam on a water film falling down a rotating surface. *Heat Transfer—Sov. Res.*, **16**(3), 132–7.

ROBINSON, A.F. (1950). Mechanical performance of the T3 water-cooled turbine. NGTE Memo. M83.

ROETZEL, W. (1977). Laminar film surface evaporation with uniform heat flux in a fast rotating drum. *Int. J. Heat Mass Transfer*, **20,** 549–53.

ROHANI, A.R. and TIEN, C.L. (1973*a*). Steady two-dimensional heat and mass transfer in the vapour gas region of a gas-loaded heat pipe. *J. Heat Transfer*. (Aug.), **95C**(3), 377–82.

ROHANI, A.R. and TIEN, C.L. (1973*b*). Analysis of the effects of vapour pressure drop on heat pipe performance. *Int. J. Heat Mass Transfer*, **17**, 61–7.

ROHSENOW, W.M. (1952). A method of correlating heat transfer data for surface boiling liquids. *Trans. ASME*, **74**, 969–76.

ROKICKI, H. (1981). Study of heat pipe application to recovery of waste heat of flue gases of boilers (in Polish). Ph.D. thesis. Politechnika, Gdańska.

ROKICKI, H. (1983). Two-phase closed thermosyphon (in Polish). *Gospodarka paliwami i energia*, **31**, 16–18.

ROWBOTHAM, G.F., HAIGH, A.L., and LESLIE, W.G. (1959). Cooling cannula for use in the treatment of cerebral neoplasms. *Lancet*, **1**, 12–15.

RUCKENSTEIN, E. and METIU, H. (1965). On dropwise condensation on a solid surface. *Chem. Eng. Sci.*, **20**, 173–80.

SAKHUJA, R.S. (1973). Flooding constraint in wickless heat pipes. ASME Paper #73-WA/HT-7.

SANG, Z., LI, P., HU, Z., and GUO, Y. (1988). The development and performance of a heat pipe for cooling thyristor. *Proc. 2nd Nat. Heat Pipe Conf., Dayong*.

SASIN, V.J., POLÁŠEK, F., and ŠTULC, P. (1984). Analysis of thermal and fluid characteristics of heat pipes with axial grooves. *Proc. 5 Int Heat Pipe Conf., Tsukuba*, pp. 248–58.

SAVCHENKOV, G.A. and GORBIS, Z.R. (1976). Boiling heat transfer in low-temperature evaporating thermosyphons. *Heat Transfer—Sov. Res.*, **8**, 52–6.

SAVCHENKOV, G.A. and SHEVCHENKO, V.G. (1976). Bibliography of low-temperature two-phase thermosphyons. *Heat Transfer—Sov. Res.*, **8**, 84–6.

SAVCHENKOV, G.A., CHUMAK, I.G., GLUSHENKO, N.I., and KUNAKOV, V.G. (1978). Theoretical and experimental research on low-temperature thermosyphons and their application in refrigeration engineering. AIAA Paper #78-642.

SAVCHENKOV, G.A., CHUMAK, I.G., and SHULAVSKY, T. (1982). Natural convection of two-phase boiling mixture in the thermosyphon investigation, In *Advances in heat pipe techhnology*, pp. 105–14. Pergamon, Oxford.

SCHENK, J., DIEPERINK, G.W., and DEN OUDEN, C. (1969). Sur l'écoulement et la distribution de température dans un thermosyphon clos. *Appl. Sci. Res.*, **20**, 246–50.

SCHMIDT, E. (1951). Heat transmission by natural convection at high centrifugal acceleration in water-cooled gas turbine blades. *General Discussion on Heat Transfer, Proc. Inst. Mech. Eng. (London)*, **IV**, 361–3.

SCHMIDT, E. (1960). Wärmetransport durch natürliche Konvektion in Stoffen bei kritischem Zustand. *Int. J. Heat Mass Transfer*, **1**(6), 92–101.

SCHMIDT, E. (1962). Heat transfer by natural convection. *Int. Dev. in Heat Transfer Conf., University of Colorado*, **XXXIX–XL.**

SCHMIDT, E. and BECKMAN, W. (1930). Das Temperatur und Geschwindigkeitfeld von einer Warmeabgebenden, Senkrechter Platte bei naturlicher Konvektion. *Forschung in Ingenieurwesen*, **1**, 391–406.

SCHNELLER, J., POKORNÝ, B., and POLÁSEK, F. (1984). Heat transfer in rotating co-axial and parallel heat pipes and their applications in machinery. In *Heat and mass transfer in rotating machinery*, (ed. D.E. Metzger and N.H. Afgan), pp. 669–88. Hemisphere, Washington, DC.

SCURLOCK, R.G. and THORNTON, G.K. (1977). Pool heat transfer to liquid and supercritical helium in high centrifugal acceleration fields. *Int. J. Heat Mass Transfer*, **20**(1), 31–40.

SCURRAH, N.H. (1990). Heat pipe heat recovery from boiler and furnace flues. *Proc. 7th Int. Heat Pipe Conf., Minsk*, D11.

SEKI, N., FUKUSAKO, S., and KOGUCHI, K. (1980). Single-phase heat transfer characteristics of concentric-tube thermosyphon. *Wärme-und Stoffübertragung*, **14**, 189–99.

SEMENA, M.G. and MEL'NICHUK, G.A. (1978). Mean-velocity distributions in a falling film. *Fluid Mech.—Sov. Res.*, **7**(5), 145–51.

SHAO, X., HOU, Z., and GUO, S. (1988). Experimental study and analysis of the operation reliability of heat pipe in rotating conditions. *Proc. 2nd Nat. Heat Pipe Conf., Dayong.*

SHEKRILADZE, I.G. and RUSISHVILI, D.G. (1987). Evaporation and condensation on grooved capillary surfaces. *Proc. 6th Int. Heat Pipe Conf., Grenoble.*

SHIRAISHI, M., KIKUCHI, K., and YAMANISHI, T. (1982). Investigation of heat transfer characteristics of a two-phase closed thermosyphon. In *Advances in heat pipe technology*, pp. 95–104. Pergamon, Oxford.

SHIRAISHI, M., YONEYA, M., and YABE, A. (1984). Visual study of operating limit in the two-phase closed thermosyphon. *Proc. 5th Int. Heat Pipe Conf., Tsukuba*, pp. 10–17.

SCHIROKY, G. H. and ROSENBERGER, F. (1984). Free convection of gases in a horizontal cylinder with differentially-heated endwalls. *Int. J. Heat Mass Transfer*, **27**, 587–98.

SIMPKINS, P.G. and CHEN, K.S. (1986). Convection in horizontal cavities. *J. Fluid Mech.*, **166**, 21–39.

SIMPSON, G.A. (1986). Geometrical effects in the closed tube thermosyphon. M.Sc. thesis. University of Alberta, Edmonton.

SMITH, A. and PEARSON, R. (1950). The cooled gas turbine. *Proc. Inst. Mech. Eng. (London)*, **163**, 221.

SOFRATA, H. (1980). Theoretical study of filmwise condensation considering wave initiation. *Wärme-and Stoffübertragung*, **14**, 201–10.

SOMOGYI, D. and YEN, H.H. (1973). An approximate analysis of the diffusing flow in a self-controlled heat pipe. *J. Heat Transfer* (Feb.,), **95**(pt 2), 93–106.

SPARROW, E.M. and ECKERT, E.R.G. (1961). Effects of superheated vapour and noncondensable gas on laminar film condensation. *AIChEJ.*, **7**, 473–7.

SPARROW, E.M. and GREGG, J. (1959). A boundary layer treatment of laminar film condensation. *J. Heat Transfer*, **81C**, 13–18.

SPARROW, E.M. and GREGG, J. (1960). The effect of vapour drag on rotating condensation. *J. Heat Transfer*, **82**, 71–2.

SPARROW, E.M. and LIN, S.H. (1964). Condensation heat transfer in the presence of noncondensable gas. *J. Heat Transfer*, **86C**, 430–6.

SPARROW, E.M., CHRYSLER, G.M., and AZEVEDO, L.F. (1984). Observed flow reversals and measured-predicted Nusselt numbers for natural convection in a one-sided heated vertical channel. *J. Heat Transfer*, **106**, 325–32.

SPENDEL, TH. (1982). Laminar film condensation heat transfer in closed two-phase, thermosyphons. In *Advances in heat pipe technology*, pp. 163–73. Pergamon, Oxford.

SPENDEL, TH. (1984). Laminar film condensation heat transfer in closed, two-phase thermosyphons. *Proc. 5th Int. Heat Pipe Conf., Tsukuba*, pp. 208–13.

ŠTULC, P., KISELEV, V.G., and MATVEYEV, J.N. (1984). Heat pipe heat exchangers for low-potential heat utilization (in Russian). *Proc. 7th Heat and Mass Transfer Conf., Minsk*, **8**, 169–74.

SUBBOTIN, V.I., SOROKIN, D.N., OVECHIN, D.M., and KUDRYUAVTSEV, A.P. (1969). *Heat exchange on boiling of metals under natural convection conditions*. Nauka, Moscow.

SUEMATSU, H., HARADA, K., INOUE, S., FUJITA, J., and WAKIYAMA, Y. (1978). Heat transfer characteristics of heat pipe. *Heat Transfer— Jn. Res.*, **1**, 1–22.

SUGAWARA, S. and KATSUKA, K. (1966). Fundamental study of dropwise condensation. *Proc. 3rd Int. Heat Transfer Conf., Chicago*, **2**, 354–61.

SUN, K.H., DIX, G.E., and TIEN, C.L. (1975). The effect of precursory cooling on falling film rewetting. *J. Heat Transfer*, **96**, 126–31.

SUN, Q. and ZHU, G. (1988). The application of a rotor heat pipe for a regulating volt and speed a.c. motor. *Proc. 2nd Nat. Heat Pipe Conf., Dayong*.

SUNDELL, R.E., GOODWIN, W.W., KERCHER, D.M., DUDLEY, J.C., and TRIANDAFYLLIS, J. (1980). Boiling heat transfer in turbine bucket cooling passages. ASME Paper #80–HT-14.

SUZUKI, H., HAGIWARA, Y., MIWATA, T., and SATO, T. (1983). Analysis of film condensation heat transfer considering the effect of interfacial waves (in Japanese). *Proc. 20th Nat. Heat Transfer Symp.*, pp. 253.

SYMONS, J.G. and PECK, M.K. (1984). Natural convection heat transfer through inclined longitudinal slots. *J. Heat Transfer*, **106**, 824–9.

TAKUMA, M., MAEZAWA, S. and TSUCHIDA, A. (1984). Effects of turbulent diffusion on thermal characteristics of a gas-loaded, two-phase closed thermosyphon. *Proc. 5th Int. Heat Pipe Conf., Tsukuba*, pp. 232–9.

TANAKA, O., YAMAKAGE, H., OGUSHI, T., MURAKAMI, M., and TANAKA, Y. (1982). Snow melting using heat pipes, *Advances in Heat Pipe Technology*, Pergamon Press, Oxford, *Proc. 4th Int. Heat Pipe Conf., London:* 11–23.

TERDTOON, P., SHIRAISHI, M., and MURAKAMI, M. (1990). Investigation of effect of inclination angle on heat transfer characterics of closed two-phase thermosyphon. *Proc. 6th Int. Heat Pipe Conf., Minsk*, B9P.

THOMAS, L.C. and FAN, L.T. (1971). A model for turbulent transfer in the wall region. *ASCE. J., Eng. Mech. Div.*, **97**, 169–73.

THOMAS, L.C. and WOOD, M.L. (1979). A new approach to the analysis of turbulent free convection heat transfer. *Int. J. Heat Fluid Flow*, **1**(2), 93–6.

THOMSON, J. (1881–2). On a changing tesselated structure in certain liquids. *Proc. Glasgow Philos. Soc.*

THOREN, F. (1984). Heat pipe cooled induction motors. *Proc. 5th Int. Heat Pipe Conf., Tsukuba*, **1**, 365–71.

TIEN, C.L. (ed.) (1973). Heat pipes. *AIAA Selected Reprint Series*, **16**.

TIEN, C.L. (1975). Fluid mechanics of heat pipes. *Ann. Rev. Fluid Mech., Am. Inst. Phys.*, **7**, 167–85.

TIEN, C.L. (1985). Heat pipes. In *Handbook of heat transfer*. McGraw-Hill, New York.

TIEN, C.L. (1985). Heat pipes. In *Handbook of heat transfer applications*, (ed. W.M. Rohsenow), Ch. 5. McGraw-Hill, New York.

TIEN, C.L. and CHEN, S.J. (1984). Non-condensable gases in heat pipes. *Proc. 5th Int. Heat Pipe Conf., Tsukuba*, pp. 97–101.

TIEN, C.L. and SUN, K.H. (1971). Minimum meniscus radius of heat pipe wicking material. *Int. J. Heat Mass Transfer*, **14**, 1853–5.

TODA, S. and UCHIDA, H. (1972). Study of liquid film cooling with evaporation and boiling. *Trans. JSME*, **38**, 1830–48.

TODO, I. (1970). Dynamic response of bayonet-type heat exchanger, Part 1. Response to inlet temperature changes. *Bull. JSME*, **19**(136), 1135–40.

TODO, I. (1978). Dynamic response of a bayonet-type heat exchange, Part 2. Response to flow rate changes. *Bull. JSME*, **21**(154), 644–51.

TONG, M., XIN, M., and CHEN, Y. (1983). Pool boiling heat transfer over a heating surface separated by a performated plate (in Chinese). *J. Eng. Thermophys.*, **4**(2), 159–64.

TONG, M., SHI, C., and XIN, M. (1984). Heat transfer enhancement in a two-phase closed thermosyphon, (in Chinese). *J. Eng. Thermophysics*, **5**(4), 371–3.

TSUCHIYA, F., RYOKAI, K., and MOCHIZUKI, S. (1989). Utilization of heat pipes for artificial permafrost storage. *Proc. 2nd Int. Symp. on Cold Regions Heat Transfer, Sapporo*, pp. 279–83.

TURNER, J.S. (1973). *Buoyancy effects in fluids*. Cambridge University Press, Cambridge.

UEHARA, H., NAKOKA, T., KUSUDA, H., and NAKAJIMA, S. (1982). Wavy film condensation heat transfer on a vertical surface (in Japanese). *Trans. JSME Sec. B*, **48**(433), 1751–60.

UMUR, A. and GRIFFITH, P. (1964). Mechanism of dropwise condensation. ASME Paper #64–WA/HT-3.

VASILIEV, L.L. (1985). Low temperature heat pipes. *Heat Recovery Syst.*, **5**, 203–16.

VASILIEV, L.L. (1987*a*). *Heat pipe heat exchangers*. Nauka i Tekhnica, Minsk.

VASILIEV, L.L. (1987*b*) Heat pipes for heating and cooling the ground. *J. Eng. Phys.*, **51**, 676–87.

VASILIEV, L.L. (1989). Heat pipe research and development in the USSR. *Heat Recovery Syst. CHP*, **9**(4), 313–33.

VASILIEV, L.L. and KHROLENOK, V.V. (1976). Centrifugal coaxial heat pipes. *Proc. 2nd Int. Heat Pipe Conf., Bologna*, pp. 293–302.

VASILIEV, L.L. and KHROLENOK, V.V. (1983). Study of a heat transfer process in the condensation zone of rotating pipes. *Heat Recovery Syst.*, **3**(4), 281–90.

VASILIEV, L.L. and KHROLENOK, V.V. (1987). *Heat and mass transfer in centrifugal heat pipes*, Preprint 42. Heat and Mass Transfer Institute, Minsk.

VASILIEV, L.L. and KHROLENOK, V.V. (1990). Heat transfer in rotating heat pipes. *Proc. 7th Int. Heat Pipe Conf., Minsk.*

VASILIEV, L.L. and KONEV, S.V. (1972). *Heat pipes*. Nauka i Tekhnica, Minsk.

VASILIEV, L.L. and KONEV, S.V. (1974*a*). *Heat pipes*. Scripta Technica, Washington, DC.

VASILIEV, L.L. and KONEV, S.V. (1974*b*). Heat pipes. *Heat Transfer—Sov. Res.*, **6**(1), 1–97.

VASILIEV, L.L., VAAZ, S.L., KISILEV, V.G., KONEV, S.V., and GRAKOVICH, L.P. (1976). *Low temperature heat pipes*. Nauka i Tekhnica, Minsk.

VASILIEV, L.L., KONEV, S.V., and KHROLENOK, V.V. (1983). *Heat transfer enhancement in heat pipes*. Nauka i Tekhnica, Minsk.

VASILIEV, L.L., RABETSKY, M.I., and KISELEV, V.G. (1987). A heat transfer device for cooling elongated horizontal objects. *J. Eng. Phys.*, **52**, 62–6.

VERKIN, B.I., KIRICHENKO, Y.A., KOZLOV, S.M., and LEVCHENKO, N.M. (1978). Studies of cryogenic liquid boiling in centrifugal acceleration fields. *Proc. 5th Int. Heat Transfer Conf., Toronto*, **2**, 117–22.

VEST, C.M. and ARPACI, V.S. (1969). Stability of natural convection in a vertical slot. *J. Fluid Mech.*, **36**, 1–15.

VINZ, P. and BUSSE, C.A. (1973). Axial heat transfer limits of cylindrical sodium heat pipes between 25 W cm^{-2} and 15.5 kW cm^{-2}. *Int. Heat Pipe Conf., Stuttgart*, Paper 2-1.

VORONIN, V.G. (1976). *Low temperature heat pipes for aircraft*. Mashinostroyeniye Press, Moscow.

WADOWSKI, T., AKBARZADEH, A., and JOHNSON, P. (1990). Characteristics of a gravity-assisted heat pipe-based heat exchanger. *Proc. 7th Int. Heat Pipe Conf., Minsk*, D-2P.

WALLIS, G.B. (1969). *One dimensional two-phase flow*, Ch. 8. McGraw-Hill, New York.

WALLIS, G.B. (1974). The terminal speed of single drops or bubbles in an infinite medium. *Int. J. Multiphase Flow.*, **1**, 491–511.

WANG, C.S. and GREIF, R. (1981). The effect of impingement on heat transfer in rotating condenstation. *Int. J. Heat Mass Transfer*, **24**(7), 1097–104.

WATERS, E.D. (1974*a*). Heat pipes to stabilize pilings on elevated Alaska pipeline sections. *Pipeline Gas J.*, 201(10), 46–58.

WATERS, E.D. (1974*b*). Arctic tundra kept frozen by heat pipes. *The Oil Gas J.*, pp. 122–25 (August 26).

WATERS, E.D. (1976). Heat pipes for the Trans-Alaska pipeline. *Proc. 2nd Int. Heat Pipe Conf., Bologna*, **2**, 803–14.

WEBB, R.L. (1981). The evolution of enhanced surface geometries for nucleate boiling. *Heat Transfer Eng.*, **2**(3–4), 46–9.

WEBER-CARSTANJEN, C.T. and BUSSE, C.A. (1984). Liquid–vapour interaction in vertical gravity-assisted heat pipes. *Proc. 5th Int. Heat Pipe Conf., Tsukuba*, pp. 102–7.

WHALLEY, P.B. (1987). *Boiling, condensation and gas-liquid flow*, Ch. 11. Oxford University Press, Oxford.

WILSON, C.H., POPE, D.H., CUNDY, V.A., NYDAHL, J.E., and PELL, K.M. (1978). A demonstration project for de-icing of bridge decks. *Bridge Eng.*, **1**, 189–97.

WINTER, E.R.F. and BARSCH, W.O. (1971). The heat pipe. In *Advances in heat transfer*, **7**, 219–320. Academic Press, New York.

WINTERS, K.H. and BROWN, T.W. (1985). Benard convection in a three-dimensional cavity: the effects of aspect ratio and tilt. AERE Harwell, Theoretical Physic Division, T.P. 1121, 10 pp.

WU, C., RU, C., HU, Y., and GAO, Z. (1988). Application of heat pipe type air-preheater for boiler in power stations. *Proc. 2nd Nat. Heat Pipe Conf., Dayong*.

XIA, X., WANG, J.C.Y., LIU, J., and TAN, H. (1990). The heat transfer characteristics of phase change heat storage medium used in a heat pipe type solar collector. *Proc. 7th Int. Heat Pipe Conf., Minsk*.

XIN, M. and XIA, J. (1986). Heat transfer by dropwise condensation in the two-phase closed thermosyphon. *Proc. 8th Int. Heat Transfer Conf., San Francisco*, **4,** 1683–8. Hemisphere, Washington, DC.

XIN, M., XIE, H., and CHEN, Y. (1984). Boiling heat transfer on heating surfaces covered by metal screens. *Proc. 5th Int. Heat Pipe Conf., Tsukuba*, pp. 151–5.

XIN, M., CHEN, G., and CHEN, Y. (1987). Flow and heat transfer in two-phase closed thermosyphons. *Proc. 6th Int. Heat Pipe Conf., Grenoble*.

YABE, A., SHIRAISHI, M., TAKETANI, T., KASHIZAWA, H., and KIKUCHI, K. (1984). Studies of enhanced boiling surfaces for two-phase closed thermosyphons. *Proc. 5th Int. Heat Pipe Conf. Tsukuba*, pp. 121–6.

YAMAGASHI, H., ISHIGURO, R., KUMADA, T., and SUGIYAMA, H. (1989). Numerical simulation of natural convection in an open circular thermosyphon for ethylene glycol. *Proc. Int. Symp. on Cold Regions Heat Transfer, Hokkaido*, pp. 189–96.

YANG, H.Q., YANG, K.T., and LLOYD, J.R. (1986). Flow transition in a three-dimensional tilted rectangular enclosure. *Proc. 6th Int. Heat Transfer Conf., San Francisco*, **4**, 1495–500. Hemisphere Pub. Corp., Washington.

YANG, H.Q., YANG, K.T., and LLOYD, J.R. (1987). Laminar natural convection flow transitions in tilted three-dimensional longitudinal rectangular enclosure. *Int. J. Heat Mass Transfer*, **3**, 1637–44.

YANG, H.Q., YANG, K.T., and LLOYD, J.R. (1988). Three dimensional bimodal flow transitions in tilted enclosures. *Int. J. Heat Mass Transfer*, **9**, 90–7.

YANG, K.T. (1987). Natural convection in enclosures. In *Handbook of single-phase convection heat transfer*, (eds S. Kakac *et al.*), Ch. 13. Wiley-Interscience, New York.

YANG, K.T. (1988). Transitions and bifurcations in laminar buoyant flows in confined enclosures. *J. Heat Transfer*, **110**, 1191–204.

YANG, Q. and SONG, X. (1988). The design and testing of the rotating heat pipe. *Proc. 2nd Nat. Heat Pipe Conf., Dayong*.

YANG, S.L. and HSIEH, C.K. (1988). Fluid flow and heat transfer in a single-pass, return-flow, bayonet heat exchanger. *Proc. ASME–AIChE. Nat. Heat Transfer Conf., Houston*, **3**, 355–64.

YANG, W. (ed.) (1987). *Heat transfer and fluid flow in rotating machinery*. Hemisphere, Washington, DC.

YUAN, S.W. and FINKELSTEIN, A.B. (1956). Laminar pipe flow with injection and suction through a porous wall. *Trans. ASME*, **78**, 719–24.

ZARLING, J.P., HAYNES, F.D., and DALY, S.F. (1988). On the application of thermosyphons in cold regions. *Proc. 7th Int. Conf. on Offshore Mechanics and Arctic Engineering*, **IV**, 280–6.

ZHANG, Y.M. and ZHANG, H.Q. (1988). Experimental investigation of flow characteristics about tip heat transfer in bayonet tube (in Chinese). *Huagong Jixie*, **15**(2), 96–102.

ZHAO, LITONG (1992). Rotating thermosyphons: a numerical study of laminar behaviour. Ph.D. thesis. University of Alberta.

ZHAO, LITONG, and LOCK, G.S.H. (1992*a*). Numerical analysis of impeded flow in an eccentric rotating thermosyphon. *Proc. 4th Int. Symp. on Transport Phenomena and Dynamics of Rotating Machinery, Hawaii.*

ZHAO, LITONG and LOCK, G.S.H. (1992*b*). Impeded flow in a rotating concentric thermosyphon. *Proc. 4th Int. Symp. on Transport Phenomena and Dynamics of Rotating Machinery, Hawaii.*

INDEX